# Basic Plane Surveying

ARTHUR SIEGLE

VAN NOSTRAND REINHOLD COMPANY
NEW YORK CINCINNATI TORONTO LONDON MELBOURNE

Library of Congress Catalog Card Number 78-27291
ISBN 0-442-27939-6



Printed in the United States of America

Published in 1979 by Van Nostrand Reinhold Company
A division of Litton Educational Publishing, Inc.
135 West 50th Street, New York, NY 10020, U.S.A.

Van Nostrand Reinhold Limited
1410 Birchmount Road
Scarborough, Ontario MIP 2E7, Canada

Van Nostrand Reinhold Australia Pty. Ltd.
17 Queen Street
Mitcham, Victoria 3132, Australia

Van Nostrand Reinhold Company Limited
Molly Millars Lane
Wokingham, Berkshire, England

16 15 14 13 12 11 10 9 8 7 6 5 4 3 2 1

**Library of Congress Cataloging in Publication Data**

Siegle, Arthur G date
Basic plane surveying.

Includes index.
1. Surveying. I. Title.
TA545.S517 526.9 78-27291
ISBN 0-442-27939-6

# PREFACE

Surveying is the art and science of determining the area of a portion of the earth's surface, the lengths and directions of boundary lines, and the contour of the earth's surface. These three-dimensional characteristics are measured by a surveyor, who often records them on a two-dimensional drawing surface. Because of the high degree of accuracy required, many years of education and practice are necessary to earn the title "surveyor."

The need to divide land into specific areas with established boundaries has existed since the earliest records of civilizations. This need increases with the importance of exact land boundaries, increasing land values, and the expanding construction industries.

This book is written with the emphasis on basics. The sequence of subjects is carefully chosen, reasons are explained, the scope is broad, and the presentation is concise.

If you have not previously studied trigonometry, drafting, slide rule, or calculator operation, you can manage, but you are encouraged to build up your knowledge of these subjects with work and practice. Field work with basic surveying equipment is essential.

Instruction on the adjustment of specific surveying instruments is not covered. The best instruction is found in the equipment manufacturers' manuals furnished with their products.

This book contains the following major features:

- Performance objectives precede each chapter telling you what is to be learned
- Summaries follow each chapter to review and emphasize important points
- Activities give you "hands on" experience in the field doing actual surveying
- All the material is up to date, presented in logical sequence to eliminate confusion

- The material is based upon the author's actual use of the information over many years of teaching surveying
- Review questions at the end of each chapter evaluate your comprehension of the material. Practical problems are carried through from one chapter to another to demonstrate different methods of solving them
- Technical terms are explained in the text when they are introduced

The author, Mr. Arthur Siegle, had been a highway engineer for thirty-eight years with the United States Federal Highway Administration before teaching surveying at the Tri-County Community College in Murphy, North Carolina. He is a graduate of the University of Pittsburg, with a degree in civil engineering.

# CONTENTS

# CHAPTER 1

# Introduction

## OBJECTIVES

After studying this chapter, the student will be able to:

- define the purpose and scope of plane surveying.
- explain the nature of the work surveyors perform.
- discuss the kinds of basic tools, training, and skills surveyors need.

## THE MEANING OF SURVEYING

*Surveying* is the occupation of defining the relative position of points on or near the earth's surface. It consists of making linear and angular measurements, and applying certain principles of geometry and trigonometry.

*Plane surveying* involves ordinary measurements where neglecting the earth's curvature causes no significant error. Where linear distances are so great that the earth's curvature requires consideration, *geodetic surveying* is involved.

Plane surveying meets a variety of needs. The location of land boundaries is probably the best known. Construction needs for design data and the layout of facilities actually place the major demand upon surveyors' time. A third use of plane surveying is the definition of the shape of the earth's surface and location of details for mapping. This third branch of plane surveying is called *topographic surveying.*

## SURVEYING CREWS

Surveying crews not only perform outdoor measurement work, but also do substantial amounts of indoor work. This indoor work includes:

- studying deeds for properties being surveyed and adjoining properties.
- consultations with clients.
- computing.

- drawing maps.
- drafting plats.
- summarizing data for construction design.
- preparing field notes for construction layout.
- researching nearby control points.
- maintaining adequate control data.

Surveying crew members are assigned jobs in accordance with their skills, which are developed from experience and training. However, quality depends upon responsible and cooperative effort by novices as well as by more experienced personnel. Teamwork is essential. Each person must always be alert and raise questions whenever observations or results seem incorrect.

Land-surveying crews are usually supervised by a registered land surveyor or a professional engineer. To become registered, a person must pass a written examination given by a state board of registration. A person is allowed to take the examination only after gaining sufficient land-surveying experience. Experience requirements are prescribed by state law and state boards. Some surveying courses offered by engineering schools and community colleges are acceptable substitutes for a year or more of experience. The time a person practices before being permitted to take the examination is thereby shortened. The trend is toward stricter requirements. College degrees may someday be required.

Crews engaged in construction or topographic surveying are not necessarily under the direction of registered surveyors or engineers. Many government agencies, and some private organizations, do not require registered personnel.

Often people with neither training nor knowledge of surveying find surveying work. Some, beginning as helpers to clear brush, advance to skilled positions. For rapid advancement, and to achieve one's highest potential level, a person should have academic surveying training. Greater job satisfaction and assurance resulting from a thorough understanding of the work are further benefits of study. An applicant's opportunities are greatly improved by completing a surveying course before seeking employment.

This book is written to prepare surveying students to do their best in this occupation. Proficiency and training in mathematics are assets. For those who lack geometry and trigonometry, essential elements of both subjects are introduced as needed.

## UNITS OF MEASUREMENT

Dimensions for most purposes are measured horizontally or vertically. When distances are measured along sloping surfaces they are also converted to horizontal or vertical measurements.

Horizontal linear measurements are in units of feet, rods (poles), chains, or stations. A *pole* is the same length as a *rod,* that is, sixteen and one-half feet long. Lengths of rods smaller than one unit are measured in either decimals or fractions. A *station* is one hundred feet. For portions of a station, feet are expressed in whole numbers and decimals. A portion of a station is referred to in terms of a *plus*. For example, a distance of 1764.31 feet is referred to as station 17+64.31. At one time, surveyors used a *Gunter's chain* to measure horizontal distances. It was sixty-six feet long, made up of one hundred links. Land surveys and deeds record lengths in any of the above terms. Assume that the line in figure 1-1 is 357.41 feet long. Notice the different ways of expressing this length.

The most common units for both horizontal and vertical measurement today are feet. Lengths smaller than one foot are expressed in decimals. Inches are used only for certain kinds of construction layout for linear distances in plane surveying.

Angular measurements are ordinarily made in units of degrees (°) and minutes ( ′ ). Seconds ( ″ ) are used where that amount of precision is needed. There are 360 degrees in a complete circle. There are 60 minutes in one degree, and 60 seconds in one minute. Angles are occasionally expressed in decimals of degrees, minutes, or seconds.

For example, the minute hand of a clock rotates 360 degrees each hour as it travels completely around. In fifteen minutes it rotates 90 degrees. It rotates 6 degrees each minute. Minutes and seconds of angles are not to be confused with minutes and seconds of time. A second hand that travels through 5.2 seconds of time actually travels through an angle of 31° 12′ (31 degrees 12 minutes). A pie cut into portions of one minute each would contain 21,600 portions. A pie cut into portions of one second each would contain 1,296,000 portions.

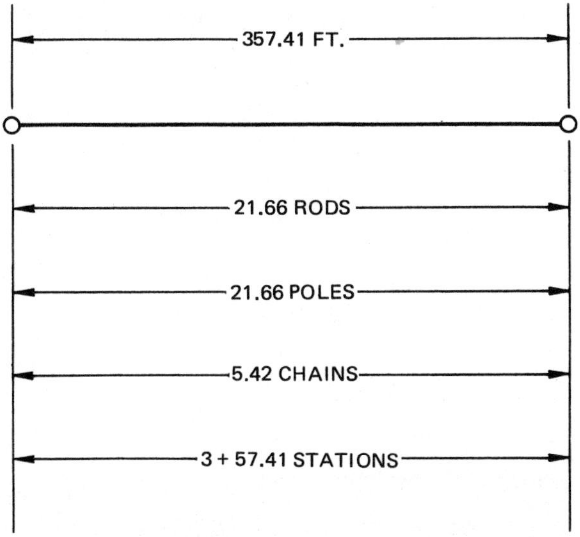


**Fig. 1-1 Same distance expressed in different units**

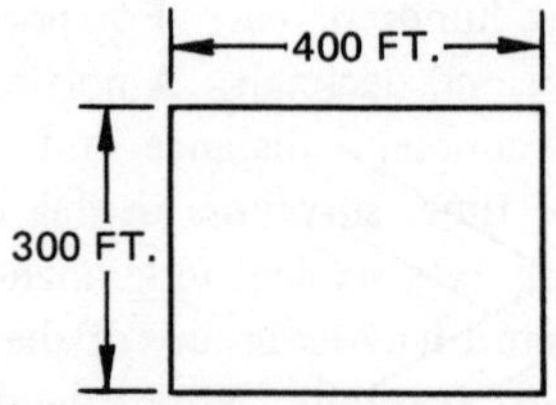


| | feet | yards | rods | chains | stations | acres |
|---|---|---|---|---|---|---|
| Length | 400 | 133.3 | 24.2 | 6 | 4 + 00 | |
| Width | 300 | 100 | 18.2 | 4.5 | 3 + 00 | |
| Area | 120,000 sq ft | 13330 sq yds | 440 sq rods | | | 2.8 |

**Fig. 1-2 Same dimensions and area expressed in different units.**

Areas computed from linear distances and angles are in units of square feet, square yards, or acres. Nine square feet equal *one square yard,* and 43,560 square feet equal *one acre.* Land areas are computed on the basis of horizontal dimensions.

Figure 1-2 shows a rectangular piece of land 300 feet by 400 feet. The lengths and area of the land are listed in different systems to show the many ways area can be expressed.

Volumes computed from linear distances and angles are in units of cubic feet or cubic yards. Twenty-seven cubic feet equal *one cubic yard.*

A typical problem may be to find the amount of dirt that has to be excavated before the construction of a house foundation can begin. In figure 1-3, the dimensions of a hole to be dug are shown. How many cubic yards must be excavated?

Different units are not combined in measurements of length, area, or volume. A distance of 145.06 feet is never reported as 1 station, 2 rods, and 12.06 feet. It is however, customary to combine units in angular measurement. An angle of $60^\circ$ 45′ 36″ is so recorded. It is also acceptable to express this same angle as $60.760^\circ$, or $60^\circ$ 45.6′.

## CONSIDERATIONS PERTAINING TO ACCURACY

No measurement is exact. How close a measurement is to the true value depends upon:

- the choice of equipment.
- the precision of the equipment chosen.
- the quality, condition, and adjustment of that equipment.
- the refinements and care that are used in making the measurements.

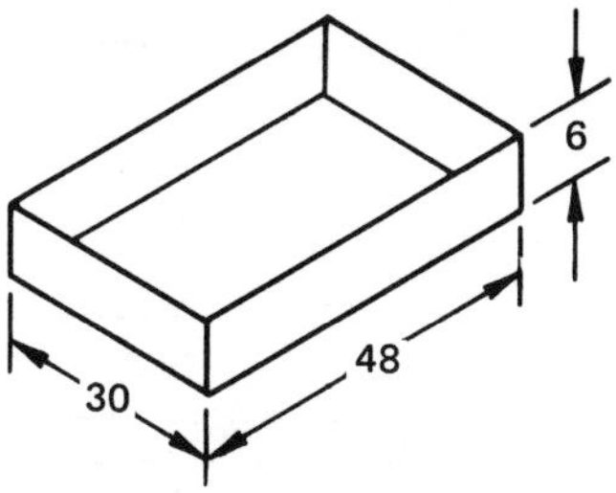


Volume = length x width x height
= 48 ft. x 30 ft. x 6 ft.
= 8640 cubic feet

Since there are 27 cubic feet in a cubic yard:

$$\text{Volume} = \frac{8640 \text{ cubic feet}}{27 \text{ cubic feet}} = 320 \text{ cubic yards}$$

**Fig. 1-3 Units for expressing volume**

The degree of precision used for a measurement depends upon what is required of the measurement. The boundaries of high-cost land are defined with higher degrees of precision than are the boundaries of low-cost land. A rail on a new railroad is located to the nearest 0.01 foot for construction. It might be located only to the nearest foot for mapping. Measurements made with more precision than necessary waste time and money, and should be avoided.

Errors in measurement occur because of faulty equipment, human limitations, and natural causes. Some are avoided by methods which cancel out their effects. Some errors are so small that they can be tolerated. Some errors can be reduced as the skill and proficiency of the surveyor increases. Suppose faulty instrument adjustment makes the first measurement of an angle too small by one minute. The instrument can be manipulated during a second measurement to force that faulty adjustment to indicate a reading one minute too large. Accepting the average of the two readings cancels the error. If the wind prevents a crew member from holding the tape directly over a point, a reading too long by 0.01 feet can occur. The required precision of the survey may be such that an excess of 0.01 feet is tolerable. In another instance, a tape is found to be 0.02 feet short. In that case, all observed measurements are corrected by the appropriate amounts. Whatever their cause, errors must be handled in a way that achieves the required precision.

*Mistakes* are blunders. They are not errors. Mistakes include numerals recorded incorrectly, or an angle turned to one point wrongly recorded as having been turned to another. It is important that precautions be taken to minimize such occurrences. To avoid wasted time and effort, detect and correct mistakes when they occur. Another visit to a job site merely to find or correct mistakes is costly. Regardless of what must be done, the final job must be free from mistakes.

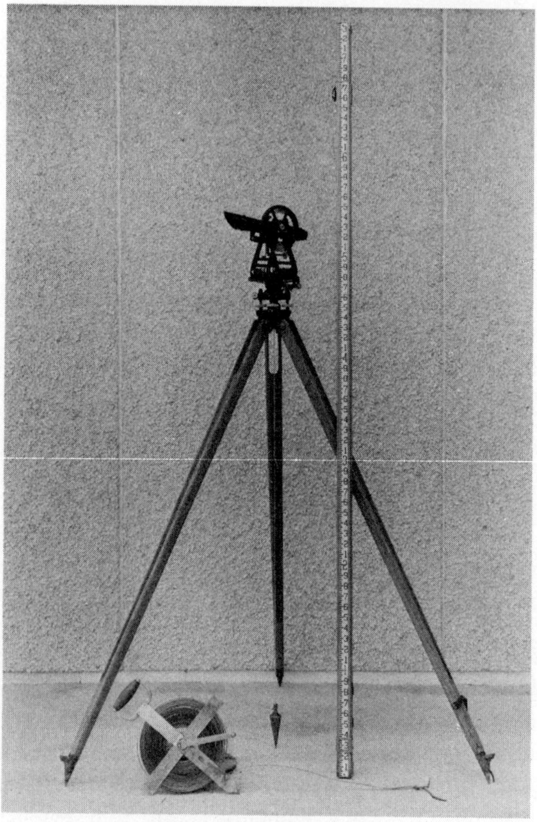

**Fig. 1-4 Transit, tape, and level rod**

## EQUIPMENT

Basic surveying equipment consists of a transit, tape, and level rod, figure 1-4. A person skilled in the operation of this equipment can easily adapt to more sophisticated surveying equipment. A *transit* measures horizontal and vertical angles and indicates the direction of a line by a built-in compass. Vertical distances are measured with a transit and a level rod. A *level rod* is a long wooden ruler graduated in feet and hundredths of a foot. A *tape* is used for horizontal and slope distances. This book explains the basic operation and functioning of the equipment. Because there are many manufacturers, specific references to the care and adjustment of equipment are left to the manuals supplied with the equipment.

For computation, surveyors must be able to work with slide rules, calculators, and printed reference tables.

For mapping and drawing, skill with drafting instruments and hand lettering is important. A client's impression of the surveyor's ability is strongly influenced by the quality of the lettering and drafting work.

## THE SURVEYOR'S FIELD NOTEBOOK

All field measurements are recorded in a *field notebook*. The entries must be legible and written to clearly express their meaning. A good test of the adequacy of notes is to see how clearly they are understood one year after being recorded.

Notebooks for surveying can be obtained from an engineering supply house. Examples of appropriate forms of notes appear in later chapters.

All entries are made at the time of their observation. Entries should never be copied from notes first recorded elsewhere. Words and numerals are hand lettered, never written in script. Improper entries should not be erased. Instead, lines are drawn through them, and correct values are entered above. Sketches are drawn as necessary. It is important that pencils with hard lead, 3H or harder, be used to avoid smearing.

It is customary to enter at the top of each pair of pages the following:

- the page number.
- the title of the survey.
- the date.
- the weather.
- the names of the crew members.
- the principal duty of each.

In large organizations, crews may not use the same equipment regularly. In that case, serial numbers of transits, tapes, and other principal items are also entered. The first few pages of the notebook are reserved for an index.

Field notebooks are carefully preserved, and can be important legal evidence in court cases.

## SUMMARY

Surveying is the occupation of defining the relative positions of points on or near the earth's surface.

Plane surveying involves ordinary measurements where neglecting the earth's curvature causes no significant errors. Its three major branches are land, construction, and topographic surveying.

Plane surveying requires both indoor and outdoor work. The teamwork of crew members is essential.

The most common unit of linear measurement used today in surveying is the foot. Smaller measurements are expressed in decimals.

Angular measurement is given in terms of degrees, minutes, and seconds.

The degree of precision of measurements depends upon the requirements of the specific job.

Errors are handled in a manner which permits achieving the required precision. Mistakes must be located and eliminated.

Basic surveying equipment consists of a transit, tape, and level rod.

Surveyors need to develop skill in using calculators, slide rules, and printed reference tables.

A client's impression of a surveyor's ability is strongly influenced by the quality of the surveyor's lettering and drafting.

All field measurements are recorded in a field notebook at the time they are made. Field notebooks are carefully preserved, and may become important legal evidence in court cases.

## ACTIVITIES

1. Make a table showing for each 0.01 foot from 0.00 feet to 1.00 foot the equivalent length in inches to the nearest 1/8 inch.
2. Record the time to the nearest second as indicated by a watch or clock. Calculate the number of degrees and angular minutes the minute hand has rotated since the hour began.

## REVIEW QUESTIONS

### A. Multiple Choice

1. The measurement of long distances requiring the consideration of the earth's curvature involves a branch of surveying called
   a. plane surveying. c. construction surveying.
   b. geodetic surveying. d. topographic surveying.
2. Surveying involving ordinary measurements where the neglect of the earth's curvature causes no significant error is called
   a. plane surveying. c. construction surveying.
   b. geodetic surveying. d. topographic surveying.
3. The type of surveying that creates the greatest demand for surveyors' time is
   a. plane surveying. c. construction surveying.
   b. geodetic surveying. d. topographic surveying.
4. A length of four stations is equivalent to
   a. 66 feet. c. 400 feet.
   b. 264 feet. d. 400 chains.

### B. Short Answer

5. What are the three basic pieces of surveying equipment?

### C. Problems

6. A rectangular field is 18.5 rods long and 14.5 rods wide. Express its dimensions in feet. What is its area in acres?

7. Excavation for a house is 27.0 feet x 38.0 feet x 5.6 feet. What is the volume in cubic yards?

8. A ditch for a sewer averages 3.0 feet in width and 6.0 feet in depth. How many cubic yards are excavated between stations 16 + 40 and 20 + 87?

9. How many degrees and minutes does a watch's second hand rotate during 18 1/2 seconds? How many degrees and minutes does the second hand rotate during 2 minutes 5.2 seconds?

# CHAPTER 2

# Measuring Distances

## OBJECTIVES

After studying this chapter, the student will be able to:

- read a surveyor's steel and woven tape.
- apply corrections to steel tape measurements.
- tape horizontally and on slopes by conventional methods.
- use tapes for establishing perpendiculars, measuring around obstacles, and measuring the area of a field.

## METHODS

Most people are accustomed to measuring short distances with rulers, yardsticks, six-foot rules, and tapes. They read them in units of inches and fractions of inches. *Pacing* is a means of approximating longer distances when using a short measuring device is difficult or time consuming. To determine a distance by pacing, a person multiplies the number of steps by their average length. For long distances, odometers on vehicles indicate miles and tenths of miles traveled.

A surveyor commonly measures distances between points on the earth's surface with a tape. It is read in units of feet and decimals of a foot. Refinements of the basic measurement are often made to achieve the desired degree of precision.

A surveyor also employs a transit and stadia rod for measuring distances. This method is particularly effective in locating points for mapping purposes where precise results are unnecessary.

Electronic distance measurement is becoming popular with surveyors. Electronic instruments produce a high degree of precision, but are expensive. They utilize lightwaves, microwaves, radiowaves, or laser beams. A survey crew member, having learned to measure properly with a tape, can easily learn to use electronic instruments.

**Fig. 2-1 Woven tape**

## SURVEYOR'S TAPES

Surveyors usually work with steel tapes, but also find woven tapes, figure 2-1, handy for some purposes. The woven tapes are made of cloth with metal wires or other reinforcement materials woven along their length. They are designed primarily for use in construction, and are not as reliable as steel tapes.

Steel tapes are available in lengths of 100, 200, 300, and 500 feet. A few are also available in other lengths. The most common lengths of woven tapes are 50 and 100 feet.

A steel tape is read directly to the nearest 0.01 foot. Thousandths of a foot may be estimated if needed. Values as read may need correction to achieve the required degree of precision. Such corrections compensate for one or more different effects, such as temperature, tension, or sag.

### The Effect of Temperature

A steel tape is calibrated at a temperature of 68°F. For each drop of 15°F below 68°F, a 100-foot tape becomes 0.01 feet shorter. For each rise of 15°F above 68°F, it becomes 0.01 foot longer.

For example, a 100-foot steel tape being used at a temperature of 38°F has actually shortened to 99.98 feet long. A 100-foot steel tape being used at a temperature of 83°F has actually lengthened to 100.01 feet. More specifically, steel tapes have a coefficient of expansion of approximately 0.0000065 feet for every 1°F rise in temperature. This means for every 1°F rise in temperature above 68°F, a tape increases 0.0000065 feet for every foot in its length. The actual length of a 100-foot steel tape used at a temperature of 98°F is found in the following manner:

$$\begin{aligned} \text{Length} &= 100 \text{ feet.} + (0.0000065)(100 \text{ ft.})(98^\circ - 68^\circ) \\ &= 100 \text{ ft.} + .02 \text{ ft.} \\ &= 100.02 \text{ ft.} \end{aligned}$$

The opposite is also true. A 1°F drop in temperature below 68°F decreases the tape 0.0000065 feet for every foot in its length. The actual length of a 200-foot steel tape used at a temperature of 30°F is:

$$\begin{aligned} \text{Length} &= 200 \text{ ft.} - (0.0000065)(200 \text{ ft.})(68^\circ - 30^\circ) \\ &= 200 \text{ ft.} - .05 \text{ ft.} \\ &= 199.95 \text{ ft.} \end{aligned}$$

Tape and air temperatures are not the same. For precise work one or more thermometers are attached to the tape.

**The Effect of Tension**

Steel tapes are usually calibrated when subjected to 10-pound pulls while fully supported on flat surfaces. Under greater tension, the amount that they stretch depends upon their weights. A 100-foot steel tape stretches 0.01 feet when the pounds equal 8.80 times its weight. That amount of pull also equals 3000 times the number of square inches in the tape's cross section. Widths, thicknesses, and weights of tapes are usually given by the manufacturer. A medium weight 100-foot tape weighing 1.70 pounds takes 15.0 pounds to stretch it 0.01 foot. This is 15.0 pounds over and above the basic pull of 10.0 pounds. A lightweight 100-foot tape weighing 1.02 pounds takes only 9.0 additional pounds to stretch it 0.01 foot. A heavyweight 100-foot tape weighing 2.66 pounds takes 23.4 additional pounds to stretch it the same amount.

**The Effect of Sag**

Most measurements of distance outdoors cannot be made with the tape supported throughout its entire length. A tape sags between the points where it is supported. That sag makes the distance along the tape greater than the distance between points. The difference is 0.01 feet when the center of the 100-foot tape sags 0.60 feet. The effects of sag are offset by pulling harder on the tape whenever possible. However, it becomes impractical to apply sufficient tension to offset sag effects for long unsupported lengths. The heavier the tape, the shorter is the unsupported length for which it is practical. When a significant error cannot be eliminated, correction must be made.

Figure 2-2(A) shows the effects of sag and offsetting pulls for medium-weight tapes at 68°F. Sag effects are offset by adding one pound of pull for each additional 5 feet of unsupported length. Thus, pulls of 20, 25, and 30 pounds offset sag effects for 50, 75, and 100 feet respectively.

To use figure 2-2(A), find the unsupported length of tape on the bottom line, for example, 60 feet. If the temperature is 68°F, extend a line from 60 feet

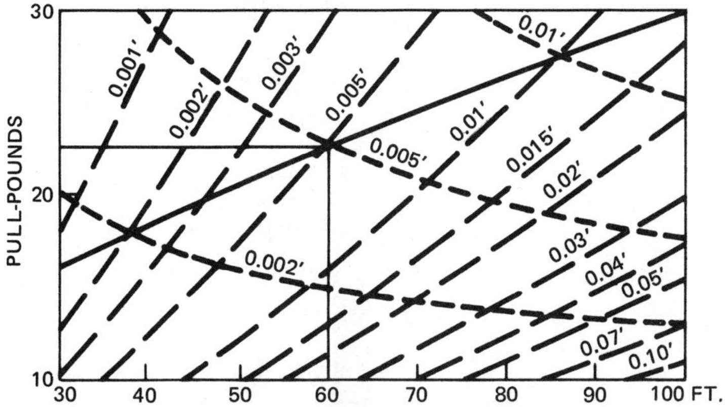


(A) UNSUPPORTED LENGTH OF MEDIUM-WEIGHT TAPE-FEET

— — AMOUNT UNSUPPORTED TAPE LENGTH IS SHORTENED BY SAG

– – – – AMOUNT UNSUPPORTED TAPE LENGTH IS STRETCHED BY APPLIED TENSION

—— POUNDS OF PULL REQUIRED TO AVOID CORRECTION AT 68°F

**Fig. 2-2(A) Effects of sag and tension at 68°F**

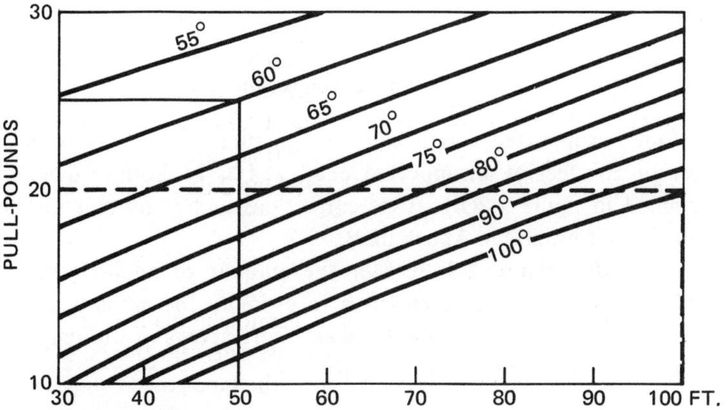


(B) UNSUPPORTED LENGTH OF MEDIUM-WEIGHT TAPE-FEET

—— POUNDS OF PULL REQUIRED TO AVOID CORRECTION AT INDICATED TEMPERATURE

**Fig. 2-2(B) Pull required to avoid correction**

straight up until it intersects the solid diagonal line. Read off the left scale 22.5 pounds, the amount of pull required to eliminate the effects of sag.

For unsupported tape lengths exceeding 100 feet, it becomes difficult or impossible to achieve precision without correction. On cold days correction is needed even for shorter unsupported lengths.

Figure 2-2(B) shows how pulls vary when simultaneously compensating for both sag and temperature. A 25-pound pull is required to avoid correction

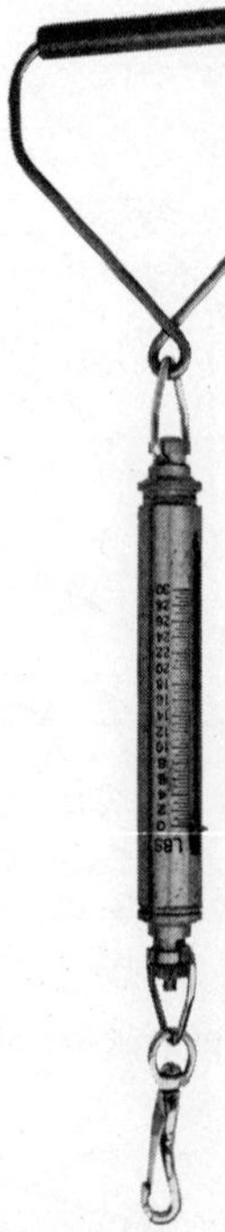

**Fig. 2-3 Tension handle**

of a 50 foot unsupported length at 60°F. A pull of only 20 pounds avoids correction of 100 feet of unsupported length at 100°F. Both of these examples are diagramed in figure 2-2(B). Tape corrections under any circumstances can be calculated by formulas in Appendix B.

*Tension handles* figure 2-3, register the amount of applied tension. Experienced survey crews train themselves to estimate the amount of pull being applied by practicing with tension handles. For surveys of ordinary precision, they then rely upon their estimating ability.

**Addition or Subtraction of Tape Corrections**

Novices experience difficulty in using the proper algebraic sign when making corrections. An excessively expanded or stretched tape gives readings too short and corrections must be added. Excessive shrinkage or sagging gives readings that are too long and corrections must be subtracted. Several examples can help to explain this concept:

> A 300-foot steel tape is continuously supported for a measurement with 10 pounds of tension at 90°F. The tape reading is 282.67 feet. The tape reading is less than the actual distance because the tape is expanded. Therefore, the correction is added to the reading to obtain the true distance:

$$
\begin{aligned}
\text{True distance} &= 282.67 \text{ ft.} + (0.0000065)(282.67 \text{ ft.})(90^\circ - 68^\circ) \\
&= 282.67 \text{ ft.} + 0.04 \text{ ft.} \\
&= 282.71 \text{ ft.}
\end{aligned}
$$

The same tape is supported at its midpoint for measuring the same distance at 68°F. The tape reading is 282.86 feet. This tape reading is 0.15 feet greater than the actual distance, sag accounting for + 0.17 feet and tension for - 0.02 feet. Therefore, the correction is subtracted from the reading:

$$
\begin{aligned}
\text{True distance} &= 282.86 \text{ ft.} - 0.15 \text{ ft.} \\
&= 282.71 \text{ ft.}
\end{aligned}
$$

## READING THE TAPE

Survey crew members should study the way the tapes they use are marked and calibrated. The zero mark is at different places on different tapes. Tapes graduated only in feet usually have the last foot at each end graduated in tenths or hundredths. On some 100-foot tapes, the foot graduated to hundredths at the zero end may be a part of the 100 feet. This is called a "minus" tape. To find the total length of a measurement, readings to hundredths are subtracted from the whole number of feet. On some tapes, there is an extra foot beyond the zero mark graduated in hundredths, figure 2-4. To find the total length of a measurement, this graduated portion of the tape is added to the whole number of feet. This is called a "plus" tape. Most steel tapes have blank sections of tape between the last graduations and loops at the ends. Woven tapes are usually graduated uniformly throughout. Zero marks are usually at the ends of the loops, figure 2-5.

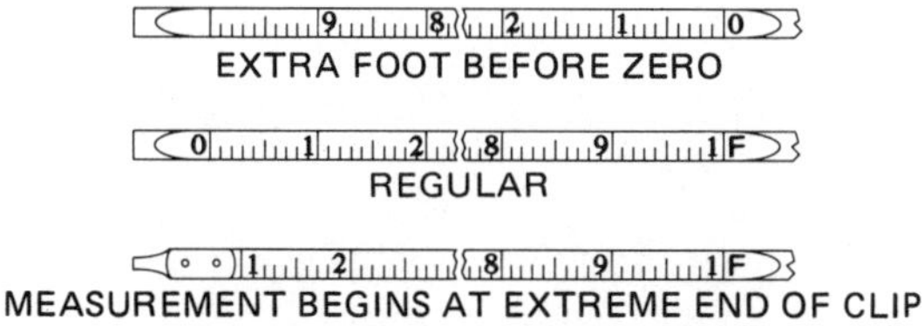


**Fig. 2-4 Styles of steel tape ends**

**Fig. 2-5 End of woven tape**

The following precautions are taken to eliminate mistakes in reading tapes. When using a steel tape having an extra graduated foot beyond the zero mark, care is taken to exclude that extra foot when measuring whole numbers of feet. When using a woven tape be sure to include the length of the unmarked loop. When verifying a foot mark on a tape, make sure the adjacent foot marks are one foot greater and smaller than the reading.

## FIELD PROCEDURE FOR TAPING

Both persons holding the tape stand at the side of the tape when stretching and reading it. The tape is held securely on or above both points between which a measurement is desired. It is stretched straight because deflection by the wind or contact with any object can introduce error. An error of 0.01 feet occurs when the center of a 100-foot tape is deflected 0.70 feet. Intermediate supports for minimizing sag effects are set at elevations aligned approximately with the elevations of the tape ends. The tape is pulled at its proper tension to avoid the necessity of correction. Or, conditions requiring corrections are noted. When a tape cannot be held and read directly on a point, a plumb bob is used. The *plumb bob,* a pointed cone-shaped weight, is suspended by a string held in contact with the correct graduation, figure 2-6. A swinging plumb bob is touched lightly to the ground repeatedly until it stops swinging. Each distance is measured at least twice, once forward and once backward.

**Fig. 2-6 Measuring with tape and plumb bob**

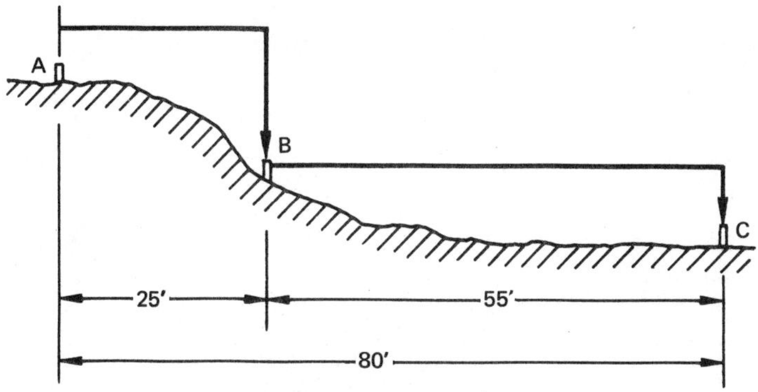


**Fig. 2-7 Breaking tape**

The elevations of two points between which a measurement is desired are seldom at the same height. The crew may choose to make a direct horizontal measurement using the plumb bob and tape. An alternative is to make a slope measurement and convert it to the horizontal distance by calculation.

## Horizontal Measurement

When direct horizontal measurements are made, sloping ground may require measured distances to be broken into segments. It is difficult to properly hold and read the tape at more than chest height. Lengths of segments should be allowed to accumulate on the tape rather than to be recorded separately. The rear person holding the tape simply holds at each intermediate point the forward tape reading for the previous segment. Measuring in segments in this manner is known as *breaking tape.*

As an example of this operation, assume that a horizontal measurement is being taken between points A and C, figure 2-7. The tape is held with the zero at A. Extending the tape toward C, the crew finds they can go no further than point B before the tape has to be held above shoulder height to keep it level. Therefore, a mark is set at point B at exactly 25 feet horizontal distance using a plumb bob. Rather than move the tape's zero end to point B, to take a second new measurement, the 25-foot mark on the tape is held at point B and the tape extended to C, which is also plumbed. The reading, 80 feet on the tape over point C, is the horizontal measurement of distance AC. This method eliminates possible error caused by adding lengths AB to BC to get the total distance.

Errors occur in horizontal measurements when crew members incorrectly judge whether a tape is horizontal. An error of 0.01 feet occurs when one end of a 100 foot tape is 1.4 feet lower than the other end. A hand level is used to make sure it is horizontal. In figure 2-8, page 18, the leveler's eye is at same level as the far end of tape. The person holding the tape in the foreground adjusts the height of the tape to that level.

**Fig. 2-8 Leveling the tape**

## Slope Measurement

Slope measurements permit longer distances to be measured in one operation. There is no need to break tape because of the slope. Opportunities generally exist for intermediate support, permitting long distances to be measured without significant sag effects. Either the tape's slope angle or the difference in elevation between the tape readings is measured. An abney hand level is used to make vertical angle measurements, figure 2-9. For slopes not exceeding 20 feet vertically per 100 feet horizontally, a formula for converting slope distance is:

$$C = \frac{h^2}{2s}$$

h is the difference in elevations

s is the slope distance

C is the amount to be subtracted from the slope distance to give the horizontal distance. All units are in feet.

Assume in figure 2-10 that point B is 13.00 feet above point A. The distance from A to B measured along the slope is 100.00 feet. The horizontal distance is found by:

$$C = \frac{h^2}{2s}$$

$$C = \frac{13.00^2}{2 \times 100.00}$$

$$C = \frac{169}{200}$$

$$C = 0.84 \text{ ft.}$$

Horizontal distance = 100.00 ft – .84 ft. = 99.16 ft.

Another formula for converting slope distance is:

$$C = 0.015\, S\,(\theta)^2$$

$\theta$ is the angle of slope in degrees.

S is the slope distance in 100 feet stations.

C is the amount in feet to be subtracted from the slope distance to give horizontal distance.

**Fig. 2-9 Abney hand level**

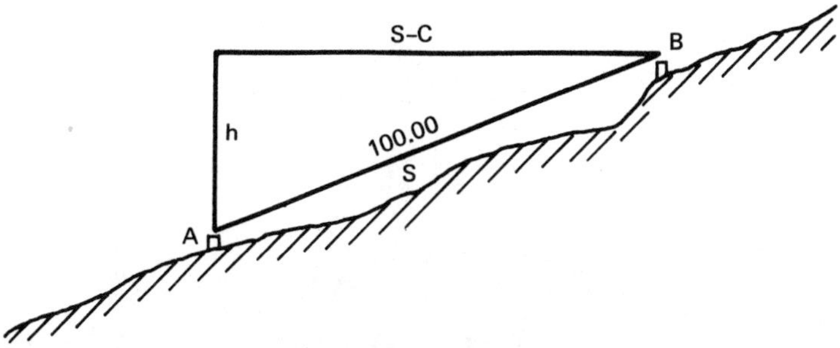


**Fig. 2-10 Slope measurement using difference in elevation**

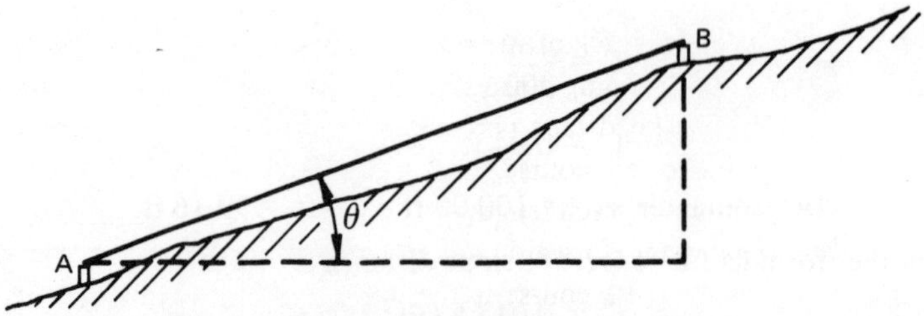


**Fig. 2-11 Slope measurement using angle of slope**

Assume in figure 2-11 that the slope measurement between A and B is 154.31 feet. The angle $\theta$ is 10°. The horizontal distance from A to B is found by the formula:

$$C = 0.015\ S\ (\theta)^2$$
$$C = 0.015 \times 1.5431 \times 10^2$$
$$C = 2.31 \text{ ft.}$$

Horizontal distance = 154.31 - 2.31 = 152.00 ft.

These formulas apply only for each individual segment. Data for a number of segments is never combined before applying them.

## Distances Longer than a Tape Length

Points set at the ends of each tape length are aligned either by transit or by eye. Range poles, figure 2-12, are usually sighted to establish alignment of these points. No significant errors result when they are lined in by eye with normal care.

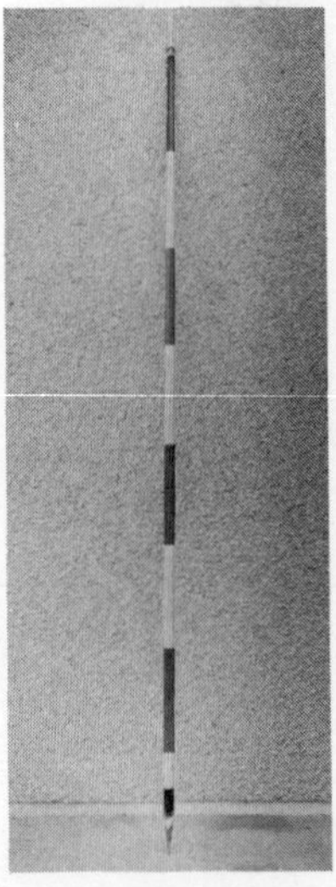

**Fig. 2-12 Range pole**

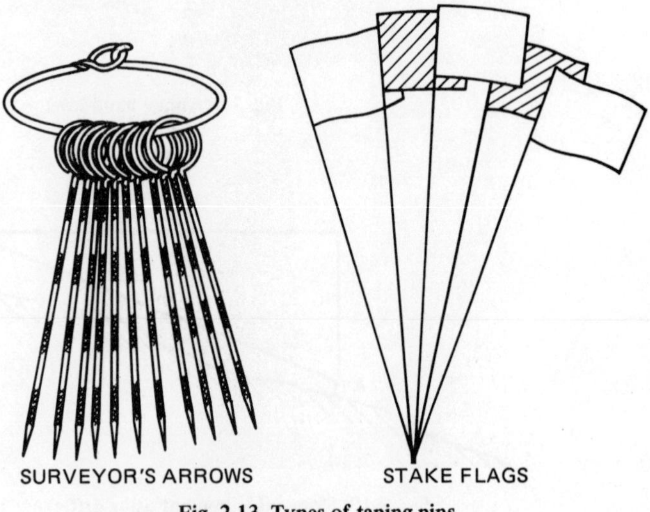


**Fig. 2-13 Types of taping pins**

A system for keeping track of the number of tape lengths must be used. One system is to use a set of taping pins, sometimes called surveyor's arrows or stake flags, figure 2-13. The head tape person sticks a taping pin in the ground at the beginning of the measured course, and at the end of each measured length. After the measurement of each length is completed, the rear person recovers the taping pin before moving forward. The number of pins in the rear person's possession at the end of the course is the number of full tape lengths. Another system is to drive stakes at the ends of each tape length, marking them by station numbers. They may be driven beside nails stuck in the ground to mark the exact points. Or, the stakes may be set on the line, and exact points marked on them by pencil lines or tacks, figure 2-14.

## SPECIAL USES OF THE TAPE

The tape alone is used in unique ways to provide solutions for a number of problems. It can be used to lay out perpendicular lines, measure around obstacles, or measure a field if all its sides are straight lines.

**Fig. 2-14 Marking points on stakes**

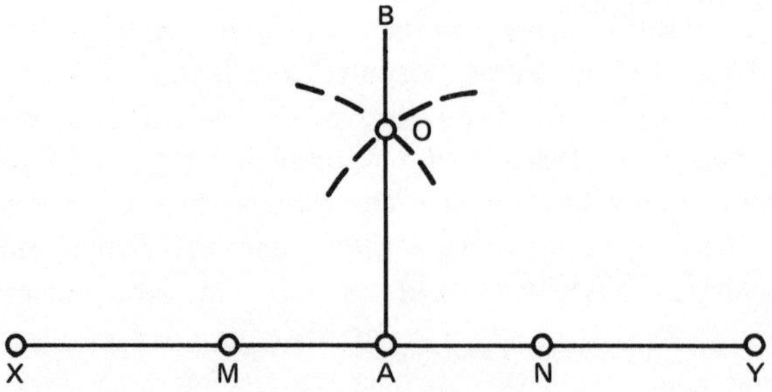


Fig. 2-15 Constructing a perpendicular using only a tape

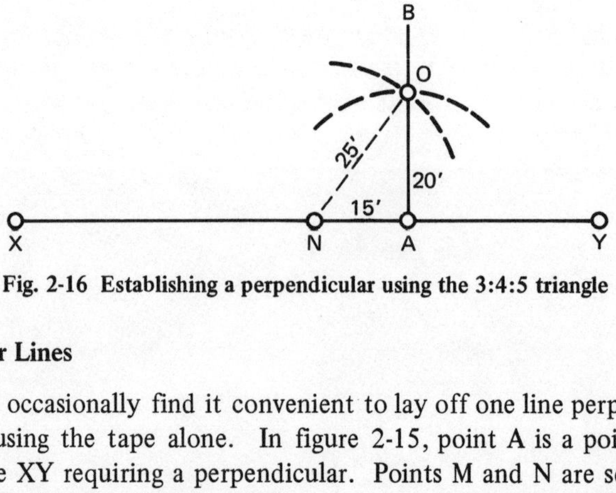


Fig. 2-16 Establishing a perpendicular using the 3:4:5 triangle

## Perpendicular Lines

Surveyors occasionally find it convenient to lay off one line perpendicular to another by using the tape alone. In figure 2-15, point A is a point on an established line XY requiring a perpendicular. Points M and N are set along line XY by making distances AM and AN equal. Arcs are struck on the ground by swinging the tape with radii MO equal to NO. Point O is the intersection of the arcs. Distance AB is measured from point A through O to establish point B. Line AB is perpendicular to line XY.

Another way to lay off one line perpendicular to another is to employ the 3:4:5 triangle method. Triangles with sides having lengths in proportion of 3, 4, and 5 have a right angle opposite the longest side. Consider a triangle with sides five times the 3:4:5 ratio; or 15, 20, and 25 feet. In figure 2-16, point N is set 15 feet from point A representing the 15 feet side. Arcs are struck swinging the tape with radius NO = 25 feet and radius AO = 20 feet. The intersection of the arcs fixes point O. Line AO is perpendicular to line XY. It is extended to point B as in the first method.

## Measurement Around Obstacles

When trees, buildings, ponds, or other objects obstruct a line, a parallel offset line can be substituted for a portion of the obstructed one. In figure 2-17,

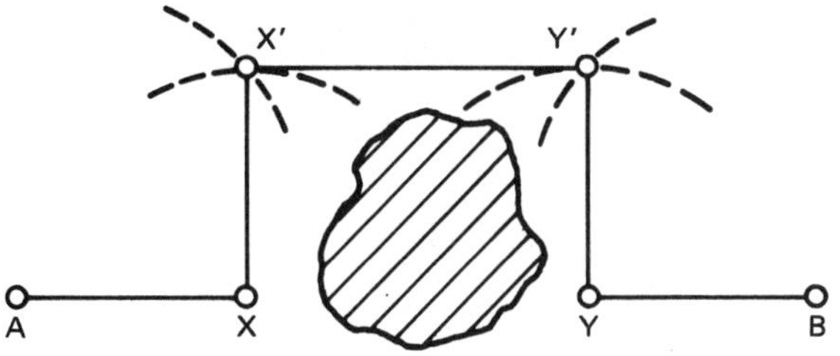


**Fig. 2-17 Parallel line measurement around an obstacle**

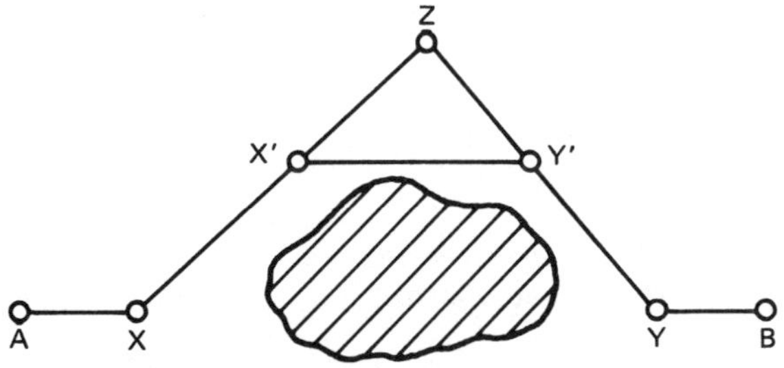


**Fig. 2-18 Triangle method around an obstacle**

an obstacle prevents measurement between points A and B. At X and Y, perpendiculars are laid off to one side. Points X′ and Y′ are set with distance XX′ and YY′ equal. Distance X′Y′ is then measured and substituted for distance XY.

Another way to determine the length of XY is shown in figure 2-18. A single point Z is located out to one side where convenient. Distance XZ is measured and point X′ is set on line XZ midway between X and Z. Distance YZ is then measured, and point Y′ is set midway between Y and Z. By the geometric principle of similar triangles, line X′Y′ is always one-half the length of line XY. Distance X′Y′ is therefore measured, and twice its length is substituted for distance XY. Extra care in measuring is advisable because errors in measuring distance X′Y′ become doubled for distance XY.

## Measurement of a Field

An entire field may be measured by taping methods. It is laid out as a network of triangles. The lengths of all three sides of each triangle are measured. When the triangles are plotted on paper, boundaries of the field are shown. The altitude of each triangle is scaled or calculated. The area is computed by the following formula:

$$A = 1/2\, bh$$

A is the area of the triangle.

b is the length of the base of the triangle.

h is the height or altitude of the triangle.

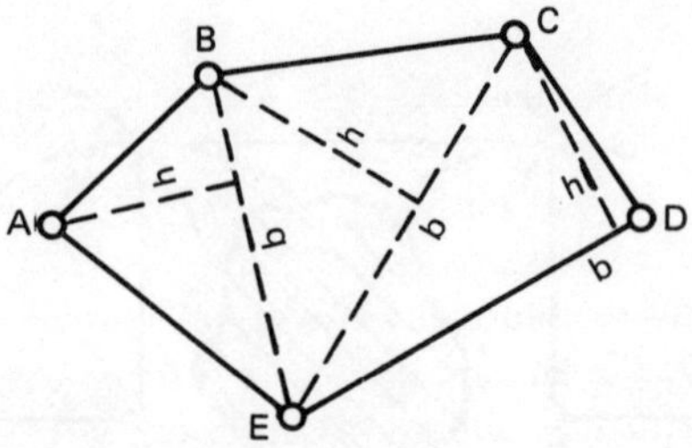


**Fig. 2-19 Triangle method of finding area for a five-sided field**

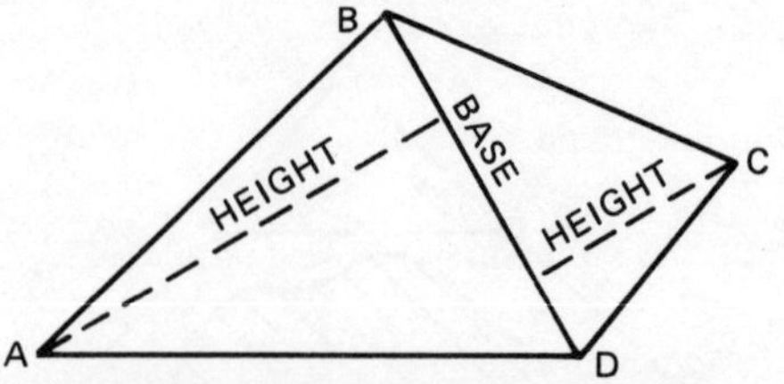


**Fig. 2-20 Triangle method of finding the area of a four-sided field**

Figure 2-19 shows how this method is laid out on a five-sided field. A four-sided field, figure 2-20, is measured by laying out only two triangles.

By tape measurement, the following dimensions are recorded:

AB = 185.00 ft.
BC = 150.00 ft.
CD = 100.00 ft.
AD = 205.00 ft.
BD = 156.00 ft. and divides the field into two triangles

From a scale drawing of the field, the heights of each triangle are approximated:

Height of triangle ABD = 177 feet
Height of triangle BCD = 93 feet

$$\text{Area of ABD} = \frac{\text{base x height}}{2} = \frac{156 \times 177}{2} = 13{,}806 \text{ square feet}$$

$$\text{Area of BCD} = \frac{\text{base x height}}{2} = \frac{156 \times 93}{2} = \underline{7{,}254} \text{ square feet}$$

$$\text{Approximate total area} = 21{,}060 \text{ square feet}$$

**SUMMARY**

Surveyors commonly measure with steel tapes calibrated in units of feet and hundredths of a foot.

A measurement with a steel tape varies with temperature, tension, and sag.

In warm weather the applied tension can compensate for sag on unsupported tape lengths up to 100 feet.

In cold weather, sufficient tension cannot be applied to compensate for tape shrinkage. Correction is necessary for precise work.

Unsupported lengths are shortened by using intermediate supports.

For ordinary work, the amount of tension being applied is usually estimated.

Become familiar with the calibration and design of each tape before using it.

The reading of feet marks on tapes is verified by noting the adjacent readings, both larger and smaller.

Tapes deflected by the wind or contact with intermediate objects produce erroneous results.

When a tape cannot be held down on a point without misalignment, plumb bobs are used.

All distances should be measured at least twice, once forward and once backward.

Where two points are at substantially different elevations, slope measurement is usually a good idea.

Slope measurements are converted to horizontal distances.

For slope measurements, either the tape's slope angle or differences in elevation are obtained.

Horizontal measurements between points on a steep slope are also made possible by breaking tape.

For long distances, a system is used for keeping track of the number of tape lengths.

One line can be laid out perpendicular to another by taping.

Taping of parallel offset lines or triangular detours offers ways of measuring distances around obstacles.

One unique application of taping methods is to determine the area of a field without using a transit.

## ACTIVITIES

For these activities, students are organized into teams. The size of each team is determined by the amount of equipment available.

1. Each team sets two points 100 feet apart by tape measurement. Each student counts paces while walking at a normal rate between these points five times. Calculate the average number of paces per hundred feet.
2. Each team sets points A and B approximately 300 feet apart by pacing. About 100 feet from each point, and in line with them, set intermediate points X and Y. These intermediate points are lined in by eye. Assume an obstacle prevents the direct measurement between X and Y. Using a tape, determine the distance between them by the parallel line method, and the similar triangle method. Measure the actual length of line XY and compare the results.

3. Each team sets a stake at each corner of a four-sided field with sides 100 to 150 feet long. Measure the length of a diagonal between the two opposite corners. With that diagonal as a base, erect and measure perpendiculars to the other two corners. Compute the area of the field to the nearest hundredth of an acre. Make the measurements to the nearest hundredth of a foot.
4. Use the same field staked for activity 3, ignoring the diagonal used for that problem. Measure the other diagonal and the lengths of all four sides. Plot to a scale of one inch equals 10 feet. Scale the altitudes of the resulting two triangles and calculate the area of the field. Compare this answer with the answer for activity 3.
5. Each team sets two points several hundred feet apart on a slope steep enough to require breaking tape. Measure the distance between them using both horizontal taping and slope taping. Convert the slope measurement to horizontal and compare the results.
6. Each team measures a course several hundred feet long in several segments. Course AXYB of activity 2 is suitable. This time, each member pulls at prescribed tensions as registered by tension handles. From A to B, pull each segment at a tension of 10 pounds. Returning from B to A, pull each segment at a tension of 20 pounds. Observe the sags. Read distances to the nearest hundredth of a foot. Note the effect of different pulls on the measurement. Compare the results of different team members. Students should then measure again at other tensions of their choice.

## REVIEW QUESTIONS

### A. Multiple Choice

1. A 100-foot steel tape is usually calibrated at
   a. 68°F, 15-pound pull, supported at quarter points.
   b. 68°F, 10-pound pull, fully supported.
   c. 70°F, 10-pound pull, supported at both ends.
   d. 68°F, 10-pound pull, supported at both ends.
2. The two factors that cause a tape to stretch are
   a. sagging between supports.
   b. rise of temperature.
   c. drop of temperature.
   d. application of tension.
3. Assume that a 100-foot steel tape has a basic pull of 10 pounds. Additional pounds equal to 3000 times the square inches of its cross section are added to the pull. Under this tension, the tape
   a. breaks.
   b. stretches to its maximum safe limit.
   c. stretches 0.01 feet.
   d. is not changed.

## B. Short Answer

4. Two students are discussing whether a 15°F rise in temperature or a 15-pound increase in pull has the most effect on a medium-weight steel tape. What is the relationship?

5. What two factors cause a tape reading to be greater than the actual distance being measured?

6. How much sag causes an error of 0.01 feet when a 100-foot tape is supported only at its ends?

7. How much deflection sideways at the center of a 100-foot tape causes an error of 0.01 feet?

8. How many feet out of level must a 100-foot tape be to cause an error of 0.01 feet?

9. What is it called when a distance is measured by accumulating the lengths of short horizontal segments on a tape?

## C. Problems

10. A right triangle has two perpendicular sides of 24 feet and 32 feet. How long is the third side?

11. The differences in elevation and slope distances for six intermediate measurements between points A and B are:

| Difference in Elevation feet | Slope Distance feet |
|---|---|
| 4.1 | 60.00 |
| 4.7 | 70.00 |
| 4.4 | 50.00 |
| 5.2 | 40.00 |
| 4.9 | 60.00 |
| 3.2 | 41.65 |

What is the horizontal distance from A to B?

12. The slope distances and angles for segments of line C to D are:

| Slope Distance feet | Slope Angle $\theta$ degrees |
|---|---|
| 100.00 | 3 1/2 |
| 40.00 | 7 |
| 60.00 | 4 |
| 80.00 | 8 1/2 |
| 100.00 | 2 |

What is the horizontal distance from C to D?

13. What pull is required to avoid correction at 68°F for unsupported tape length of 38 feet? For 84 feet? Answer to the nearest pound using values indicated in figure 2-2(A).

14. What pull is required to avoid correction at 55°F for an unsupported tape length of 38 feet? At 100°F for 84 feet? Answer to the nearest pound using values indicated in figure 2-2(B).

15. A medium-weight steel tape measurement of 66.67 feet is obtained when the temperature is 23°F. There are no intermediate supports. The pull on the tape is 24 pounds to compensate for sag. No attempt is made to increase the tension to compensate even partially for tape shrinkage due to temperature. What correction should be made for temperature? What is the corrected length?

16. A medium-weight steel tape measurement of 100.00 feet is obtained when the temperature is 98°F. There are no intermediate supports. The pull on the tape is 25 pounds. What correction should be made? What is the correct length?

17. A team is laying off a distance for precisely 100.00 feet with a medium-weight tape. The temperature is 80°F. To avoid correction they know from figure 2-2(B) that for an unsupported length of 100.00 feet they should apply how many pounds of tension? Not wishing to pull the tape that hard, they place an intermediate support at the midpoint. What tension do they need to avoid correction? Give the answers to the nearest pound.

18. Continuing question 17, the team returns the next day to check their measured distance. They again place an intermediate support at the midpoint. The temperature is now 59°F. What tension is needed to avoid correction?

19. In figure 2-19, assume the following dimensions in feet:

| Triangle | Base b | Altitude h |
|---|---|---|
| ABE | 463 | 260 |
| BCE | 580 | 330 |
| CDE | 530 | 295 |

What is the area of the field to the nearest tenth of an acre?

# CHAPTER 3

# Leveling

## OBJECTIVES

After studying this chapter, the student will be able to:

- measure vertical distances as differences in elevation.
- perform differential leveling operation, notekeeping, and calculations using conventional methods.
- perform trigonometric leveling.
- exercise the proper care and manipulation of surveying equipment used for leveling.

## METHODS

Almost everyone has measured vertical distances between two points in the home or in school. Such measurements are frequently made by holding a ruler vertically between two points. The interval between the ruler readings opposite the points is the vertical distance. That kind of direct measurement works well when both points are in the same vertical line. Surveyors can seldom use this method in plane surveying. Most of the points with which they are concerned are not directly one above the other. Also, distances often exceed ordinary ruler lengths.

A surveyor's instrument enables the operator to look out in any direction at the same level. Level lines of sight in all directions are considered to be in the same horizontal plane. Vertical distances are measured between that plane and any other point above or below it. Suppose a level line of sight is fixed in a plane 3.50 feet above point A, figure 3-1, page 30. Suppose the line of sight is 8.00 feet above point B. Then point A is 4.50 feet above point B. From a single location of the instrument, the surveyor can measure vertical distances to many points. The vertical distance between any two of those points is the difference between the measurements to them.

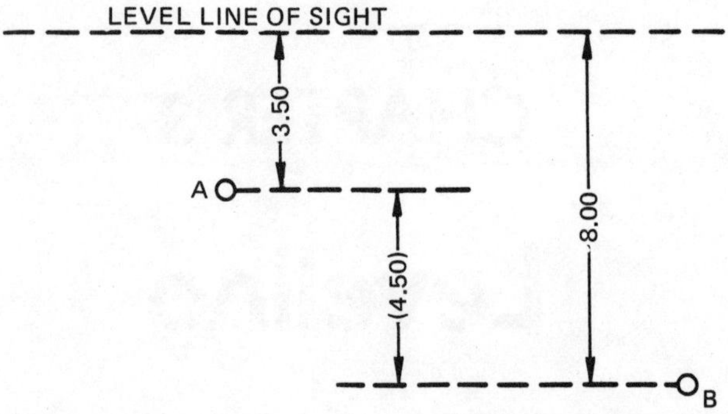


**Fig. 3-1 Measuring vertical distance by reference to a horizontal plane**

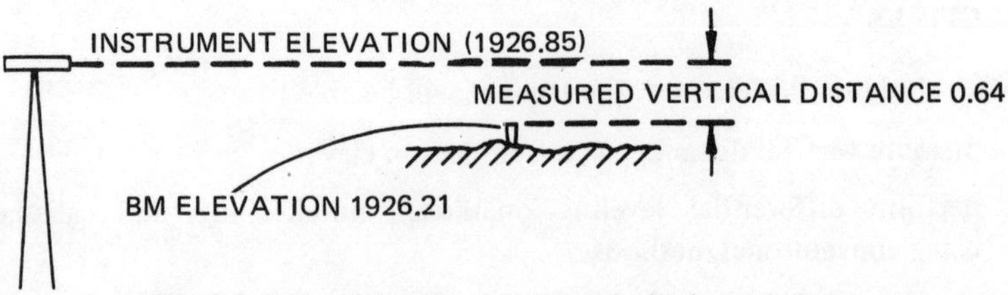


**Fig. 3-2 Determining the elevation of a horizontal plane**

## ELEVATIONS

The height of any point above mean sea level is called its *elevation.* A point 1933.21 feet above mean sea level is said to have an elevation of 1926.21 feet. Throughout the United States monuments called *bench marks* (BM) have been set at known elevations.

A surveyor sets up the instrument slightly higher than a bench mark. The vertical distance between it and the horizontal plane of the instrument is measured. The instrument's elevation is the bench mark's elevation plus the measured vertical distance, figure 3-2. Measurements are then made vertically downward from the instrument's horizontal plane to any point of interest. The elevation of these points is the instrument's elevation minus the measured vertical distances. The vertical distance between any two points is their difference in elevation, figure 3-3.

Bench marks are not always conveniently located in the immediate vicinity of a survey. In that case, a surveyor sometimes assumes some convenient point to be the zero reference level. The reference point that is said to be at an elevation of 0.00 feet is called the *datum.*

## CURVATURE AND REFRACTION

The curvature of the earth affects calculations of differences in elevation of points widely separated horizontally. At sea, figure 3-4, a horizontal plane

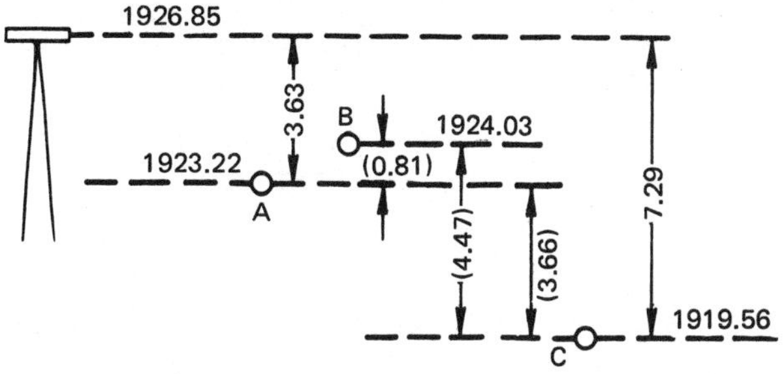


**Fig. 3-3 Vertical distances as differences in elevation**

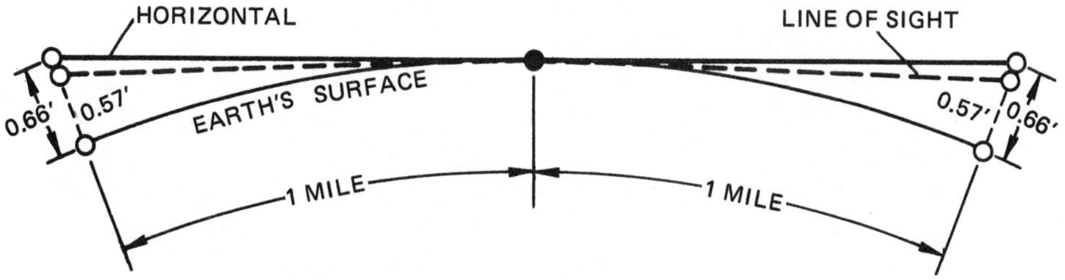


**Fig. 3-4 Divergence of horizontal lines of sight from earth's surface**

extending outward from any point of observation rises above sea level. At points one mile away, it reaches an elevation 0.66 feet higher than the point of observation. However, the atmosphere acts like a prism and bends the lines of sight downward, making the point appear to reach only 0.57 feet. At 1000 feet it appears to reach 0.021 feet, and at 300 feet only 0.002 feet. Effects of the earth's curvature can usually be ignored for points only a few hundred feet apart. Errors due to the curvature of the earth and refraction do not become significant at such short distances. It is also possible for the surveyor to eliminate the error created by the earth's curvature and refraction by setting up the instrument halfway between the two points to be leveled. In taking the readings on two points in this manner, the errors cancel themselves out. Formulas for calculating their effect at any distance are given in Appendix C.

## LEVELING EQUIPMENT

Basic leveling equipment consists of instruments used for establishing a level line of sight and a level rod.

For rough measurement, a hand or abney level is used as the instrument. These instruments are short tubes 5 to 8 inches long upon which are mounted level vials. Looking through the tube, one sees a horizontal crosswire and an image of the level vial's bubble. The line of sight is level when the cross-wire lines up with the center of the bubble, figure 3-5, page 32. An object then viewed through the tube is at the same level as the surveyor's eye. An *abney level*

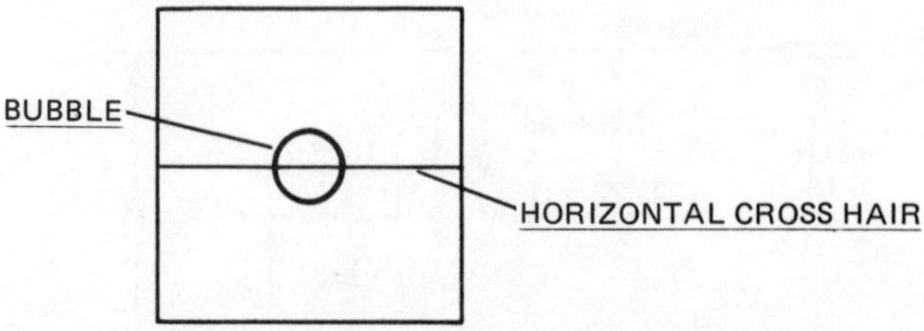


**Fig. 3-5 Lineup of hand level's horizontal cross hair and bubble**

features a level vial mounted on the axis of a rotating graduated circular arc. It allows this type of level to make rough slope measurements. Hand and abney levels have little or no magnification of the field of view. Therefore, they are used only for sighting nearby points.

For more precise work, and for points further away, a level or a transit is used, figure 3-6. Both have telescopes mounted on tripods. A level is designed specifically for leveling. Levels are available in models which give more precise results and are handier to use than transits. The telescope of a level is brought easily into a level position. Inside the telescope a horizontal and a vertical cross hair intersect at the center of the tube. Their intersection defines a point on the telescope's axis. With a level set up in its proper operating position, a surveyor looks through the telescope. The eyepiece is adjusted to get a sharp image of the cross hairs. The objective lens is adjusted to obtain a sharp image

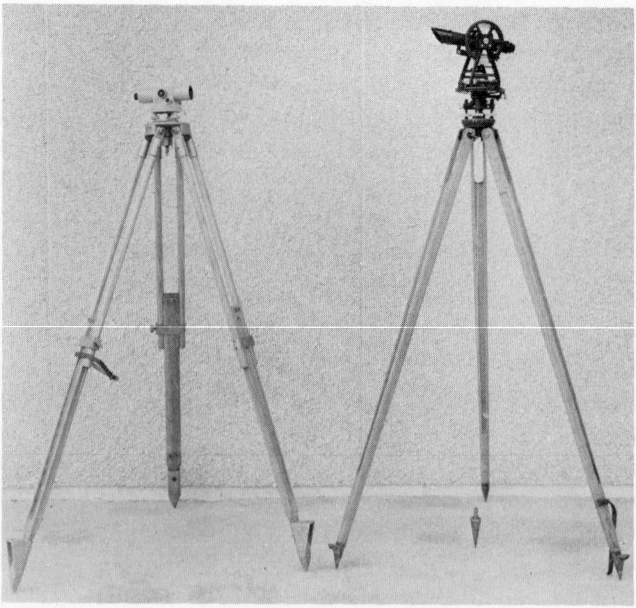

**Fig. 3-6 Level and transit**

Fig. 3-7 Philadelphia level rod

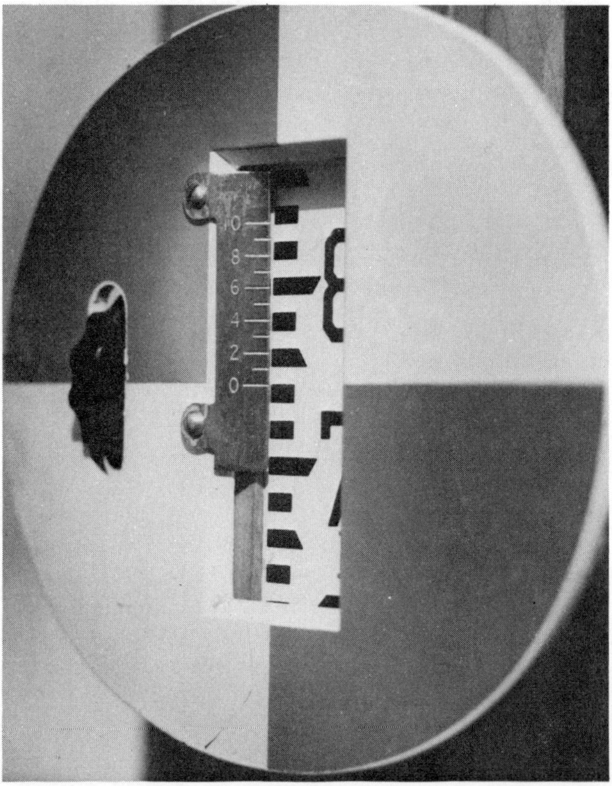


Fig. 3-8 Level rod target

of an object. Objects seen in line with the horizontal cross hair are at the same level as the axis. A transit functions in the same manner when used as a level. However, since it is built to perform other functions also, it is more difficult to use. Anyone who has learned to use a transit for leveling should also be able to handle a level.

A *level rod* is a graduated rod with zero graduation at the bottom. It is held vertically on a point. The surveyor reads the graduation in line with the horizontal cross hair of the instrument. The most popular style is the *Philadelphia rod,* figure 3-7, page 33. A Philadelphia rod is a rectangular wooden rod made in two sliding sections held together by two brass sleeves. It can be used as a short rod 7.1 feet in length. It can also be extended to its full 13.1 feet length and used as a high rod. The scale is graduated to read directly to the nearest hundredth of a foot.

Under certain conditions, readings are made easier by using a target. A *target* is a disc which slides along grooves in the rod. It can be clamped at any position, figure 3-8, page 33. Horizontal and vertical lines through its center are easy to see because the alternate quadrants are painted red and white.

Targets are useful when rod readings are made unclear by weather, foliage, or long distance. A target also permits readings to thousandths of a foot by a vernier mounted on the target. A *vernier* is a short auxiliary scale that slides along the divisions of the main graduated scale. It provides measured values for any fractional part of one of the main scale units. Target verniers have an auxiliary scale graduated to ten equal divisions within a length of 0.09 feet.

In figure 3-9, the vernier scale's zero line is opposite the 1.74 reading on the main scale. This means that the reading is also indicated to the nearest thousandth of a foot, 1.740. Note that both the zero line and the tenth line on the vernier are opposite main scale lines. Also notice that the other nine graduated lines on the vernier do not match main scale lines.

**Fig. 3-9 Position of vernier for reading of 1.740**

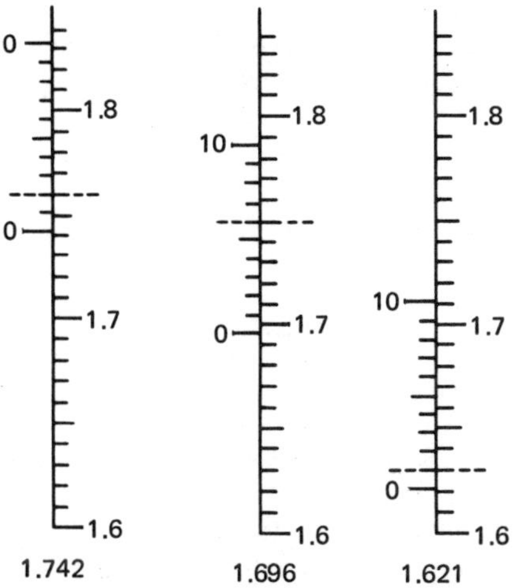


**Fig. 3-10 Several positions of vernier**

Consider the effect of raising the target 0.001 feet to a height of 1.741. The graduated line on the vernier marked 1 is opposite the level rod's 1.75 line. No other line on the vernier matches a line marking a hundredth on the level rod. Now consider the effect of raising the target another 0.001 feet to a height of 1.742, figure 3-10. The graduated line on the vernier marked 2 is opposite the level rod's 1.76 line. No other line on the vernier matches a line on the level rod.

The numbered matching vernier line is the number of thousandths of a foot added to the number of feet, tenths, and hundredths of a foot read on the level rod itself. Figure 3-10 also shows the position of a vernier for several different readings.

## FIELD USE OF LEVELING EQUIPMENT

A number of practices are similar in the proper handling of levels and transits. Before using either piece of equipment, survey crew members should become familiar with their setup and use.

Anchor the tripod legs at points where they form angles of about 60 degrees with a level ground surface, figure 3-11, page 36. Place the legs so that the tripod head is about level. On a hard surface, the legs must be kept from spreading and letting the instrument fall. Take advantage of cracks and irregularities when locating the pointed shoes of the legs on a hard surface. If necessary, tighten the wing nuts at the upper ends of the legs to increase the friction between the tripod legs and the head.

When mounting the instrument on a tripod, rotate the footplate of the instrument counterclockwise until it clicks as the threads mesh. Then rotate the footplate of the instrument clockwise until it seats firmly; but avoid overtightening.

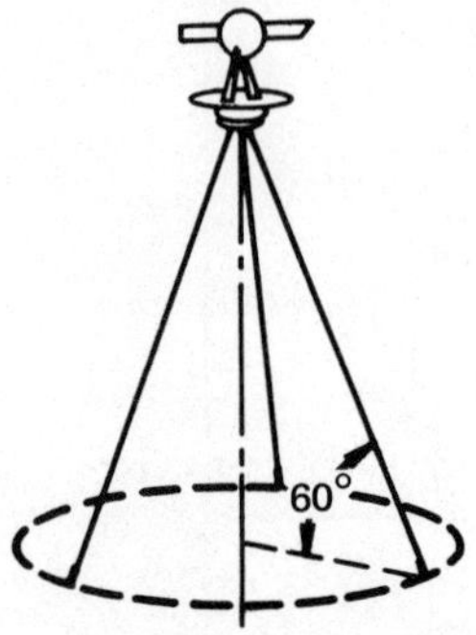


**Fig. 3-11 Proper angle of tripod legs with level ground**

After mounting the instrument on the tripod, lightly tighten the clamp screws to prevent the parts from moving freely. Parts should be able to move with a slight drag if the instrument is accidentally struck. Carry the mounted instrument over a shoulder only when the path is clear of obstacles. Carry it under the arm with the instrument forward when passing through doorways or brush. Before crossing a fence, pass the instrument over the fence and set it up on its three legs. Never carry a mounted instrument in a vehicle. The instrument should be returned to its box when being transported in vehicles and the box should be braced to prevent overturning.

Never leave the instrument unattended. It may be accidently overturned.

Keep the wing nuts at the top of the tripod legs in good adjustment. The best adjustment is usually achieved when the legs barely fall from a horizontal position by their own weight.

When setting an instrument up over a point, face uphill if the ground is not level. Straddle the point over which the instrument is to be located. Stick one leg of the tripod into the ground about two feet uphill beyond the point. Step backward and place one hand on each of the other two legs. Spread them sideways and stick them into the ground where necessary to keep the tripod head approximately level. Adjust its position by picking the tripod up and resetting it without disturbing the angles to which the legs are set. Otherwise the tripod head does not remain level. Set each tripod leg firmly in the ground by stepping on the lug on the pointed metal shoe. If more pressure is needed, push in the direction of the leg, not vertically downward.

If the instrument has four leveling screws, rotate the telescope horizontally until it is in line with two opposite screws. Manipulate these screws to bring the bubble in the level vial to its approximate center, figure 3-12. At this stage, leveling screws should support the instrument on its base just enough to prevent rocking. Rotate the telescope 90 degrees to align it with the other pair of leveling screws. Manipulate those screws to level the bubble. Repeat this process several times until the bubbles in the level vials are at the exact center for both positions. While making final adjustments, tighten the leveling screws to make them bear firmly on the base, but avoid overtightening.

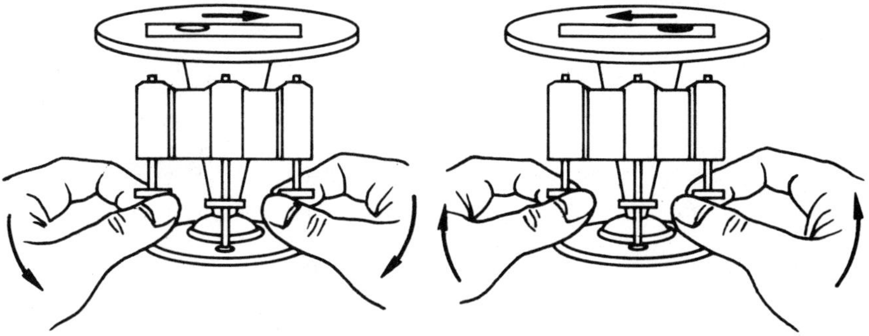

**Fig. 3-12 Manipulation of the leveling screws**

When a transit is used instead of a level, the plate bubbles are centered without rotating the telescope. However, it is also necessary to bring the level bubble on the telescope to its exact center. This requires tightening the telescope clamp screw when the bubble approaches the vial's center, and then operating the telescope tangent screw.

If the instrument has three leveling screws, rotate the telescope horizontally until it is parallel with two of these screws. Manipulate the screws to bring the bubble to its approximate center for this position. Then rotate the telescope 90 degrees and manipulate the third leveling screw to again center the bubble. Repeat this process several times until the bubble is at the exact center for both positions.

After an instrument is leveled, the line of sight is easily disturbed by unnecessary handling. Once set up and leveled, an instrument should only be touched when necessary for its operation. Avoid stepping closer than one foot to any tripod leg. The surveyor's hands should never rest on the tripod legs or the instrument.

Focus the eyepiece to gain a sharp black image of the cross hairs. Because of differences in individuals' eyes, each operator of the instrument requires individual settings of the focus adjustment ring on the eyepiece.

Aim the telescope at the level rod. Focus the objective lens to get a sharp image of the level rod's graduations. Make an approximate preliminary reading of the level rod while focusing.

Reexamine the bubble in the level vial attached to the telescope. If it has drifted, return it to the exact center before taking each final reading of the level rod. When using a transit, disregard any drift in the plate bubbles. Concentrate only on keeping the telescope bubble in the exact center for each level rod reading.

Make certain that the level rod is plumb for final readings. If the level rod and vertical cross hair are not parallel, the level rod is leaning sideways, figure 3-13, page 38. To check if the level rod is leaning toward or away from instrument, request that the rod be *waved.* This means that the rod is leaned

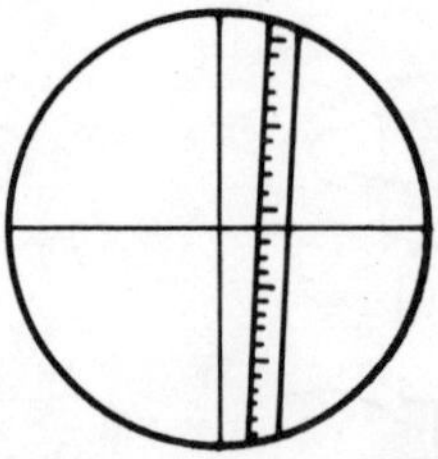

**Fig. 3-13 Level rod leaning sideways is not parallel with vertical cross hair**

slightly toward the instrument, and then slightly backward, figure 3-14. The rod is moved slowly back and forth until signaled that a reading has been made. As the rod waves, the lowest rod reading as viewed through the telescope occurs when the rod is plumb. Reexamine the level bubble on the telescope after the reading. For precise work, the whole process is repeated if a slight bubble drift requires another centering adjustment.

Methods described in the last two paragraphs are relaxed when work of low precision is performed. In some work, rod readings are needed only to the nearest tenth of a foot.

The level rod must be handled with as great care as the transit or level to assure accurate measurements. Some important points to note while handling the rod are mentioned below.

Tighten the clamp whenever the rod is in use, whether at the 7.1 or the 13.1 foot length. Be sure that the rod is fully extended when using it as a high rod. Listen for the snapping of the retaining spring for evidence that it is fully extended. Loosen the clamp when storing the rod to prevent warping that results from changes in temperature and humidity.

Hold the rod plumb. A rod level, figure 3-15, fitted to the rear side face of the level rod is helpful in plumbing. A good technique is to stand behind the rod facing the instrument, and carefully balance the rod. In the absence of wind, acceptable results are often obtained by merely balancing the rod. When winds prevent this, dependence is placed on adjustments requested by the operator of the instrument who verifies that the rod is plumb.

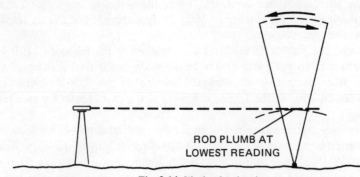

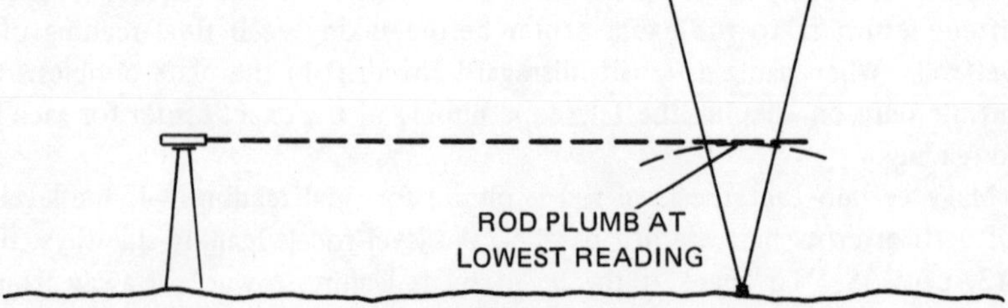


**Fig. 3-14 Waving level rod**

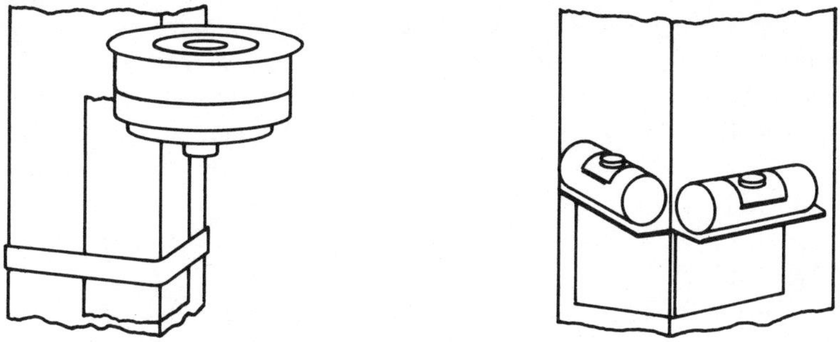

**Fig. 3-15 Rod level**

Keep the bottom of the brass plate at the foot of the rod clean. For precise readings, wipe off the plate before mounting the rod on a point.

When using a target with a high rod, set the zero line of the target vernier exactly on the 7.00 line. Verify the adjustment of the vernier on the back of the level rod by checking to make sure it agrees with the target vernier.

## FIELD OPERATIONS

The instrument is set up no higher than a high rod length above a bench mark. The ideal height depends upon whether points of desired elevation are above or below the bench mark. If above the bench mark, the anticipated rod reading should be within a foot of the top. If below the bench mark, the anticipated rod reading should be within a foot of the bottom.

The *height of instrument* (HI) is its height above the selected datum. The surveyor calculates the height of the instrument after reading the level rod held on the bench mark. Rod readings taken on points of known elevation are called backsights (BS) or plus sights (+). The height of the instrument is the elevation of the bench mark added to the backsight. In figure 3-16, the height of the instrument is the elevation of the bench mark, 1604.91 feet, added to the backsight rod reading of 0.83 feet. The height of the instrument is 1605.74 feet.

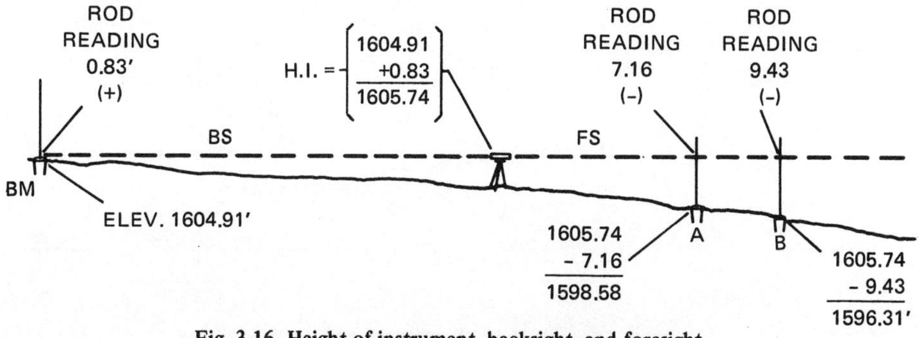


**Fig. 3-16 Height of instrument, backsight, and foresight**

The surveyor calculates the elevation of any other point after reading the rod held on that point. Rod readings taken on points of unknown elevation are called *foresights* (FS) or *minus sights* (-). The elevation of any point is the height of the instrument minus the foresight. In figure 3-16, the elevation of point A is found by subtracting the foresight rod reading of 7.16 feet from the height of the instrument, 1605.74 feet. This puts point A at an elevation of 1598.58 feet. Point B is found by the same method. Its foresight of 9.43 feet is subtracted from the height of the instrument, 1605.74 feet, to get its elevation of 1596.31 feet.

Differences in elevation between a bench mark and points where elevations are needed often exceed the length of the rod. To determine the elevation of these points, a *turning point* (TP) approximately a rod length above or below the bench mark is selected for the rod. Its elevation is determined by subtracting a foresight taken on it from the height of the instrument. The instrument is then moved to where it is used to take readings over another span of the rod. A backsight is taken on the turning point and the height of the instrument is calculated for the new instrument location. A sufficient number of turning points and instrument setups are made to determine the elevations wherever needed, figure 3-17.

The turning point must be a firm surface with an elevated area smaller than the bottom of the rod. It is important that there be no vertical movement of the turning point between the foresight and the backsight readings. It should be easy to place the rod on the turning point's highest spot for each reading. Where suitable existing objects cannot be found for a turning point, a stake or spike is driven into the ground. Care should be taken that marking crayons, if used on the turning point's surface, do not cause error. A *guard stake* is often driven beside the turning point and marked with its identifying number.

It is a good practice to equalize the distance between the instrument and the rod when taking backsights and foresights. This reduces the error caused by faulty instrument adjustment. For example, an instrument reads 0.01 foot

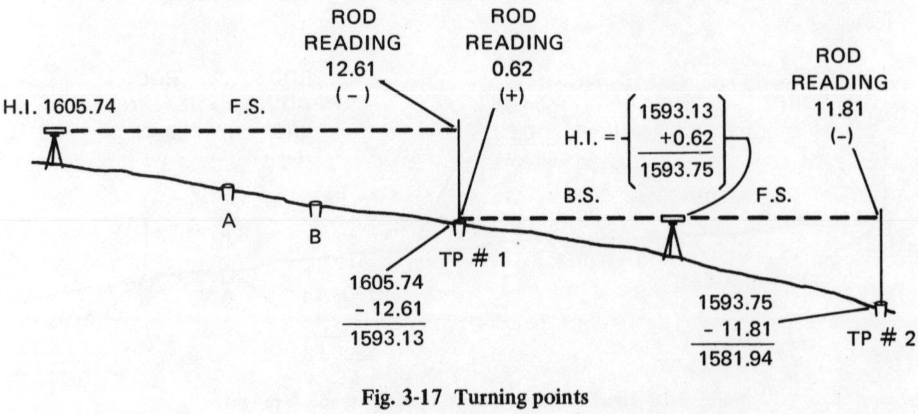


**Fig. 3-17 Turning points**

too high at a distance of 200 feet. It produces the correct difference in elevation between two points if each is 200 feet away. A 0.01 foot error in backsight compensates for a 0.01 foot error in foresight. However, a reading for a backsight only 100 feet away is only 0.005 feet too high. With a foresight 200 feet away reading 0.01 feet too high, a 0.005 foot error results. Errors resulting from neglecting the earth's curvature are also eliminated by equalizing backsight and foresight distances.

Determining differences in elevation by a series of turning points and instrument setups is called *differential leveling.* It is customary to check readings and calculations by also measuring the difference in elevation on a return trip. In returning to the bench mark, a different set of turning points is used. The round trip constitutes a circuit. Maximum acceptable error for ordinary leveling is given by the formula:

$$\text{Error} = \pm 0.10 \text{ ft.} \sqrt{\text{circuit distance in miles}}$$

For example, a surveyor is measuring differences in elevation of two points that are 2640 feet apart. Differential leveling is performed between the points in both directions. The total circuit distance covered is 5280 feet, or one mile. Using the formula, the error cannot exceed 0.10 feet between the height of the starting point at the beginning of the circuit, and its measured height at the end of the circuit.

The error is distributed by proportionate adjustment of elevations of all intermediate points. A turning point one-fourth of the way around the circuit is adjusted to compensate for one-fourth of the error. A turning point half way around is adjusted to compensate for one-half of the error.

The form of notekeeping for differential leveling that is almost universally accepted is shown in figure 3-18, page 42. Below the last entries, totals are shown for all backsight entries (+) and all foresight entries (-). The backsight sum (+) less the foresight sum (-) equals the difference between elevations of the bench mark and the last point listed. These sums are used as a convenient overall check to reveal the existence of error in intervening calculations. For a closed circuit, the difference between the backsight and foresight totals is the *error of closure.*

## TRIGONOMETRIC LEVELING

It is sometimes impossible or impractical to hold a level rod where an elevation is desired. Tops of poles, water tanks, radio towers, and tall free-standing chimneys are examples. In these cases, trigonometric leveling is usually the most practical method of measuring the vertical distance.

In the right triangle in figure 3-19, page 43, the altitude is side a, the base is side b. The acute angle at the base is angle A. The altitude divided by the base is the tangent (tan) of angle A. This relationship is expressed by the formula:

$$\tan A = a/b$$

Establishing T.B.M. - Project 106

| Sta. | B.S.(+) | H.I. | F.S.(−) | Elev. | 36<br>9-29-73 |
|---|---|---|---|---|---|
| B.M. | 8.73 | 1934.41 | | 1925.68 | T.V.A. HMK 48 |
| T.P. 1 | 6.89 | " 41.15 | 0.15 | " 34.26 | ⊼ McGraw |
| T.P. 2 | 12.82 | " 52.99 | 0.98 | " 40.17 | ϕ Nix |
| T.P. 3 | 9.26 | " 62.09 | 0.16 | " 52.83 | Notes Leek |
| T.P. 4 | 6.64 | " 68.36 | 0.37 | " 61.72 | |
| T.P. 5 | 0.90 | " 56.35 | 12.91 | " 55.45 | Cool Cloudy |
| T.P. 6 | 1.63 | " 46.05 | 11.93 | " 44.42 | |
| T.P. 7 | 10.76 | " 56.43 | 0.38 | " 45.67 | |
| T.P. 8 | 11.25 | " 66.42 | 1.26 | " 55.17 | |
| T.P. 9 | 13.03 | " 79.33 | 0.12 | " 66.30 | |
| T.B.M. 265 | | | 0.22 | 1979.11 | Highest Point on Stump 40' L. Sta. 26+50 |
| (1925.68 + 81.91 | | − | 28.48 | = 1979.11) | |
| | | | | | |
| T.B.M. 265 | 0.87 | 79.98 | | 1979.11 | 9-30-73 |
| T.P. 10 | 0.64 | 68.20 | 12.42 | " 67.56 | ⊼ Kelly |
| T.P. 11 | 0.43 | 56.62 | 12.01 | " 56.19 | ϕ Pullium |
| T.P. 12 | 5.42 | 50.18 | 11.86 | " 44.76 | Notes Stalcup |
| T.P. 13 | 12.34 | 55.39 | 7.13 | " 43.05 | |
| T.P. 14 | 9.69 | 63.87 | 1.21 | " 54.18 | |
| T.P. 15 | 2.13 | 64.84 | 1.16 | " 62.71 | |
| T.P. 16 | 0.86 | 53.37 | 12.33 | " 52.51 | |
| T.P. 17 | 1.03 | 41.93 | 12.47 | " 40.90 | |
| T.P. 18 | 3.27 | 36.79 | 8.41 | " 33.52 | T.V.A. HMK 48 |
| B.M. | | | 11.15 | 1925.64 | |
| (1979.11 + 36.68 | | − | 90.15 | = 1925.64) | Adjusted Elev. T.B.M. 265 = 1979.11 + $\left(\frac{0.04}{2}\right)$ = 1979.13 |
| page totals | 118.59 | − | 118.63 | = − 0.04 | |

**Fig. 3-18 Conventional form of notes for differential leveling**

Assume side a is 40.00 feet high and side b is 80.00 feet long. The tangent of angle A is then 0.50000. Appendix D lists the values of tangents for all angles from 0 to 90 degrees. The tangent table shows that an angle of 26°33′ has a tangent of 0.49967, and an angle of 26°34′ has a tangent of 0.50004. The angle chosen with a tangent nearest the value of 0.50000 is 26°34′.

Consider a situation where the altitude of the triangle is difficult to measure. By measuring the length of its base and the angle A, the altitude can be computed by expressing the previous formula in terms of:

$$a = b(\tan A)$$

This means that the altitude is equal to the length of the base multiplied by the tangent of angle A. Assume that with a transit set up at A, angle A is 26°34′.

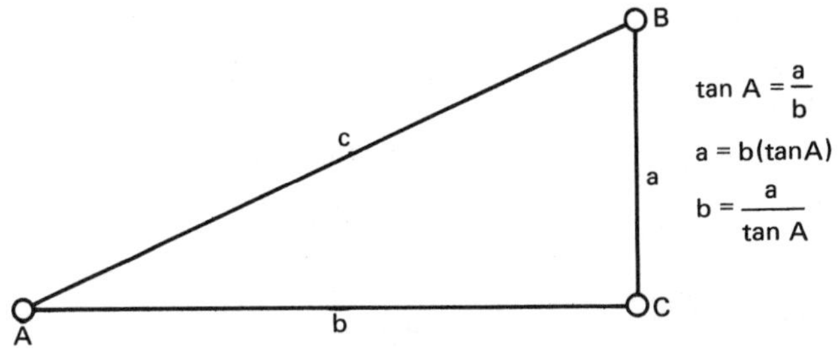


**Fig. 3-19 Tangent functions of angle A**

A tape measurement from point A shows that b is 80.00 feet. Multiplying the value for the tangent of 26°34′ (0.50004) by 80.00 gives the answer 40.00 feet for side a, the altitude.

For determining the heights of flag poles, transits are set up at any convenient distance. Where possible, the height of the instrument is at a higher elevation than the base of the flagpole, figure 3-20. The telescope is leveled and aimed at the flagpole. Point C is marked where the level line of sight strikes the pole. Measuring with the level rod, the difference in elevation between the ground (G) and point (C) is found. The instrument is elevated, bringing the telescope's horizontal cross hair to the level of the top of the pole. As the horizontal cross hair approaches this level, the clamp screw is tightened. The telescope is brought into the exact position by operating the tangent screw. Vertical angle A is read by observing the scale on the rotating vertical circle and the vertical vernier.

The scale on the vertical rotating circle reads zero when the instrument is level. On most transits the scale permits readings of vertical angles to one-half of a degree. The vertical vernier is in a fixed position. It has 30 divisions which equal 29 divisions on the rotating vertical circle. Angles read on the rotating circle appear opposite or just short of the vernier's zero line. Vertical angle A is the sum of that reading and the minute reading on the vernier scale.

Most transits have a double vernier readable in both directions from the zero index line. The right half of this vernier is read when the telescope is elevated

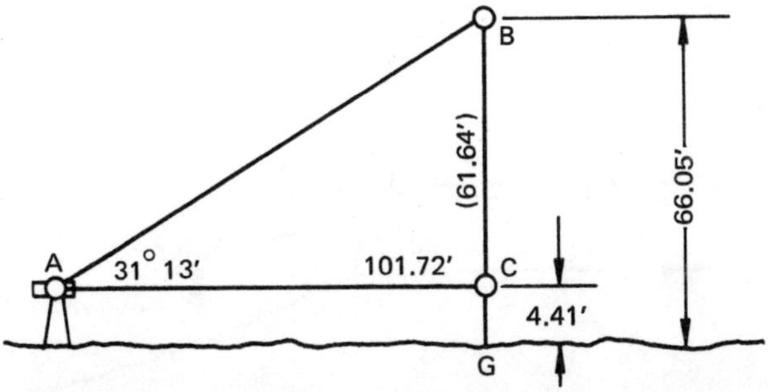


**Fig. 3-20 Determining the height of a flagpole on level ground**

above a level line. The left half is read when it is depressed below a level line.

The horizontal distance between the instrument and the center of the pole is read with a tape. The tape is held at the level at which point C is marked. The person holding the tape stands at the side of the pole as a measurement is made to the point on the pole's surface. That point is at right angles to a line between the instrument and the center of the pole. The other end of the tape is held close to the end of the hub of the telescope's transverse axis. Be careful that nothing comes in contact with the instrument.

Using figure 3-20 as an example, assume the vertical angle measured at point A is 31°13′. Assume that the horizontal distance b from the instrument to the center of the pole is 101.72 feet. The height of the pole above a level line at instrument height is:

$$a = b\,(\tan A)$$

$$a = 101.72\,(\tan 31^\circ 13')$$

$$a = 101.72 \times 0.60602$$

$$a = 61.64 \text{ feet}$$

Assume the level rod reading for distance GC is 4.41 feet. The top of the pole is then 61.64 + 4.41, or 66.05 feet above the ground from its base.

If the flagpole or other object being measured is not plumb, a modified procedure is followed. The horizontal measurement is taken to a point directly under the top of the pole. To test vertical alignment, visualize a line from the object approximately 90 degrees from line AC, figure 3-21. Set up the transit at a convenient point (D) on this line. Sight the telescope at the top of the pole, clamp the horizontal motions, and depress the telescope. If point C is out of line, another point (C′) is set for the horizontal measurement b. The height is then found using the method demonstrated in figure 3-20.

Another method is followed when a flagpole or other object is located on a hilltop. This method is different because a level line of sight from an instrument on the hillside strikes the ground before reaching the flagpole. In this case, trigonometric functions other than tangents are used.

In a right triangle, figure 3-22, diagonal side c is called the *hypotenuse.* When its length and one acute angle are known, both altitude and the base length can

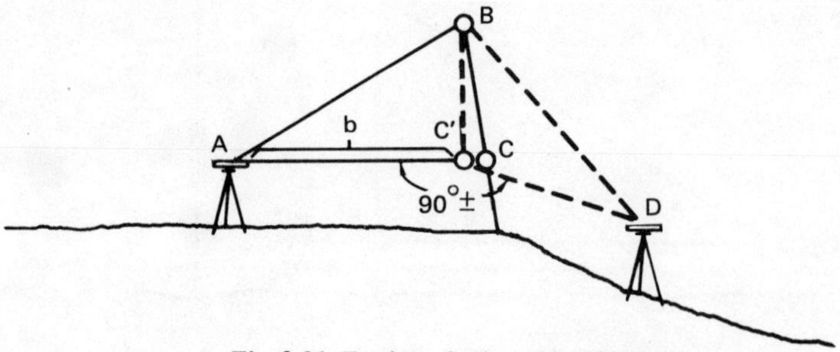


**Fig. 3-21 Testing whether object is plumb**

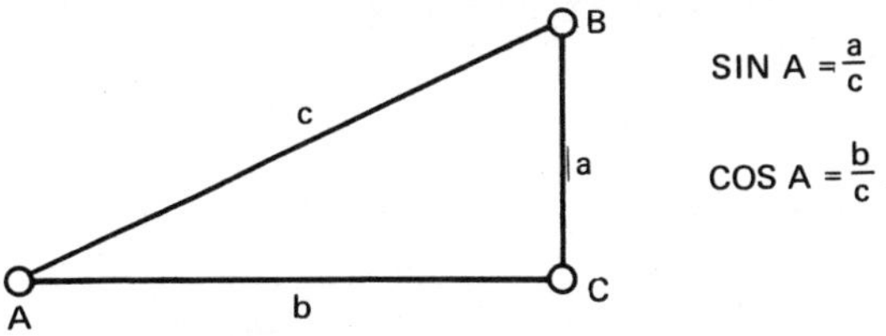


**Fig. 3-22 Sine and cosine functions of angle A**

be calculated. In trigonometry, relationships between sides a, b, and c are expressed as functions of the acute angles. These relationships for angle A in figure 3-22 are listed below. They are true only if triangle ABC is a right triangle.

$$\text{Sin A} = \frac{a}{c}$$

$$a = c\,(\sin A)$$

$$c = \frac{a}{\sin A}$$

$$\cos A = \frac{b}{c}$$

$$b = c\,(\cos A)$$

$$c = \frac{b}{\cos A}$$

Appendix D lists the value of sines and cosines for angles.

For determining the height of a pole on a hilltop, field measurements are as shown in figure 3-23. They include slope distance AB′, two vertical angles CAB′ and CAB and the height of B′ above ground.

BG, the height of the flagpole, is calculated as follows:

$$B'C = AB'\,(\sin CAB')$$
$$B'C = (80.00)\,(0.17365)$$
$$B'C = 13.89 \text{ feet}$$

$$AC = AB'\,(\cos CAB')$$
$$AC = (80.00)\,(0.98481)$$
$$AC = 78.78 \text{ feet}$$

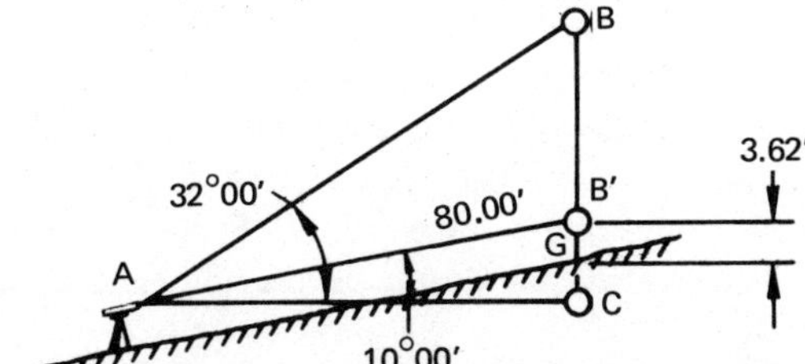


**Fig. 3-23 Determining the height of a flagpole on a hilltop**

$$AB = AC/\cos CAB$$
$$AB = 78.78/0.84805$$
$$AB = 92.90 \text{ feet}$$

$$BC = AB (\sin CAB)$$
$$BC = 92.90\ (0.52992)$$
$$BC = 49.23 \text{ feet}$$

$$BG = BC - B'C + B'G$$
$$BG = 49.23 - 13.89 + 3.62$$
$$BG = 38.96 \text{ feet}$$

## SUMMARY

Vertical distances are differences in elevation.

Vertical measurements are usually made between points that are not in the same vertical alignment.

Monuments called bench marks exist throughout the United States for elevation reference purposes.

Widely separated points with the same elevations are not in the same horizontal plane because of the earth's curvature.

Basic leveling equipment consists of an instrument for leveling and a level rod.

Close adherence to instructions for manipulating leveling equipment is necessary for measurements to be acceptable.

Bubbles in the level vials controlling level lines of sight must be in the exact center when taking readings.

Level rods must be plumb when being read.

The height of instrument is its height above the datum. The datum is usually sea level, but can be any point assumed to have zero elevation.

A backsight is a rod reading taken on a point whose elevation has been determined.

A foresight is a rod reading taken on a point whose elevation is being determined.

A firm turning point is necessary whenever a leveling instrument is moved.

The height of the instrument equals the bench mark elevation plus the backsight, or the turning point elevation plus the backsight.

The elevation of the turning point, or any other observed point, equals the height of the instrument minus the foresight.

Equalization of horizontal distances to points where a backsight and foresight are taken compensates for faulty instrument adjustment.

Differential leveling is a process of determining differences in elevation by a series of instrument setups and turning points. Such work is checked by completing a circuit. The error of closure is distributed proportionately to all intermediate points.

Trigonometric leveling allows a practical means of determining the heights of objects that are difficult to reach or inaccessible.

## ACTIVITIES

1. Each team member clamps a target on a level rod at three different settings. Other team members check the setting. Use the following settings:

| TEAM MEMBER | SETTINGS | | |
|---|---|---|---|
| 1 | 3.263 | 4.120 | 5.037 |
| 2 | 3.911 | 4.635 | 5.113 |
| 3 | 3.456 | 4.314 | 5.092 |
| 4 | 3.348 | 4.489 | 5.000 |
| 5 | 3.672 | 4.557 | 5.124 |

2. Each team is equipped with a transit or level. Each member sets up the instrument and levels it following the practices explained.
3. Each team sets up a transit or level. Team members take turns leveling it and reading a level rod to the nearest hundredth of a foot. Team members also take turns holding that level rod on a point about 100 feet away. The instrument operator takes responsibility for seeing that the level rod is plumb. After all members have had their turns, repeat the exercise reading the rod to thousandths of a foot, using the vernier of the target.
4. Each team sets up a transit. Each team member elevates the telescope to three different vertical angles. Other team members are asked to confirm the settings on the rotating vertical circle and the vertical vernier. Use the following settings:

| TEAM MEMBER | SETTINGS | | |
|---|---|---|---|
| 1 | 10°04′ | 16°41′ | 28°22′ |
| 2 | 8°11′ | 14°59′ | 23°31′ |
| 3 | 7°00′ | 11°30′ | 21°15′ |
| 4 | 9°27′ | 17°33′ | 27°57′ |
| 5 | 2°02′ | 13°43′ | 22°58′ |

5. Each team drives a stake in the ground diagonally so that one corner is higher than the others. The stake is assumed to be a BM at elevation 100.00 feet. Using a hand level or abney as an instrument, run a circuit of levels. Rod readings are to the nearest tenth of a foot. The furthest point away from the bench mark is about 100 feet. One set of turning points is used on the way out, and a different set is used on the return. Team members rotate so each gets practice with both the instrument and the rod. The error of closure is computed. Keep notes in the form shown in figure 3-18.
6. Each team begins with an assumed bench mark as in activity 5, using a transit or level. Run a circuit about one-half mile long. Rod readings are to the nearest hundredth of a foot. One set of turning points is used

for the outbound quarter mile, and a second set is used on the return. Rotate team members so each gets practice with the instrument and the rod. If error of closure exceeds the acceptable limit of ± 0.10 feet $\sqrt{\text{circuit distance in miles}}$, the circuit is rerun. Find and eliminate major errors. When error of closure is acceptable, distribute the error. Calculate the adjusted elevations at all turning points. Keep notes in the form shown in figure 3-18.

7. Each team selects a pole or other tall object suitable for an exercise in trigonometric leveling. Using trigonometric leveling as explained in the text, determine the height of the object.

## REVIEW QUESTIONS

### A. Multiple Choice

1. The height of any point above mean sea level is called its
   a. datum.
   b. height of the instrument.
   c. elevation.
   d. reference.

2. Atmospheric refraction deflects level lines of sight
   a. upward.
   b. northward.
   c. downward.
   d. southward.

3. Which of the following pieces of equipment do not have the function of establishing a level line of sight?
   a. Hand level
   b. Plumb bob
   c. Transit
   d. Level rod

4. Tripod legs are set for angles with the ground of about
   a. 30 degrees.
   b. 45 degrees.
   c. 60 degrees.
   d. 75 degrees.

5. A Philadelphia level rod when extended to its full length is approximately
   a. 7 feet long.
   b. 10 feet long.
   c. 13 feet long.
   d. 16 feet long.

6. The scale of a Philadelphia level rod is graduated to read directly to the nearest
   a. tenth of an inch.
   b. hundredth of a foot.
   c. one-eighth inch.
   d. tenth of a foot.

7. The height of the instrument equals the bench mark elevation plus the
   a. backsight.
   b. foresight.
   c. turning point.
   d. datum.

8. In plane surveying the effects of the earth's curvature are ignored for points separated by less than
   a. 200 feet.
   b. 1000 feet.
   c. one-half mile.
   d. one mile.

9. A surveyor adjusts the objective lens of a transit or level to gain a sharp image of the
   a. bubble.
   b. object.
   c. cross hairs.
   d. telescope.

10. A transit or level mounted on a tripod may be carried in a vehicle
    a. if it is well braced and protected from road shock.
    b. if it is turned sideways and held on the lap of a passenger.
    c. under no circumstances.
    d. to protect it from rain.

11. A surveyor ready to read a level rod notices that the vertical cross hair and the rod are not parallel. The most likely cause is that the
    a. telescope is not level.
    b. level rod is on the side of a hill.
    c. level rod is not plumb.
    d. metal plate on the bottom of the rod is bent.

12. The elevation of a point where a backsight is taken is
    a. higher than the height of the instrument.
    b. known.
    c. undetermined.
    d. unnecessary.

13. In a right triangle, the side opposite acute angle A divided by the hypotenuse is called the
    a. sine.
    b. cosine.
    c. tangent.
    d. datum.

**B. Short Answer**

14. Vertical distances are commonly measured by holding a ruler vertically between two points. How often can a surveyor use this method?

15. On a hillside, how are the legs of a tripod set?

16. How tight are the wing nuts at the tops of the tripod legs kept?

17. A rod reading is taken when a level rod is not plumb. How does this affect the reading?

18. How do the spaces between graduations on the adjacent main scale compare to the spaces between graduations on the vernier?

19. What is another name for a backsight?

20. In a conventional set of differential leveling notes, is the backsight on a point entered on the same line as the elevation of the point? If not, where?

21. Anyone who has learned to use a transit for leveling should be able to easily handle what other type of instrument?

22. For a closed circuit, what is the difference between the backsight totals and the foresight totals called?

23. What is helpful when rod readings are made indistinct by weather, foliage, or long distance?

24. In selecting points for instrument setups and for turning points, what can be done to minimize errors from faulty instrument adjustment and earth curvature?

## C. Problems

25. Copy and complete the following notes:

| STATION | BS | HI | FS | ELEVATION |
|---|---|---|---|---|
| BM 1 | 12.36 | | | 1916.41 |
| TP A | 11.89 | | 0.82 | |
| TP B | 12.03 | | 0.96 | |
| TP C | 6.41 | | 1.13 | |
| BM 2 | | | 2.16 | |

Total the backsights and foresights and use their sums as a check on the elevation at BM 2.

26. Copy and complete the following level notes:

| STATION | BS | HI | FS | ELEVATION |
|---|---|---|---|---|
| BM 1 | 0.64 | | | 1916.41 |
| TP 1 | 0.89 | | 12.11 | |
| TP 2 | 0.21 | | 12.85 | |
| BM 2 | 1.42 | | 9.69 | |
| TP 3 | 1.11 | | 11.30 | |
| TP 4 | 11.03 | | 10.16 | |
| TP 5 | 12.36 | | 1.24 | |
| BM 3 | 11.97 | | 0.73 | |
| TP 6 | 12.72 | | 1.09 | |
| TP 7 | 11.80 | | 0.32 | |
| BM 1 | | | 4.63 | |

Calculate and check the error of closure of the circuit and adjust the elevations of bench mark 2 and bench mark 3.

27. Points A and B are each 1000.00 feet from point C, by horizontal measurement. At point C, the vertical angle to point A is $+ 4^\circ 16'$. Also at point C, the vertical angle to point B is $+ 1^\circ 03'$. What is the difference in elevation between points A and B?

28. Indicate which of the following circuits have acceptable errors of closure for ordinary leveling:

| | Circuit length (miles) | Error of closure (feet) |
|---|---|---|
| a. | 2.5 | 0.17 |
| b. | 1.25 | 0.11 |
| c. | 0.5 | 0.07 |
| d. | 3.7 | 0.19 |
| e. | 4.0 | 0.21 |

# CHAPTER 4

# Determining Direction

## OBJECTIVES

After studying this chapter, the student will be able to:

- define the location of one point in relation to another.
- explain the behavior and limitations of magnetic compasses.
- refer to direction of lines by conventional methods.

## DIRECTION, AZIMUTH, AND BEARING

The distance between two points does not alone define the location of one point relative to another. The other necessary factor is the direction to be followed. In figure 4-1, points A, B, and C are all 100.00 feet from 0. To completely fix their location, their direction from 0 must also be known.

Surveyors define direction in terms of horizontal angles measured from north-south lines through occupied points. The horizontal angle is expressed either as an azimuth or a bearing.

An *azimuth* is a horizontal angle measured clockwise between a zero line of reference and an object. Azimuths have values between 0 and 360 degrees. The zero line of reference ordinarily extends either northward or southward from the occupied point.

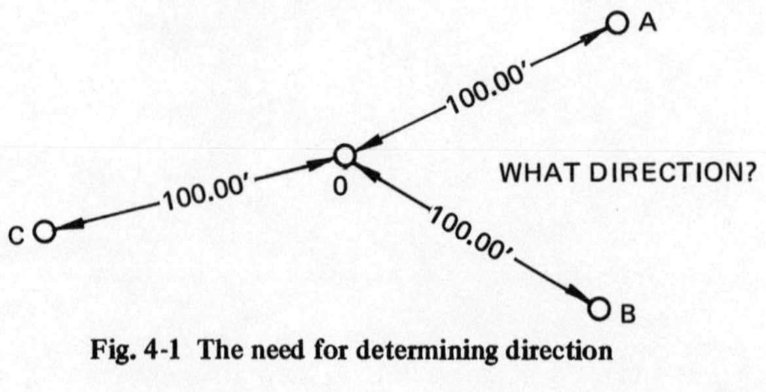


**Fig. 4-1 The need for determining direction**

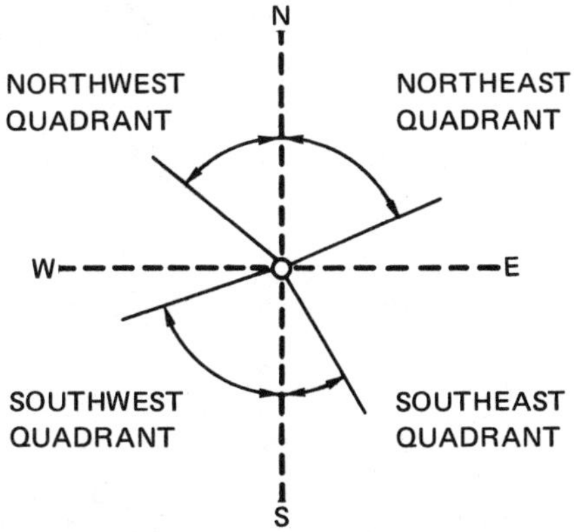


**Fig. 4-2 Reckoning direction within quadrants**

A *bearing* is an angle defined as being in one of the four quadrants of a circle. That circle is centered on a point which an observer occupies. Lines dividing the quadrants extend from north to south and from east to west. Zero lines of reference extend both northward and southward from the occupied point. Angles are reckoned clockwise in the northeast and southwest quadrants, and counterclockwise in the northwest and southeast quadrants, figure 4-2. Since each quadrant covers an arc of 90 degrees, no bearing exceeds 90 degrees. Figure 4-3 shows how the same three directions are expressed as azimuths and bearings.

A line between two points has two directions 180 degrees apart depending upon which end of the line is occupied. In figure 4-4, page 54, the bearing of line AB is N 45 degrees E, when viewed from A. Viewed from point B, the bearing of line AB is S 45 degrees W. Expressed in terms of an azimuth with the zero reference line extending southward, the azimuth from point A is 225 degrees. From point B it is 45 degrees. Thus a difference in bearing of 180 degrees is defined by changing the letters identifying diametrically opposite quadrants. A 180 degree difference in the azimuth is defined by a numerical change of 180.

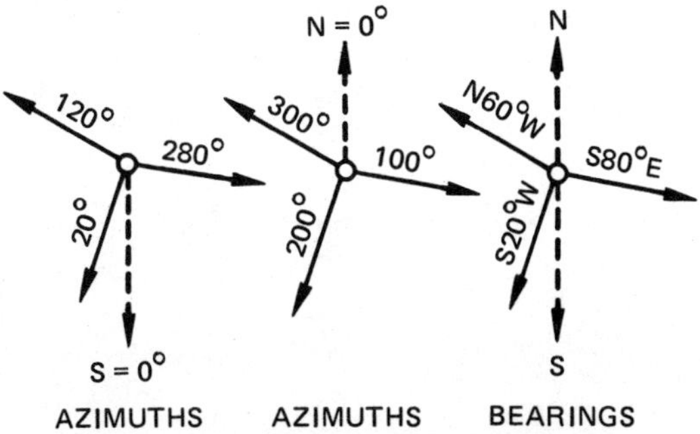


**Fig. 4-3 A comparison of azimuths and bearings**

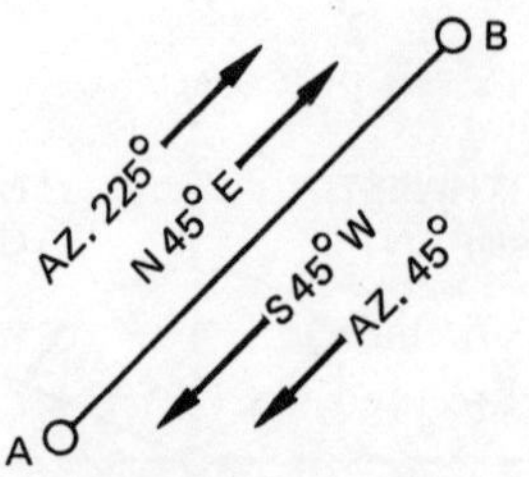


**Fig. 4-4 Direction depends upon point of view**

A transit is equipped with a compass box to indicate in which direction the telescope is pointing, figure 4-5. Its horizontal circular dial is graduated from 0 degrees to 90 degrees both ways from north and south marks. Positions of east and west at 90-degree graduations are reversed. A compass needle pivots at the center of the dial. The dial and the telescope rotate together, while the compass needle remains aligned with the earth's magnetic poles. The circular scale reading gives the angle at which the telescope's line of sight departs from magnetic north or south. The two letters at the ends of the 90-degree arc in which a reading falls identify the quadrant.

In figure 4-5, the needle's north end points to N 30 degrees E. That reading is the forward bearing of the telescope's line of sight. The needle's south end points to S 30 degrees W. That reading is the backward bearing of the telescope's

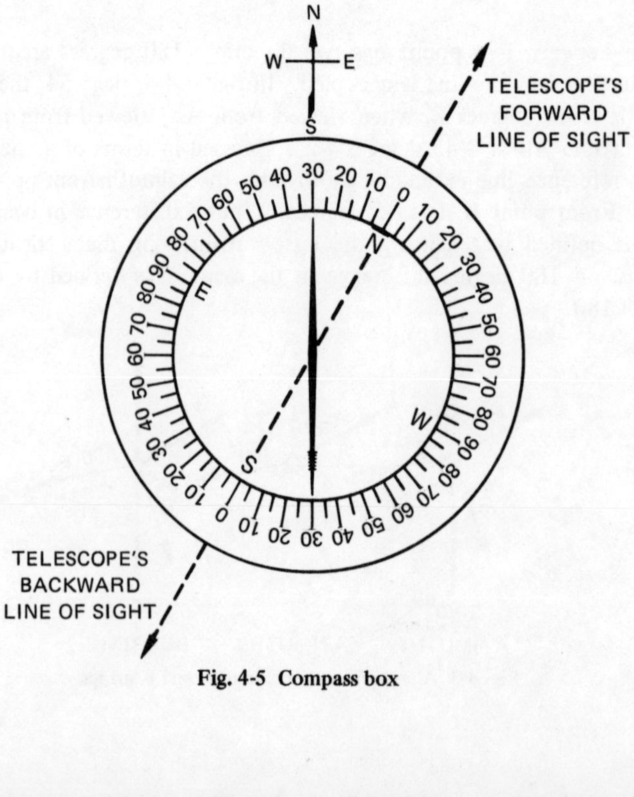


**Fig. 4-5 Compass box**

line of sight. Because the magnetic north pole pulls the needle downward as well as northward, the south half of a needle has a balancing weight attached to keep the needle level. By noting which half has the weight, a surveyor distinguishes between the needle's north and south ends.

A compass needle is kept in the locked position except when a compass observation is being made. A thumbscrew at one side of the compass box locks the needle. That screw is also used to dampen the needle's swinging motion to hasten its coming to rest.

## MAGNETIC DECLINATION

True *geographic poles* are located at the earth's axis of rotation. The earth's magnetic poles are 1000 miles or more away from them. A line from an observer's position extending through the magnetic poles is a *magnetic meridian.* A line through the true geographic poles is a *true* or *geographic meridian.* The angular difference between magnetic and true meridians at any point is the *magnetic declination.* Magnetic declinations now vary from 21 degrees west in Maine to 22 degrees east in the state of Washington, figure 4-6. Magnetic declinations are constantly changing by a few minutes per year in most places. These changes occur because of the shifting locations of the magnetic poles. On many transits, zero degree readings at north points are adjustable to compensate for magnetic declination. Such adjustment permits orientation of compasses on transits to true rather than magnetic north.

## THE LIMITATIONS OF COMPASSES

Surveyors once depended solely upon compass readings for the directions of lines. Today they depend upon horizontal angles turned with a transit or other more sophisticated equipment. Compasses are read directly only to the nearest

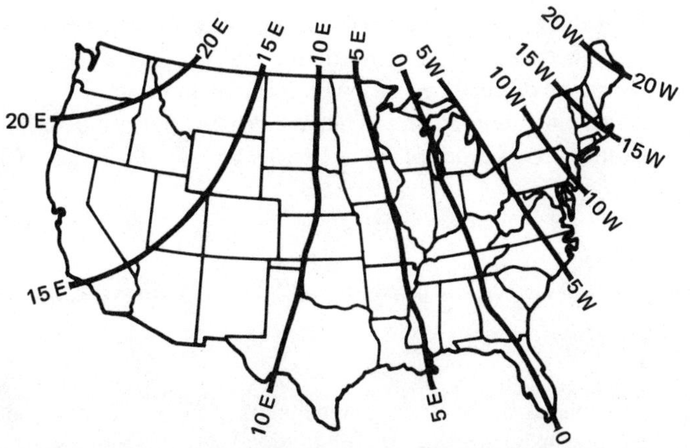


**Fig. 4-6 Magnetic declinations**

degree or half degree. The behavior of compass needles is sometimes erratic. Local magnetic fields and metallic objects deflect compass needles substantially from magnetic meridians. Local magnetic fields exist in the vicinity of electric power lines and electric stock fences. Examples of metallic objects include mineral deposits, vehicles, and iron posts. Even small objects, such as a metallic pencil or magnifying glass, interfere with accurate readings. Although they have limitations, compass readings are used effectively as supplemental information.

Approximate angles are calculated by taking the difference in compass readings for two intersecting lines. Such calculated angles are helpful in providing rough checks on more dependable angular measurements made with transits. The calculation of angles is made easier by first drawing sketches showing intersecting lines plotted to the observed compass readings.

The quality of a survey is determined largely by how precisely angles are measured. However, current practice allows considerable tolerance in orientation of a survey with respect to north-south meridians. Compass readings, although only approximate, are still accepted for relating surveys to reference meridians. There is a growing requirement for precise locations of reference meridians as calculated from celestial observations or ties to geodetic monuments.

## SUMMARY

Both distance and direction are needed to define the location of one point in relation to another.

Direction is expressed either as an azimuth or a bearing. Azimuths are always measured clockwise. Some bearings are measured clockwise and some counterclockwise from a north-south meridian.

Any straight line has two diametrically opposite directions.

The two azimuths of a line differ in value by 180 degrees.

The two bearings of a line have the same numerical values, but differ in quadrant designation.

A compass reading reveals the angle of departure from magnetic north or south. The angle between magnetic north and true north is called the magnetic declination. Magnetic declinations vary with location and time.

Compasses give only approximate values and can be erratic in behavior.

Precise direction is determined by celestial observation or ties to geodetic monuments.

Observed compass bearings provide rough checks on angular measurements made with transits.

Compass bearings are still accepted for relating surveys to reference meridians.

## ACTIVITIES

1. Each team places a transit mounted on its storage plate on a desk top. Level the transit by bringing the plate bubbles to the centers of the vials. Release the compass needle lock. Each student points the telescope in

three different directions. For each direction record the compass readings at both the north and south ends of the needle. Relock the compass needle before picking up the transit to return it to the box.

2. Using the transit on the desk top as in activity 1 above, study the disturbance of the needle by small metal objects. Each student tests the pulling effect of several metal objects held off center from the line of needle. Test with objects held at various radial distances and various off center positions.

3. Each team sets up a transit on its tripod outdoors. Level it by bringing the plate bubbles to the centers of the vials. Release the compass needle lock. Each student points the vertical cross hair of the telescope at an object about a hundred feet away. Read both the north and south ends of the compass needle. Rotate the telescope clockwise to compass readings at right angles to the observed readings. With the telescope thus aimed, have another team member place a stake about 100 feet from the instrument. Have the team member place a pencil mark on top of the stake where the vertical cross hair strikes. Variations in locations of stakes and pencil lines as placed by different students show the variations caused by using compass bearings alone to locate points. At the completion of this exercise, relock the compass needle before removing the transit from its tripod.

## REVIEW QUESTIONS

### A. Multiple Choice

1. A zero line of reference for measuring an azimuth or bearing is
   a. the datum. c. north-south.
   b. east-west. d. a compass needle.
2. Azimuths have numerical values between 0 degrees and
   a. 90 degrees. c. 270 degrees.
   b. 180 degrees. d. 360 degrees.
3. Bearings have numerical values between 0 degrees and
   a. 90 degrees. c. 270 degrees.
   b. 180 degrees. d. 360 degrees.
4. Magnetic declination
   a. is a fixed point on the earth's surface.
   b. does not depend upon the location of the point of observation.
   c. is affected by small metal objects.
   d. is the angular difference between magnetic and true meridians.

### B. Short Answer

5. What is changed to define a 180-degree difference in bearing?
6. What makes the south end of the compass needle easy to recognize?

7. What deflects the compass needle from magnetic north besides nearby metal objects?

C. **Problems**

8. Two bearings, N 8° 30′ E and N 14° W, intersect. At what acute angle do the lines cross?

9. Bearing N 80° E crosses bearing S 70° E. At what acute angle do the lines intersect?

10. A line with an azimuth of 61°30′ intersects a line with an azimuth of 94°15′. At what acute angle do the lines cross?

11. Convert the following compass bearings to azimuths with 0-degree azimuths pointing to magnetic south:

| | | |
|---|---|---|
| N 14°30′E | S 76°45′E | N 75°45′E |
| S 81°15′W | S 1°15′E | S 45°30′W |
| N 14°30′W | S 1°15′W | |

12. If magnetic declination is 3 degrees east, what are the true bearings and azimuths of courses listed in question 11?

13. Since bearings and azimuths were listed in question 12, magnetic declination has shifted 1 degree west. What are the true bearings and azimuths now?

14. What are the clockwise angles between each successive listing of bearings in question 11? What are the clockwise angles between each successive listing of bearings in question 12?

15. From figure 4-6, estimate the magnetic declination in your vicinity to the nearest degree for the year 1970.

16. The compass reading along one side of the equilateral triangle in figure 4-7 is N 45°E. Proceeding clockwise around the triangle, calculate the compass bearings of the other two sides.

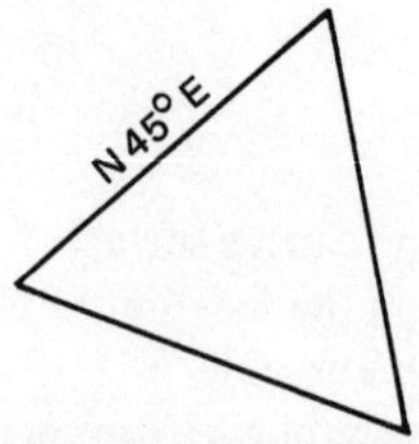


**Fig. 4-7 What are the bearings of the other sides?**

17. The compass reading along one side of a rectangular field is N 13 30′E, figure 4-8. Proceeding clockwise around the field, calculate the compass bearings of the other three sides.

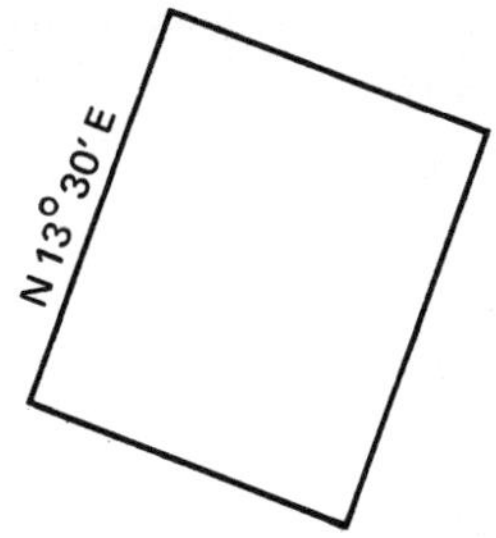


**Fig. 4-8 What are the bearings of the other sides?**

18. The azimuth along one side of a regular hexagon is 39 degrees, figure 4-9. Proceeding clockwise, calculate the azimuths for the other five sides.

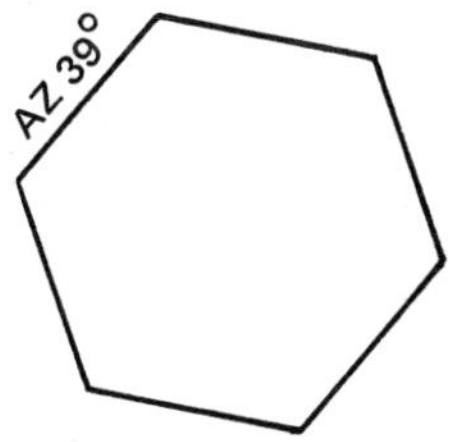


**Fig. 4-9 What are the azimuths of the other sides?**

19. List all the compass bearings from problem 16 in a counterclockwise direction. Start with the original bearing, which becomes S 45°W.
20. List all the compass bearings from problem 17 in a counterclockwise direction. Start with the original bearing, which becomes S 13°30′ W.
21. List all the azimuths from problem 18 in a counterclockwise direction. Start with the original azimuth, which becomes 219 degrees.

# CHAPTER 5

# Measuring Horizontal Angles

## OBJECTIVES

After studying this chapter, the student will be able to:

- measure horizontal angles with a transit.
- mark points appropriately.
- give backsights and foresights on points.
- prolong a straight line.
- measure horizontal angles using only a tape.

## TRANSIT SETUP

Setting up a transit for measuring horizontal angles is similar to setting up a transit for leveling. The important difference is that it is now centered over an exact point. A plumb bob is suspended by a string from a chain that hangs down through the tripod head, figure 5-1. A slipknot in the plumb bob string permits adjustments of the plumb bob to the desired height.

Tripod legs are maneuvered to positions which bring the suspended plumb bob close to the point. When it is within a few hundredths of a foot, the leveling

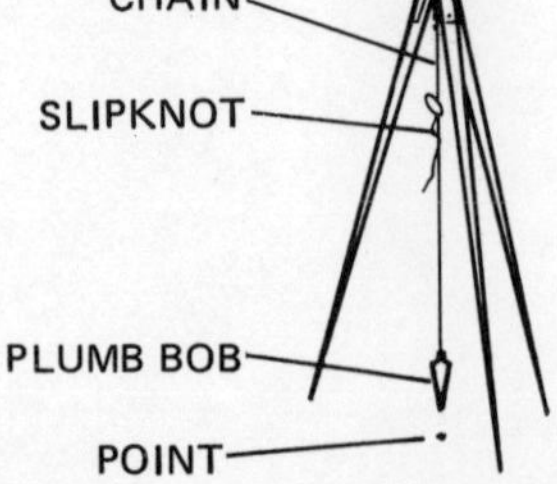


**Fig. 5-1 Suspension of the plumb bob over a point**

of the instrument begins. As the plate bubbles approach the centers of the vials, the leveling screws are loosened to allow the lateral shifting of the movable head. It is shifted as necessary to bring the plumb bob exactly over the point. Finally, the plate bubbles are centered. Attention is given to the bubble on the telescope only if leveling is also to be performed.

**Marking a Point**

There are many ways of marking points over which transits are set, figure 5-2. One way is to use nails punched through brightly colored plastic tape and pressed into the ground. Another way to mark points over which transits are set is to use surveyor's tacks or nails set into the tops of hubs driven into the ground. (A *hub* is a short square wooden stake driven flush with the surface of the ground.) The locations of either nails or hubs are protected by driving guard stakes beside them. Guard stakes are pieces of wood marked with brightly colored paint or tape for easy recognition. Numbers or letters identifying the points are written on the guard stakes. Points in asphalt pavement are marked by driving nails through tin discs or bottle-caps into the pavement. Points in concrete are marked by cutting crosses or by drilling and setting metal plugs.

**Fig. 5-2 Methods of marking points**

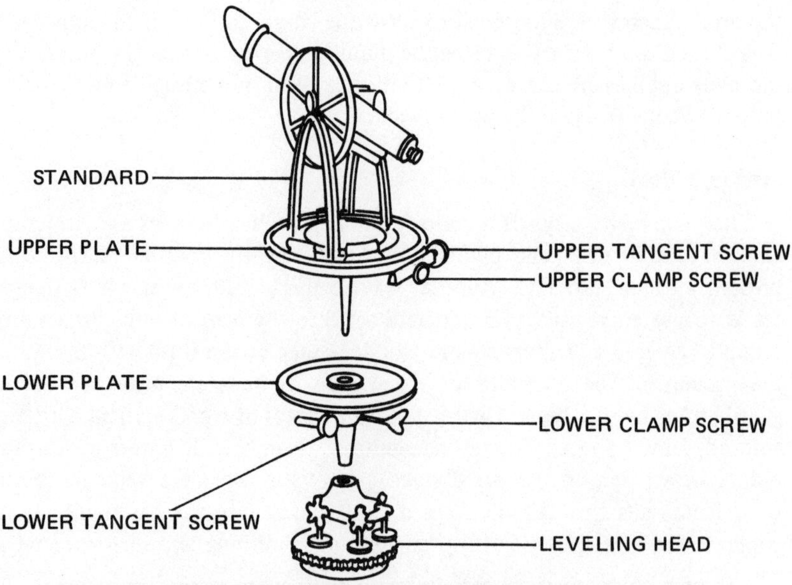


**Fig. 5-3 Transit components**

Points in steel are set by punch marks or by nicks in flanges. Permanent points are preserved by driving long pipes or setting metal plates in concrete.

## USING A TRANSIT

A conventional transit has upper and lower circular plates which rotate in the same horizontal plane, figure 5-3. The telescope is fixed to the *upper plate.* Zero index lines on the upper plate verniers are located on opposite sides of the plate's perimeter. The *lower plate* has a graduated circle, usually reading to half degrees.

Each plate has a *clamp screw,* figure 5-3, which is tightened to hold its plate at any desired setting. *Tangent screws* are turned to make small changes in settings after the clamp screws are tightened.

In turning angles, the sighting can be in either the direct or inverted position, figure 5-4. In the direct position the eyepiece is over the upper plate clamp screw and the vernier marked A. A telescope is in the inverted position when the eyepiece is over the vernier marked B.

Twin verniers A and B are each double verniers reading in either direction from their zero index lines. There are usually 30 vernier graduations in each direction lying opposite 29 half-degree divisions on the lower plate. The zero index line of vernier A is set opposite the zero reference point on the lower scale when preparing to turn an angle, figure 5-5. Figure 5-6 shows the position of vernier A after an angle of 14°47′ has been turned counterclockwise. An angle on the circle is read in the same direction that the telescope is turned.

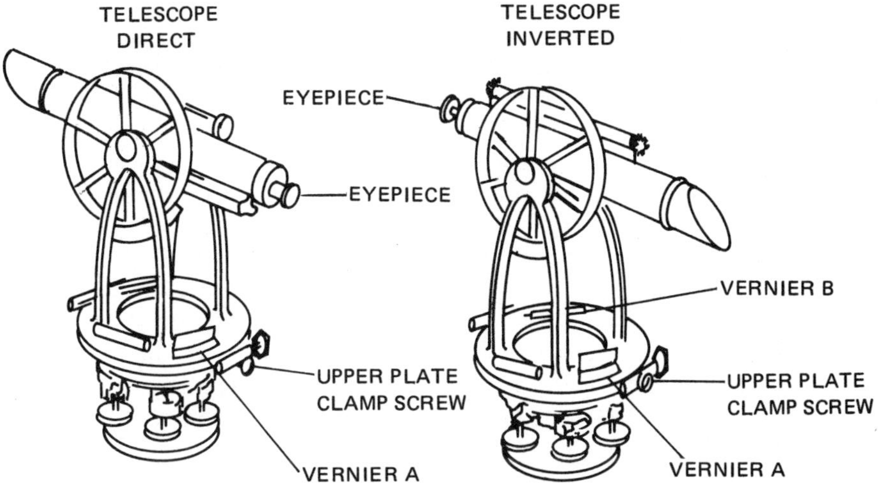


**Fig. 5-4 Telescope positions**

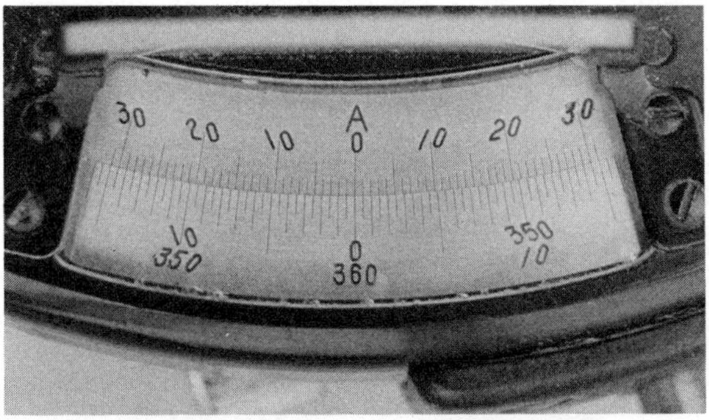


**Fig. 5-5 Plate positions before turning angle**

**Fig. 5-6 Plate positions after turning angle**

The degree or half degree on the lower plate is read opposite or just short of the vernier's zero index line. The number of minutes on the vernier scale which line up with a lower plate division is added to that reading. The vernier must be read in the same direction that the telescope has been turned; that is, to the left when clockwise, to the right when counterclockwise. To minimize mistakes, the number of minutes is estimated before the vernier is read. When a transit is properly adjusted, the angle readings at A and B verniers are 180 degrees apart.

## TURNING AN ANGLE

With the transit set up over a point and leveled, the procedure for turning an angle requires four steps: setting the upper and lower plates to zero, figure 5-7(A); taking a backsight, figure 5-7(B); taking a foresight, figure 5-7(C); and reading the angle.

### Setting the Plates

Set the A vernier zero index line exactly opposite the zero reading on the lower scale. To make this setting, loosen both the upper and lower clamp screws. Rotate the standards to bring the upper clamp and vernier in front of the operator. Revolve the lower plate by pressing lightly on its underside until the zero points of the vernier and the main scales almost match. Tighten the upper clamp and finish matching the zero points of the vernier and the main scales exactly by turning the upper tangent screw. Use a pocket magnifying glass to confirm that the zero points on both scales are exactly opposite each other.

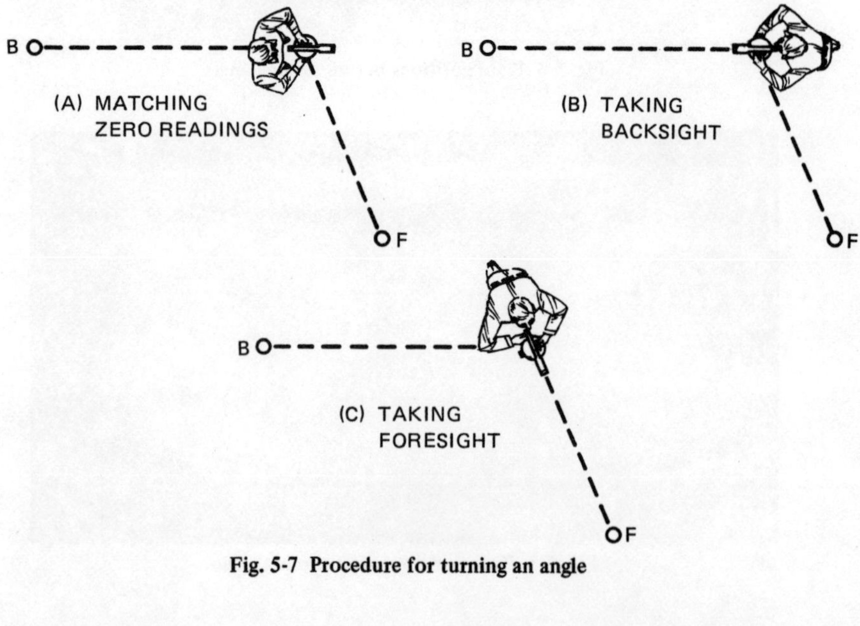


**Fig. 5-7 Procedure for turning an angle**

Leave the lower clamp loose so the telescope can be pointed in any direction without disturbing the matched zero readings.

### Taking a Backsight

Step around the transit and look over it toward the backsight station. A *backsight station* is a point from which an angle is to be turned. Aim the telescope toward the backsight station by eye. Look through the telescope and focus the eyepiece to give a sharp image of the cross hairs. Focus the objective lens to give a clear image of the point. Move the telescope until the vertical cross hair is close to the point and tighten the lower clamp. Raise or lower the telescope until the horizontal cross hair is also near the point. Bring the vertical cross hair exactly on the point by turning the lower tangent screw. Because the upper clamp remains tight and the upper tangent screw is untouched during this procedure, the A vernier still reads zero.

### Taking a Foresight

Loosen the upper clamp. Step around the transit and look over it toward the foresight station. A *foresight station* is a point to which an angle is to be turned. Aim the telescope at the foresight station and focus the objective lens. Tighten the upper clamp when the cross hairs almost cover the point. Bring the vertical cross hair exactly on the point by turning the upper tangent screw.

### Reading the Angle

Read the turned angle on the A vernier using a pocket magnifying glass. Before deciding which divisions on the vernier and lower scales match, observe the division lines adjacent to the apparent matching lines. When adjacent vernier lines to the left and right are equally offset from adjacent divisions on the lower scale, read the minute which is matched. When two adjoining vernier divisions match adjoining divisions on the lower scale equally well, the reading lies halfway between. In that case, record the angle to the intervening half minute.

## SIGHTS ON POINTS

Low obstructions often prevent clear fields of view between the instrument and the backsights or foresights.

The operator of the instrument depends upon other members of the survey crew to give sights on points. They hold range poles on points or suspend plumb bobs over them. A crew member can hold a pencil on a point when the line of sight is unobstructed. It is important that a range pole or pencil be plumb. The operator of the instrument must make sure that their centers are strictly in line with the vertical cross hair. The operator directs necessary changes in position to make them plumb. A plumb bob is steadied by tapping it lightly on the point.

Another way to steady the plumb bob is to suspend it from a leaning pole, figure 5-8. The visibility of the plumb bob string is improved by holding an object of a contrasting color behind it. At night a flashlight is held behind the plumb bob string.

## REFINEMENTS

Refinements in technique come with experience. In addition to practices already explained, there are several other suggestions to help prevent errors.

When setting or reading an angle, the surveyor's eye should be in a plane through the transit's vertical axis and the vernier's zero index.

Frequently, angles are turned to a number of different points from the backsight. It the instrument needs to be leveled again during this process, a fresh backsight is taken before continuing.

Immediately after backsighting, the surveyor searches for some clearly defined object on line with the backsight. This is used to recheck the backsight instead of sending someone back to give another sight.

Too much turning of the tangent screw is avoided by bringing the vertical cross hair fairly close to the point before clamping. The last turn of the tangent screw is clockwise. This keeps the resisting spring compressed.

**Fig. 5-8 Leaning pole steadies plumb bob**

## PROLONGING A STRAIGHT LINE

A straight line is not prolonged by turning an angle of 180°00'. Turning the telescope over on its transverse axis in the opposite direction gives better results. This method is called *plunging* the telescope.

Errors due to incorrect adjustment of the instrument can be eliminated. The telescope is plunged twice; direct for the first backsight, and inverted for the second backsight. If the lines of sight do not coincide at the set-forward point, each is marked by a pencil line. A tack is then set midway between the two marks. This process is called *double-centering*.

## DOUBLE ANGLES

For precise results, angles are turned at least twice. This procedure avoids errors due to poor adjustment of the instrument, and incorrect readings of an angle. The telescope is direct for the first backsight and inverted for the second backsight, or vice-versa. After the first turning, the lower clamp is loosened before aiming the telescope for the second backsight. A value of twice the angle is read on the plates after the second turning. Half of that value is recorded.

When angles are being laid off, marks for each line of sight are made at the point being set. A tack is set midway between the marks if they do not coincide. This process is also called double-centering.

Turning an angle twice between two points already established is called *double-sighting*. If the second reading departs more than one minute from twice the first reading, the results are rejected and the operation is repeated.

For still greater precision, the turning of the angles is repeated more than twice. Many repetitions are made in high precision geodetic surveying.

## MEASURING HORIZONTAL ANGLES BY TAPE ALONE

Angles can be measured by tape alone. Results are more accurate for small angles than for large angles.

Angle A, figure 5-9, is measured by the application of the relationship:

$$\tan A = a/b$$
$$\text{and} \qquad a = b \tan A$$

$$\text{TAN } A = \frac{a}{b}$$
$$a = b \text{ (TAN A)}$$
$$\text{SIN } A = \frac{a}{c}$$
$$a = c \text{ (SIN A)}$$

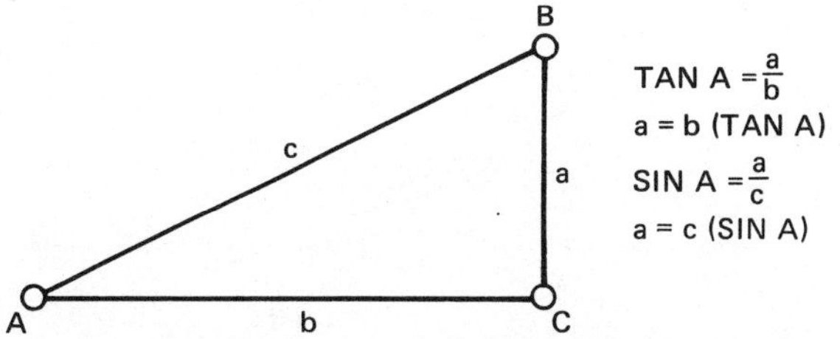

Fig. 5-9 Trigonometric relationships

Angle A is measured off line AC by measuring a perpendicular distance CB of length b tan A. For example, Assume that angle A is to be 40°00′. Line AC is 100.00 feet long. The length of perpendicular BC is found by multiplying the tangent of 40°00′ by length AC:

$$a = b \tan A$$

$$a = 100.00 \times 0.83910$$

$$a = 83.91 \text{ feet}$$

If point B is set 83.91 feet from point C, angle A is 40°00′.

Horizontal angles are also measured by using the sine formulas. In figure 5-10, angle BAC is unknown. To find it, a convenient length is selected and measured out so that AD and AE are equal. Distance DE is measured, and divided in half at F. Triangles DAF and EAF are equal right triangles. Angles DAF and EAF are equal. When they are added together they total the angle DAE. Using the values shown in figure 5-10, DAE is found by:

$$\sin a = \frac{DF}{AD}$$

$$\sin a = \frac{40 \text{ ft.}}{100 \text{ ft.}}$$

$$\sin a = .40000$$

$$\text{Angle } a = 23°35'$$

$$\text{Angle DAE} = 2a = 2(23°35') = 47°10'$$

This method is also used when the angle is known, and line AC is to be laid out from line AB using only a tape. A convenient length for AD is chosen. The length of DF is found by the formula:

$$DF = AD \sin a$$

An arc is swung from A the length of AD. Another arc equal to twice the length of DF is swung from D. The intersection of the two arcs establishes point E.

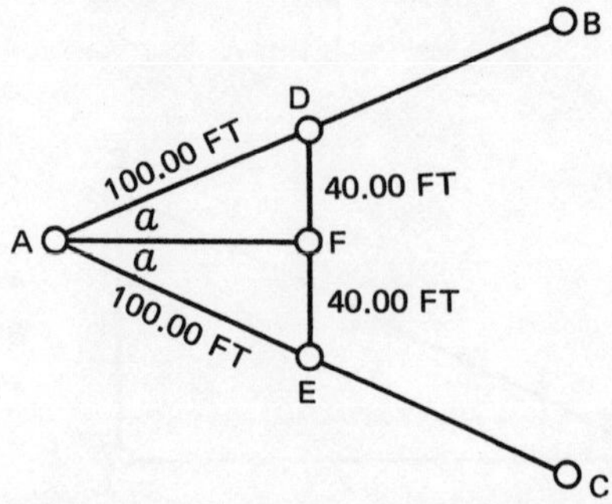


**Fig. 5-10 Measuring an angle by tape**

## SUMMARY

A plumb bob suspended from the center of the transit allows it to be set up over a point.

Points are marked by a variety of methods depending upon their location and the need for preservation.

The essential parts of the transit used for turning angles are the two rotating circular horizontal plates. They are set to read zero when the telescope is aimed at the first point, and the lower plate is clamped and set. The telescope is then aimed at the second point and the upper plate is clamped and set. The angle between the points is read from the graduated divisions on the plates.

Plunging the telescope prolongs a straight line better than turning an angle of 180°00′. Angles are commonly turned two or more times.

Double-centering and double-sighting are procedures for averaging results of doubling an angle.

Angles can be laid out or measured by tape alone.

## ACTIVITIES

1. With the transit at point 0, figure 5-11, set stakes at points A, B, C, and D, each about 100 feet from 0. Each stake should be in a different quadrant. Allow a tack on top of each stake to protrude to make sighting easier. Turn angles AOB, BOC, COD, and DOA independently. Read compass bearings of rays OA, OB, OC, and OD. The total of the four turned angles should be 360° ± 1′ or 2′. Compute the angles from differences in compass bearings and compare these measurements with the turned angles.

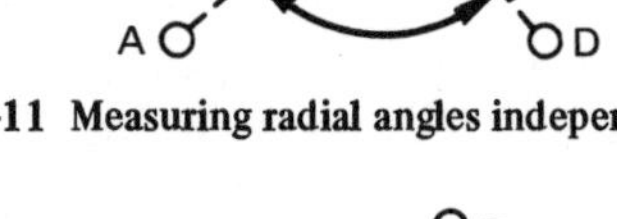

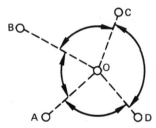

**Fig. 5-11 Measuring radial angles independently**

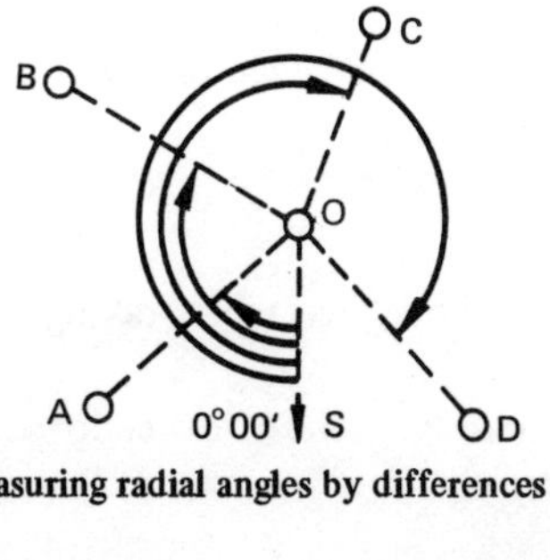


**Fig. 5-12 Measuring radial angles by differences in azimuth**

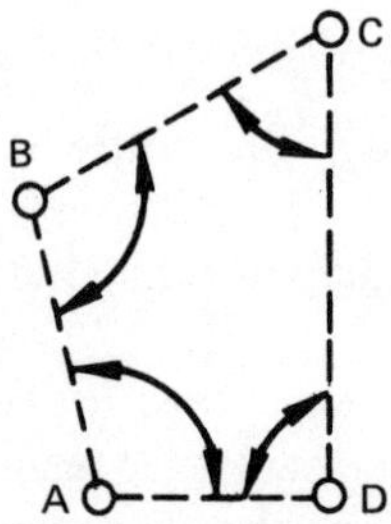


Fig. 5-13 Measuring interior angles

2. Set up the transit at 0, figure 5-12, page 69, and use the same stakes as activity 1. Fix the plates to read 0°00′ with the transit pointed south as indicated by the compass. Turn the azimuths to A, B, C, and D and read the compass bearings to each. Compute the angles from differences in the azimuths and compare with the angles turned independently in activity 1. Compute the angles from differences in compass bearings and compare.
3. With stakes A, B, C, and D from activity 1, set the transit up at each stake, figure 5-13. Measure the interior angles DAB, ABC, BCD, and CDA and note both backward and forward compass bearings. The total of the four angles should be 360°00′ ± 1′ or 2′. Compute the interior angles from differences in compass bearings and compare them with the turned angles.
4. Using the layout of stakes in activity 1, measure angles AOB, BOC, COD, and DOA using tape only. Compare the results with angles turned using a transit.
5. Prolong a straight line by double-centering with the backsight and set-forward point both exceeding 300 feet.

## REVIEW QUESTIONS

### Multiple Choice

1. Measuring an angle with a transit requires manipulation of
   a. the upper plate.
   b. the lower plate.
   c. both plates.
   d. neither plate.
2. The final setting of plates at zero is achieved by using the
   a. upper clamp screw.
   b. upper tangent screw.
   c. lower clamp screw.
   d. lower tangent screw.
3. The final setting of the plates when taking a foresight is achieved by using the
   a. upper clamp screw.
   b. upper tangent screw.
   c. lower clamp screw.
   d. lower tangent screw.
4. The final setting of the plates when taking a backsight is achieved by using the
   a. upper clamp screw.
   b. upper tangent screw.
   c. lower clamp screw.
   d. lower tangent screw.

5. Two points are marked where lines of sight strike the top of a stake as an angle is turned twice. A tack is set midway between the points. This process is called
   a. double-sighting.
   b. plunging.
   c. double-centering.
   d. midpointing.

## B. Short Answer

6. How do the setups of a transit for leveling and for measuring horizontal angles compare?
7. Which circular plate is the telescope attached to?
8. In which direction is the number of minutes read on the vernier after the telescope is turned?
9. One step in turning an angle is to clamp the lower plate while taking what type of sight, a backsight or a foresight?
10. Which plate readings are necessary in order to obtain the value of a turned angle?
11. List five ways of marking points where transits are set up.

## C. Problems

12. A transit is set up at point 0. Vernier A is set at 0°00′. The compass needle is released and the telescope in a direct position is pointed south. The lower clamp and the lower tangent screw are used to bring 0° on the compass dial under the needle. The upper clamp is loosened and a foresight is taken on point A, where an angle of 30°12′ is read. The upper clamp is again loosened and another foresight is taken on point B where an angle of 86°48′ is read. What is the value of angle AOB? What are the azimuths of lines OB and OA?
13. Angle BOC is 54°00′. B and C are each 200.00 feet from 0. What is the length of a line from B to C?
14. Two sides of a triangle are each 100.00 feet long. The third side is 80.00 feet long. What is the value of the angle opposite the 80.00 foot side?
15. Six angles are turned by double-sighting, with the following results. Copy and list in the third column, the value of the acceptable angles that should be recorded:

| FIRST READING | SECOND READING | VALUE |
|---|---|---|
| 30°03′ | 60°06′ | |
| 39°54′ | 79°47′ | |
| 44°02′ | 88°06′ | |
| 48°28′ | 96°57′ | |
| 55°41′ | 111°22′ | |
| 61°36′ | 123°09′ | |

16. After turning a clockwise angle, assume the reading on the plates to be as shown in figure 5-6. What is the value of the angle?

# CHAPTER 6

# Survey Methods

## OBJECTIVES

After studying this chapter, the student will be able to:

- locate subsidiary points from transit stations and transit lines.
- locate a point by a random line and perpendicular offset.
- go around obstacles when prolonging a straight line.
- find the point of intersection of two transit lines.
- reference a transit station.
- run a traverse.

## POINTS LOCATED FROM A SINGLE TRANSIT STATION

Measurements can be made from a single convenient point to define the positions of any other points visible to it. A transit is set up at a selected point called a *transit station*. From this station, a horizontal angle and distance are measured to each of the other points. Angles are turned from a suitable reference which can be a true, magnetic, or assumed meridian. This method of locating points is called *radiation*, figure 6-1.

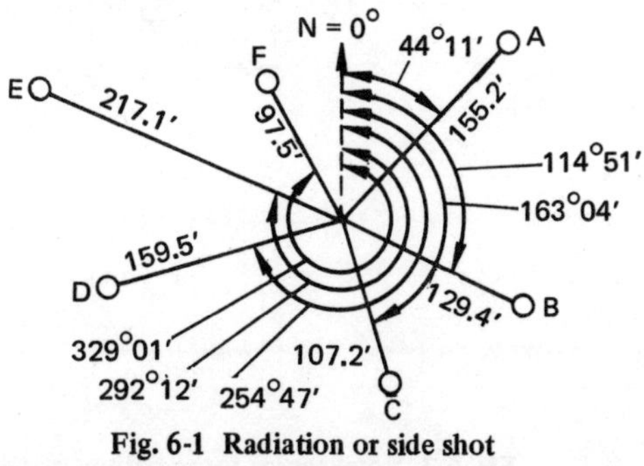


**Fig. 6-1 Radiation or side shot**

Measuring an angle and distance to a point is also referred to as taking a *side shot.* This means that all points located this way are regarded as subsidiary points to a transit station. Some subsidiary points are fence posts, trees, or terrain features which cannot be occupied as transit stations.

In notekeeping, a transit station is identified by a capital letter. Subsidiary points are identified by numbers. Sketches of objects being located are usually made on the right-hand page. On the left-hand page, measurements are entered, figure 6-2.

## POINTS LOCATED FROM TRANSIT LINES

In areas too large for all the points to be visible from one transit station, additional transit stations are needed. The lines between transit stations are called *transit lines.* When only two transit stations are needed, a line between them is called a *base line.* A base line may also be composed of more than two transit stations on a straight line. A base line composed of more than two transit stations not on a straight line is called *traverse.* A traverse may be a closed polygon or open ended.

Distances along transit lines can be measured in units of stations and pluses. The term station, when used to mean 100.00 feet, is not to be confused with the term transit station. The latter term refers to a point wherever a transit is set up. A transit station can have a station and plus designation, but often does not.

Survey of A. Ledford Property by Radiation

Transit @ A. 0° Az. is South

| Point | Az. | Dist. |
|---|---|---|
| 1 | 348°15' | 78.6 |
| 2 | 354°30' | 82.2 |
| 3 | 35°57' | 193.1 |
| 4 | 38°23' | 67.5 |
| 5 | 58°44' | 53.2 |
| 6 | 80°30' | 135.5 |
| 7 | 94°32' | 214.7 |
| 8 | 122°25' | 84.0 |
| 9 | 131°51' | 79.3 |
| 10 | 176°30' | 140 |
| 11 | 199°19' | 196.6 |
| 12 | 217°00' | 68 |
| 13 | 229°00' | 152 |
| 14 | 265°04' | 166.5 |

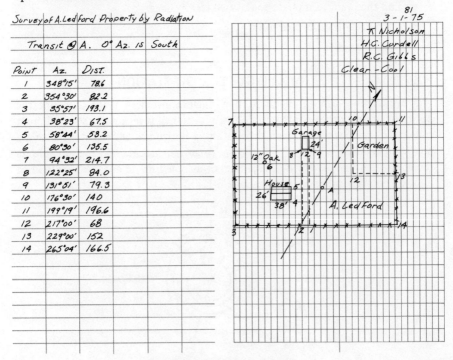

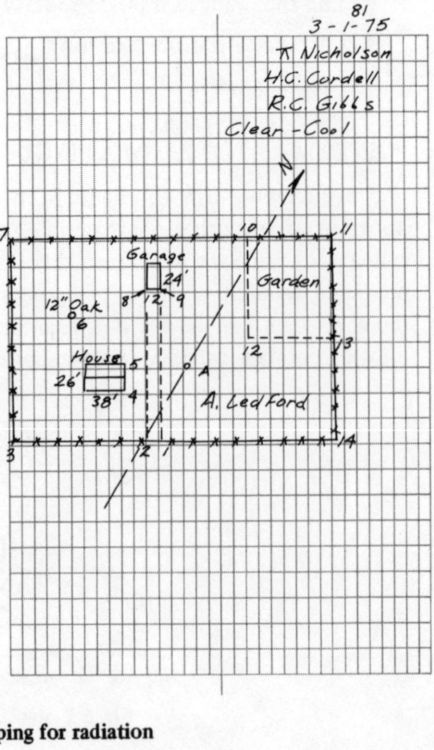


**Fig. 6-2 Notekeeping for radiation**

Once transit lines are established, methods other than radiation can be used for locating nearby subsidiary points.

**Intersection**

An angle is turned from each of two transit stations to a subsidiary point, figure 6-3. Distance c between transit stations is known. Distances from each transit station to the subsidiary point are calculated by the trigonometric law of sines. This law states that:

$$\frac{a}{\sin A} = \frac{b}{\sin B} = \frac{c}{\sin C}$$

$$\text{or } a = \frac{c \sin A}{\sin C}$$

$$\text{or } b = \frac{c \sin B}{\sin C}$$

Assume in figure 6-3 that angle A measures 35°00′ and angle B measures 40°00′. Therefore, angle C is 105°00′. Also, station 14 + 32 minus station 11 + 42 equals 290 feet. To find the lengths of a and b:

$$a = \frac{c \sin A}{\sin C} \qquad b = \frac{c \sin B}{\sin C}$$

$$a = \frac{290 \sin 35^\circ}{\sin 105^\circ} \qquad b = \frac{290 \sin 40^\circ}{\sin 105^\circ}$$

$$a = \frac{290 \times .57358}{.96593} \qquad b = \frac{290 \times .64279}{.96593}$$

$$a = 172 \text{ feet} \qquad b = 193 \text{ feet}$$

This method becomes inaccurate when intersecting lines of sight at point C form a small angle.

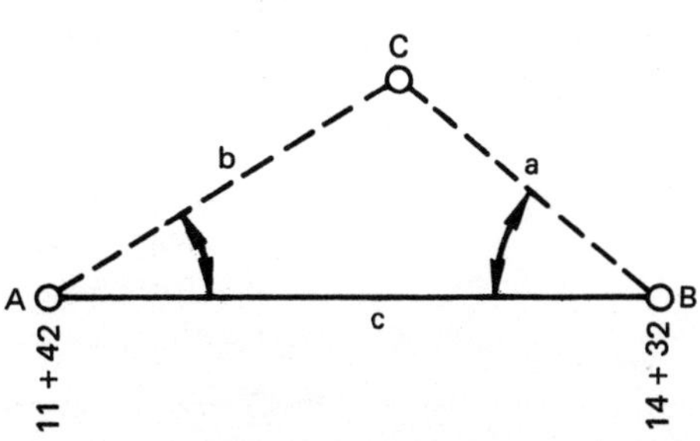


**Fig. 6-3 Intersection**

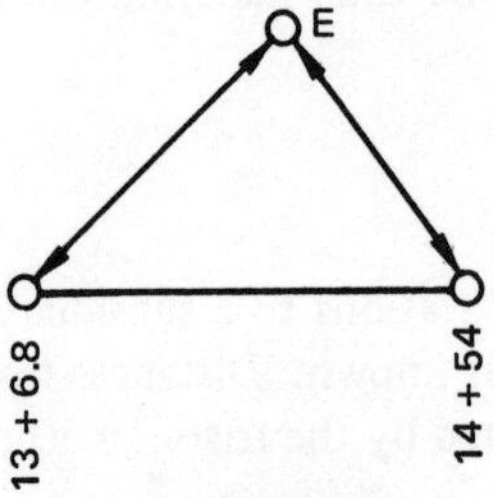


Fig. 6-4 Ties

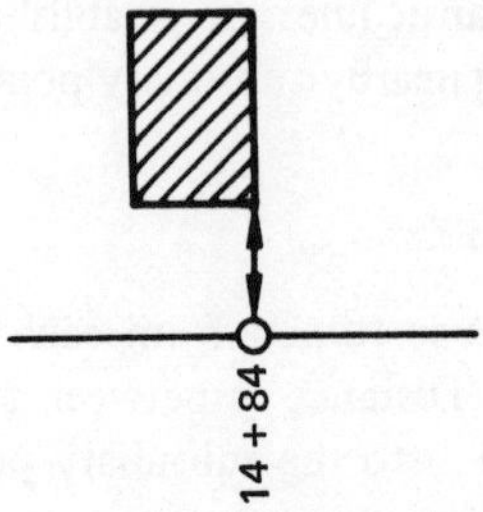


Fig. 6-5 Range tie

## Ties

A *tie* is a distance measured from a subsidiary point to any point on a transit line. The distance of that point from a transit line station must be measured if it is not already known. Two ties, preferably nearly perpendicular to each other, locate an object, figure 6-4.

## Range Ties

A line extending outward in line with one side of a building is a *range line.* When extended to a point some measured distance from the corner of the building, it is a *range tie,* figure 6-5. A range line is sighted by eye. Two intersecting range ties from points on a transit line determine the location of one corner of a building. They also fix the direction of the building's sides. One range tie can be used in combination with other methods. The dimensions of the building are obtained to permit the plotting of the building on a map.

## Perpendicular Offsets

A subsidiary point is located by measuring its perpendicular offset distance from the transit line. A measurement is taken along the transit line to learn the station and plus of the point where the perpendicular strikes. There are three procedures for locating the point where the perpendicular strikes the transit line.

**Procedure A.** With the zero end of the tape anchored at point B, figure 6-6, the tape is swung in an arc over transit line AX. The surveyor, sighting along the

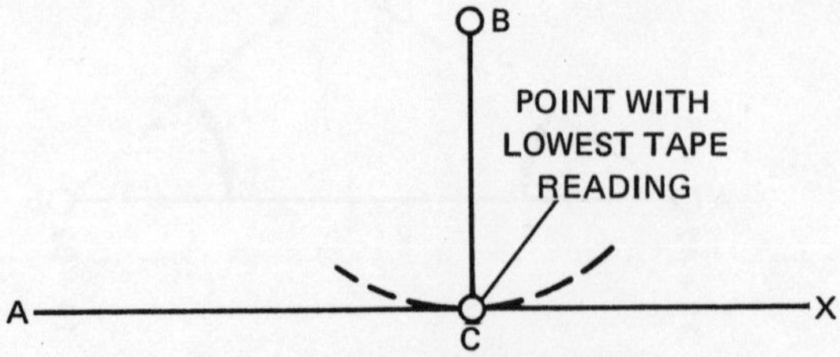


Fig. 6-6 Perpendicular offset procedure A

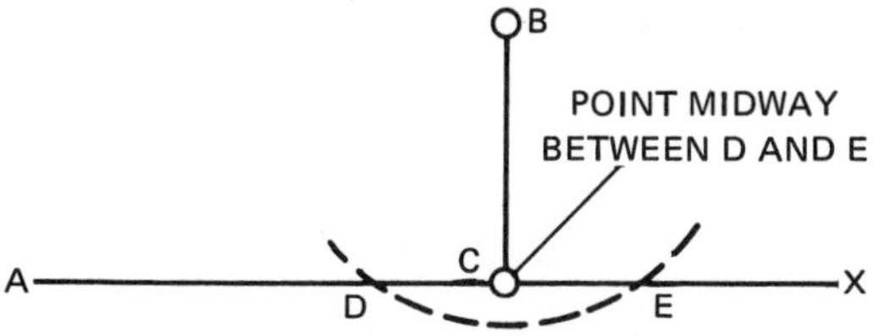


**Fig. 6-7 Perpendicular offset procedure B**

transit line, orders the tape to be stopped where the lowest reading is noted. The tape is then over point C on the transit line where the perpendicular strikes.

**Procedure B.** With the zero end of the tape anchored at point B, figure 6-7, the tape is swung in an arc over transit line AX. The radius of the arc exceeds the estimated offset distance by a number of feet. Two points, D and E, are marked where the tape reading crosses the transit line. Point C, midway between points D and E, is the point on the transit line where the perpendicular strikes.

**Procedure C.** The point where the perpendicular strikes is sometimes estimated. This is done by standing on the transit line, with the arms extended full length to the left and right along the transit line, figure 6-8. Bring the hands together and sight forward toward them, figure 6-9, page 78. Step to the left or right along the line to where the object is found to be in the line of sight. This is the point on the transit line where the perpendicular strikes.

By procedures A and B, the perpendicular offset distance BC cannot be determined precisely. For the precise location of the point where the perpendicular strikes the transit line, an additional step is necessary. Point C as established by any one of the three procedures is regarded as a trial point. With a transit at C, an angle of 90°00′ is turned from the transit line. By whatever amount the line of sight misses point B, point C is shifted to make distance AC correct.

**Fig. 6-8 Perpendicular offset procedure C step 1**

**Fig. 6-9 Perpendicular offset procedure C step 2**

Illustrations of these methods of locating points from a transit line are shown in figure 6-10. The form of notes for recording these measurements is shown in figure 6-11.

## SUPPLEMENTARY TRANSIT LINES

Where a line is run between points that are not visible to each other, the precise direction is often unknown. Even in cases where the exact direction is known, an obstacle, such as a large tree or building, may be encountered.

Where the precise direction is unknown, a random line is run from one point in the approximate direction of the other point. The random line is run past the unknown point. A perpendicular offset distance is measured. The difference between the bearings of the random line and a line between the points is designated angle $a$, figure 6-12. The perpendicular offset BC divided by the length of the random line AC is the value of tangent $a$. The length of the random line AC divided by cosine $a$ is the distance AB between points.

To get around an obstacle, alternate methods are the parallel line offset and the isosceles triangle offset.

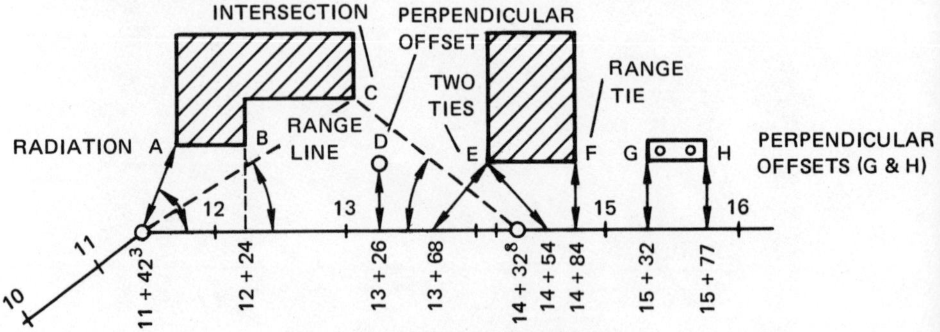


**Fig. 6-10 Application of methods**

Locating Buildings & Facilities – 3rd Ave.

| Sta. | Point | Az. | Dist | |
|---|---|---|---|---|
| $11+42^{3}$ | 16- | 162°55' | | |
| $11+42^{3}$ | A | 99°30' | 73.4 | |
| $11+42^{3}$ | C | 132°10' | | |
| $14+32^{8}$ | C | 21°15' | | |
| 13+26 | D | | 47.7 | ⊥ |
| +68 | E | | 68.7 | |
| 14+54 | E | | 66.5 | |
| +84 | F | | 50.6 | Range |
| 15+32 | G | | 51.0 | ⊥ |
| +77 | H | | 51.3 | ⊥ |
| 12+24 | B | | | Range |

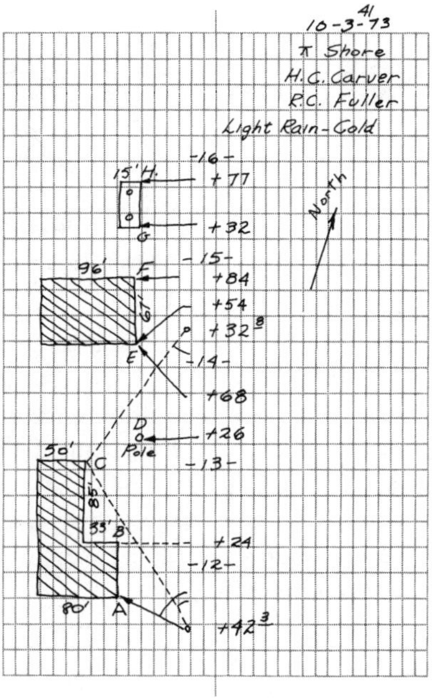


**Fig. 6-11 Notekeeping for variety of methods**

## Parallel Line Offset

An obstructing building makes it impossible to project line XA by direct line of sight beyond point A. A parallel line offset can be made, figure 6-13, page 80. An angle of 90 degrees is turned and A′ is established at a convenient perpendicular distance from A. Through A′ another transit line is established parallel to XA and projected forward beyond the building. At point B′ another 90-degree angle is turned and B is established by making BB′ = AA′. Through B, the transit line is continued forward, keeping it parallel to line X′Y′. Line BY is then a prolongation of XA.

This method is successful only when backsights of adequate length are taken when occupying points A′ and B. When distances AA′ and BB′ are only a few feet, they are too short for backsighting. When this happens, there are two ways to find backsights of adequate length.

Before moving the transit from A, find a distant fixed object in line with A′. After moving to A′, backsight on that fixed object and then turn 90 degrees to set B′. If an object suitable for a backsight cannot be sighted, stake is set approximately 100 feet from point A to be used as an adequate backsight.

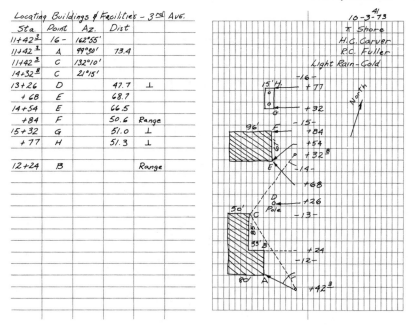

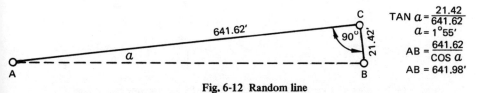

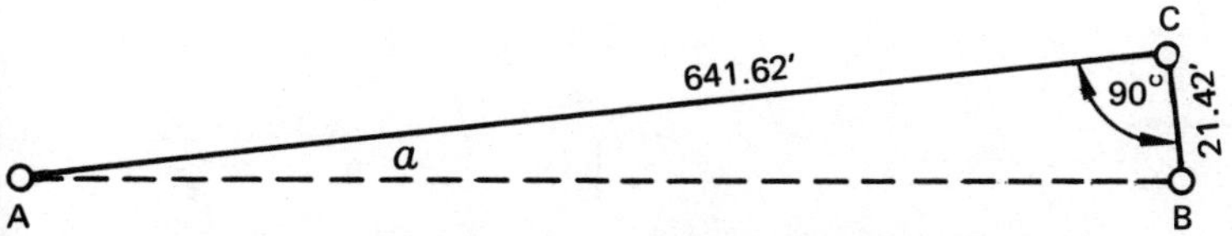


$$\text{TAN } a = \frac{21.42}{641.62}$$

$$a = 1°55'$$

$$AB = \frac{641.62}{\cos a}$$

$$AB = 641.98'$$

**Fig. 6-12 Random line**

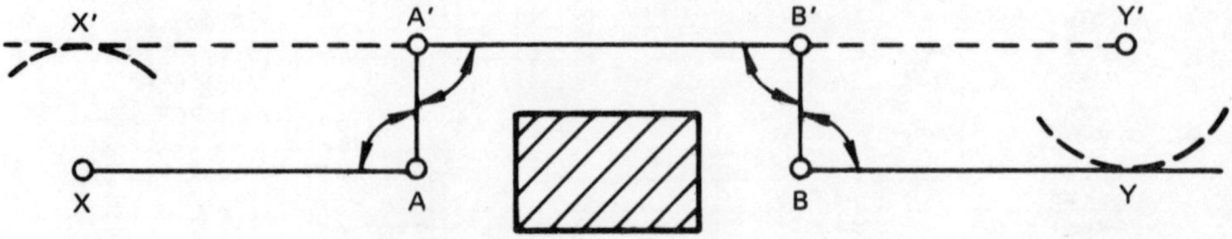


**Fig. 6-13 Parallel line offset**

Another way of establishing an adequate backsight is by plunging. With the transit at A', a backsight is taken on X', and the telescope is plunged to set B' and Y'. Point X' is established so that perpendicular XX' equals perpendicular AA'. One method of establishing the perpendicular is to swing the tape in an arc centered on point X. By turning the lower tangent screw, the tape is tracked following the reading corresponding with distance XX'. At the place where that reading is furthest from point X, the backsight is parallel to XA.

These procedures described for point A' can also be followed for establishing other points in the offset.

### Isosceles Triangle Offset

Three transit stations, A, B, and C, are established with equal acute angles $a$ at A and B, figure 6-14. The external angle at C is equal to their sum $2a$. Since angles A and B are equal, sides a and b are equal. Points X, A, B, and Y are then in line. The length of line AB is found by using the formula:

$$AB = b \cos a + a \cos a$$

Since sides a and b are equal, the formula is also written:

$$AB = 2a \cos a$$

## INTERSECTION OF TRANSIT LINES

Frequently, it is desired to prolong two transit lines to a point of intersection. To do this, two temporary points are set on one line. One of these temporary points is set on each side of the estimated point of intersection. A string is stretched between those points. The transit is moved to the other line which is then prolonged to intersect the string line, figure 6-15.

Temporary points for this purpose are usually tacks or pencil marks on tops of stakes. The stakes are removed after the point of intersection is established.

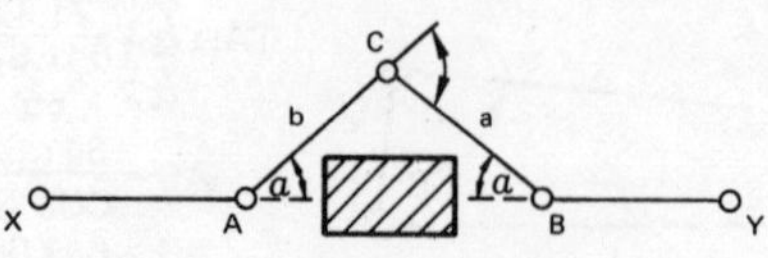


**Fig. 6-14 Isosceles triangle offset**

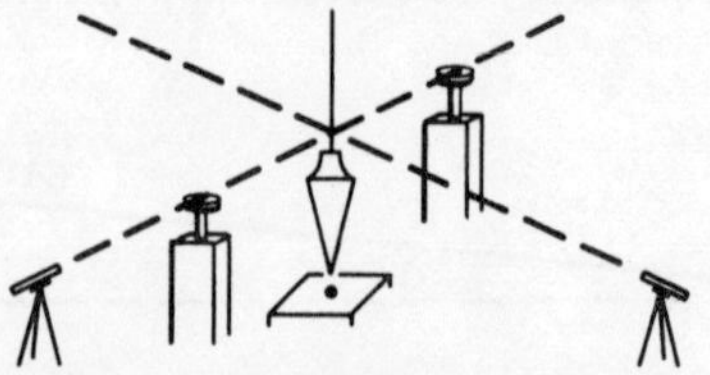

**Fig. 6-15 Intersecting transit lines**

## REFERENCING TRANSIT STATIONS

*Referencing* is a way to find or reestablish points that have been covered over, moved, or destroyed.

One method of referencing is to make two nearly perpendicular ties to identified points on fixed objects. Identification is made by placing a nail in a tree, making an X mark in concrete, or other identifying markings. Such reference points are called *witness points,* or *witnesses.* Intersecting arcs swung with a tape from witness points relocate a transit point. A tie to a third point is occasionally made as an additional safeguard.

Where suitable fixed objects are not available within 100 feet, another method is used. In construction surveying, transit stations are regularly lost by movement of the earth. To permit easy restoration, a pair of reference hubs are set along each of two nearly perpendicular radiating lines. The hubs are generally 50 feet or more apart and are protected by guard stakes. Prolonging transit lines through both pairs of hubs to a point of intersection reestablishes the transit station. Distances are occasionally measured from the transit station to each hub as additional safeguards in case a reference hub also becomes lost.

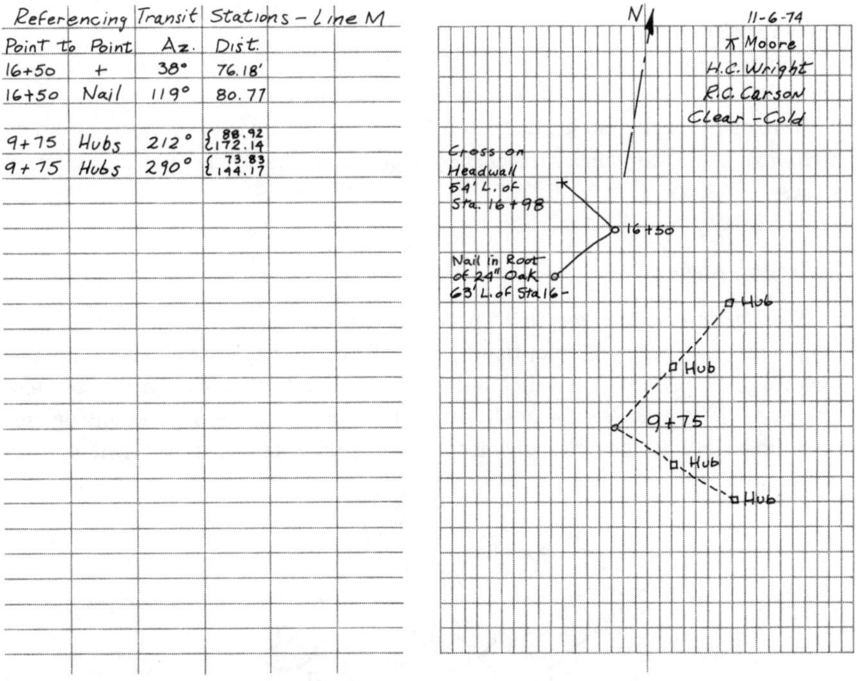


**Fig. 6-16 Notekeeping for referencing transit stations**

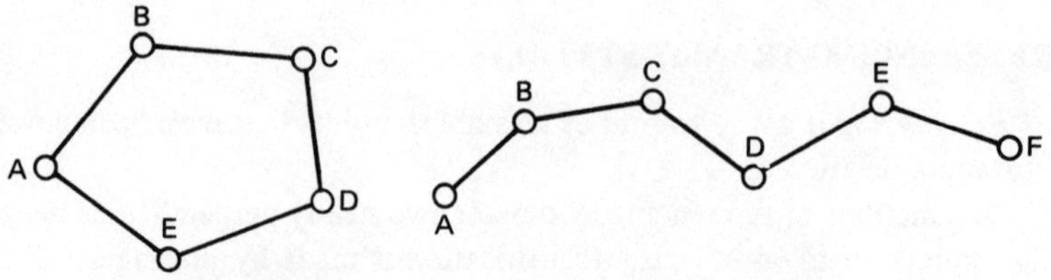


**Fig. 6-17 Traverses**

Notes for referencing transit stations should include a description and a sketch of the reference points, measured lengths of ties, and compass bearings or azimuths of witness points or reference ties, figure 6-16, page 81.

## TRAVERSES

A series of transit stations connected by transit lines leading back to the starting point is a *closed traverse.* The geometric figure they trace is a polygon. A series not returning to the starting point is an *open* or *continuous traverse,* figure 6-17.

A closed traverse allows an opportunity for checking the accuracy of the angle and distance measurements. The position of a calculated point of closure is compared with the position of the point of beginning. An open traverse affords a similar opportunity only when the end points are at monuments with known positions.

Transit stations located along a traverse are called *traverse stations.* Transit lines between them are called *traverse lines* or *courses.*

Horizontal angles at traverse stations are measured by several different methods: deflection angle, traverse by azimuth, angles to the right, and interior angles.

### Deflection Angle

A *deflection angle* is measured between a succeeding line and a forward prolongation of a preceding line, figure 6-18. When measured clockwise from the forward prolongation, it is called a *deflection angle to the right.* When counterclockwise, it is a *deflection angle to the left.* After taking a backsight on the preceding station, the telescope is plunged before taking a foresight on the

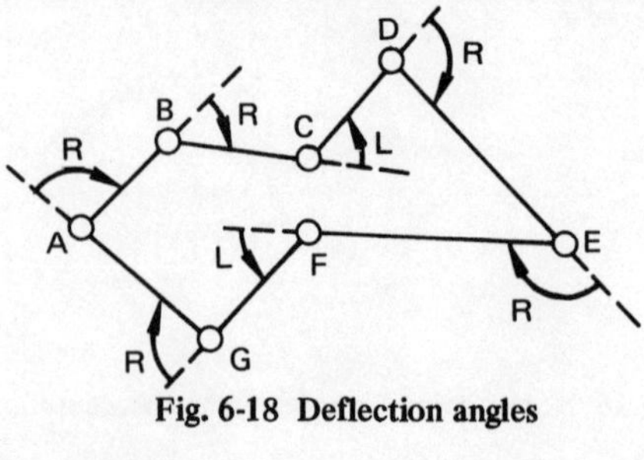


**Fig. 6-18 Deflection angles**

succeeding station. Keep the lower clamp of the instrument ahead in the direction of forward prolongation to avoid mistakes. For a closed traverse, the sum of deflection angles to the right should differ from the sum of deflection angles to the left by 360°.

### Traverse by Azimuth

Clockwise angles are accumulated at successive traverse stations. A reading of 0 degrees is obtained only at a station where the transit points in the same direction as the reference meridian. In this method the A vernier scale is not reset to 0 degrees at each station. It remains opposite the lower scale reading to which it was turned at the preceding station. After taking a backsight on the preceding station, the telescope is plunged before taking a foresight on the succeeding station, figure 6-19, page 84. At the first station, the azimuth is recorded not only to the succeeding station, but also to the preceding station. The latter should be 180 degrees opposite that recorded for that same course at the last station.

### Angles to the Right

Angles to the right are measured clockwise from the preceding line to the suceeding line, figure 6-20, page 84. Their sum depends upon whether the survey's route is clockwise or counterclockwise and the number of sides in the polygon. If the route is clockwise, the angles to the right are outside the polygon and should total (n + 2) 180 degrees, where n is the number of sides. If the survey's route is counterclockwise, the angles are inside the polygon and should total (n - 2) 180 degrees.

### Interior Angles

An interior angle is an angle inside a polygon measured from the preceding line to the succeeding line, figure 6-21, page 84. Whether the survey's route is clockwise or counterclockwise around a polygon, the interior angles should total (n - 2) 180 degrees.

## ASSURING ACCEPTABLE RESULTS

Traverses are run with reasonable assurance of acceptable results when certain practices are used.

Horizontal angles at all traverse stations are doubled for accuracy.

Distances between traverse stations are measured forward and backward along each traverse line, figure 6-22, page 85. When measuring slope distances, one end of the tape is centered at the end of the telescope's transverse axis. The other end of the tape may be held either on the adjacent traverse point or any level above it. The vertical angle is measured to a point on the tape

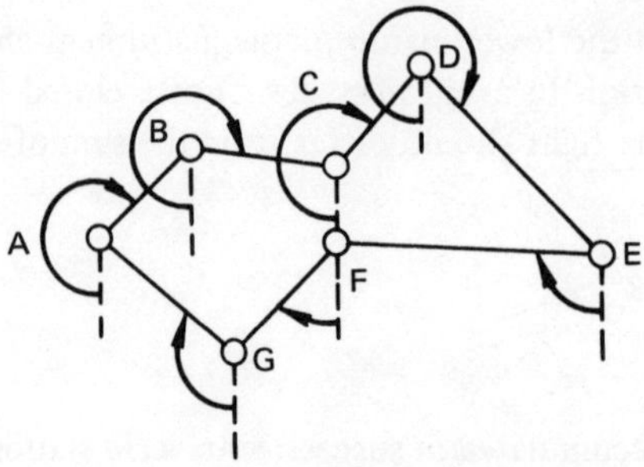


Fig. 6-19 Traverse by azimuth

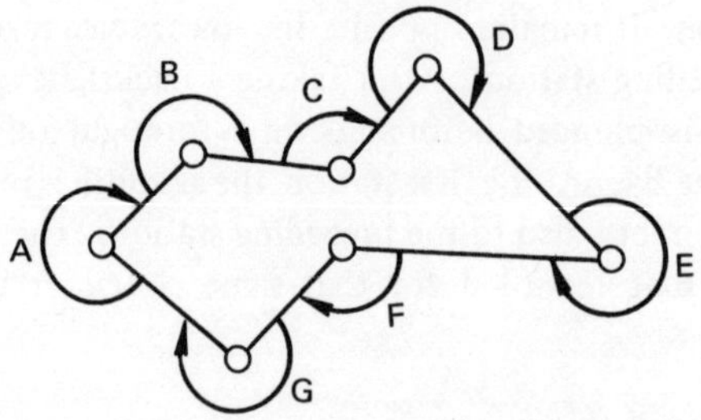


Fig. 6-20 Angle to the right

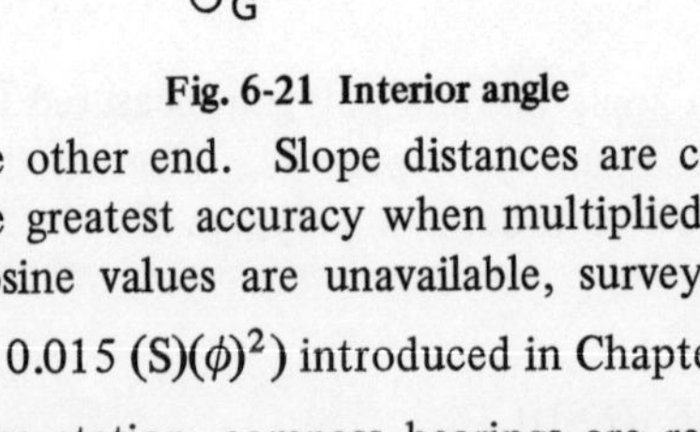

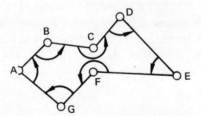

Fig. 6-21 Interior angle

where read at the other end. Slope distances are converted to horizontal distances with the greatest accuracy when multiplied by the vertical angle's cosine. When cosine values are unavailable, surveyors use the correction formulas ($\frac{h^2}{2s}$ and $0.015\,(S)(\phi)^2$) introduced in Chapter 2.

At each traverse station, compass bearings are read along both traverse lines. If the difference between these two bearings fails to be in close agreement with the turned angle, the reason is found at once. Any error is corrected before the transit is moved.

It is impractical to locate traverse stations to coincide with all property corners and other significant features. Those not occupied are located by side shots from nearby traverse stations, or in some other way.

A good system of field notes should be adopted, such as that shown in figure 6-23, page 86.

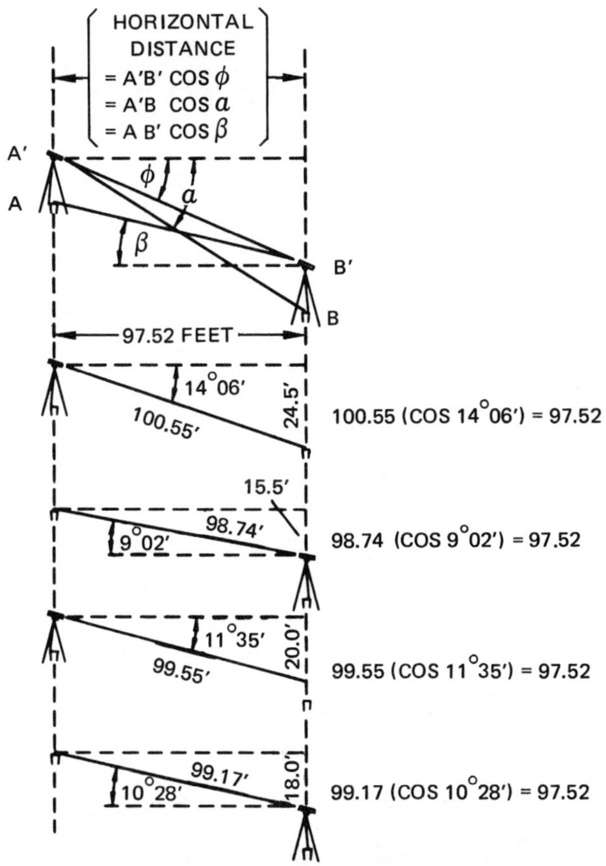


**Fig. 6-22 Ways of measuring slope distances between same two points**

Traverses are oriented to reference meridians by compass, celestial observations, or by tie-ins with geodetic monuments. The choice of method depends primarily upon the degree of precision required.

The traverse method is the most widely used plane surveying method for establishing and maintaining horizontal control.

## TRIANGULATION

*Triangulation* is a system for locating control monuments at the vertices of interconnected triangles. This system employs highly precise surveying techniques. Basically the method begins by measuring the length of a single base line and determining its direction. Horizontal angles of all the triangles are measured. The lengths and directions of the sides of all the triangles except the base line are computed.

The National Geodetic Survey uses triangulation for establishing high order horizontal control nationally. Cities use triangulation for establishing accurate control monuments locally. Agencies making maps of large areas extend the

Closed Traverse with Side-Shots @ E & F

| Sta. | Defl ∠ | Bearing | Slope D | Vert. ∠ | Hor. D |
|---|---|---|---|---|---|
| A | D 20°10'R | F N39½E | F 223.14 | -6°10' | 221.85 |
| | F 40°20'R | B S39½W | B 222.44 | +4°14' | 221.83 |
| | 20°10'R | N39½E | | | 221.84 |
| B | D 102°28'R | F N60°E | F 350.30 | +3°26' | 349.67 |
| | F 204°56'R | B S60°W | B 350.93 | -4°57' | 349.63 |
| | 102°28'R | N60°E | | | 349.65 |
| C | 67°31'L | S17½E | 287.62 | -5°11' | 286.44 |
| | 135°02'L | N17½W | 287.05 | +3°45' | 286.44 |
| | 67°31'L | S17½E | | | 286.44 |
| D | 129°23'R | S84½E | | — | 87.33 |
| | 258°46'R | N84½E | | — | 87.32 |
| | 129°23'R | S84½E | | | 87.32 |
| E | 39°54'R | S47½W | 171.61 | -2°18' | 171.47 |
| | 79°48'R | N47½E | 171.50 | -1°02' | 171.47 |
| | 39°54'R | S47½W | | | 171.47 |
| F | 58°22'R | S85W | 424.88 | +3°21' | 424.15 |
| | 116°44'R | N85E | 424.46 | -2°19' | 424.11 |
| | 58°22'R | S85W | | | 424.13 |
| G | 77°11'R | N36½W | | — | 126.70 |
| | 154°22'R | S36½E | | — | 126.72 |
| | 77°11'R | N36½W | | | 126.71 |
| A | | | | | |
| E-M | | N5°E | | | 20.4 |
| F-N | | N38½W | | | 18.6 |

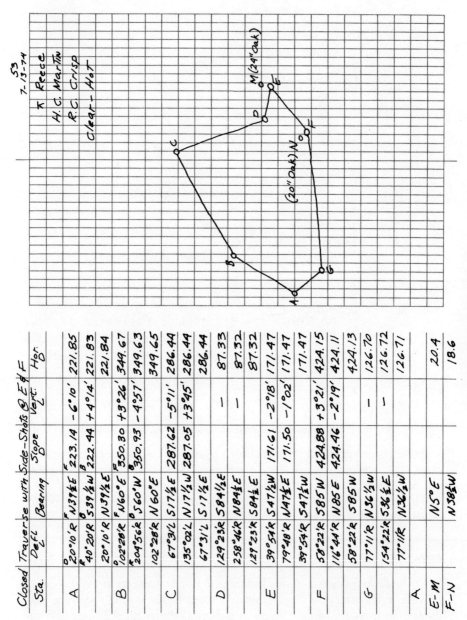

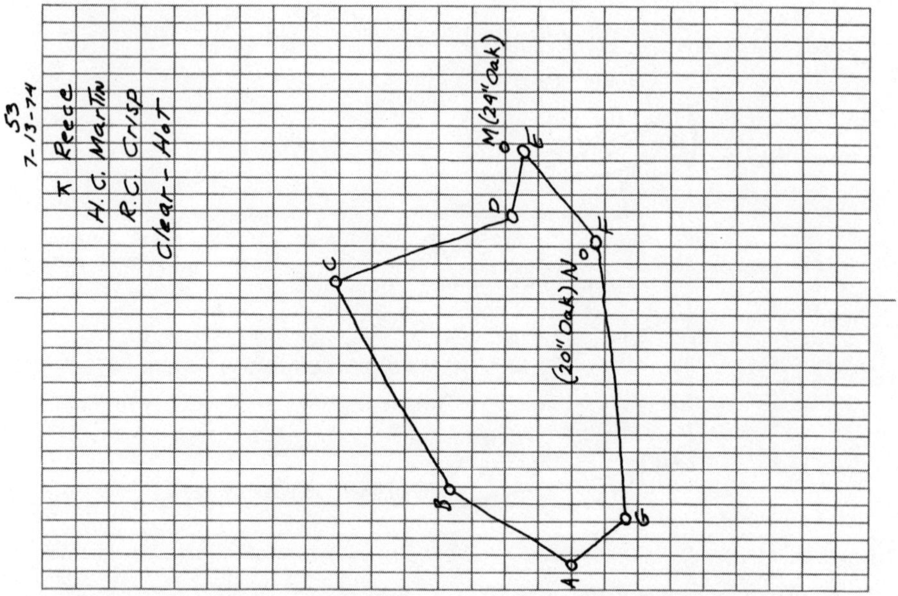


**Fig. 6-23 Notekeeping for traverse**

government triangulation networks to gain adequate horizontal control of their work. Monuments established by triangulation are located throughout the United States. Traverses and other plane surveys are tied in to them when the accuracy of the job justifies the cost.

## SUMMARY

Radiation is a method of defining positions of subsidiary points in relation to a transit station. It involves measuring a horizontal angle and distance to each point. This is called *taking side shots.*

The establishment of transit lines connecting transit stations permits the use of additional methods of defining the positions of points. Points are located by intersection, ties, range ties, or perpendicular offsets.

Where it is impractical or impossible to measure directly to a point, a random line and perpendicular offset are adequate.

Obstacles encountered when prolonging a line are bypassed by parallel line offsets or isosceles triangle offsets.

A point of intersection of two transit lines is established in two stages. A pair of points straddling the approximate point of intersection is set along one line. A string is then stretched between these two points. The second transit line is then prolonged to an intersection with the string.

One method of referencing transit stations is to make two nearly perpendicular ties to witness points. Another is to set a pair of reference hubs along each of two nearly perpendicular radiating lines.

A traverse is a series of transit stations connected by transit lines, also called traverse lines or courses. Traverses are closed when they return to the starting point. The geometric figure traced by a closed traverse is called a polygon. The open traverse does not return to its starting point.

Four methods of turning horizontal angles at traverse stations are: deflection angles, traverse by azimuth, angles to the right, and interior angles.

Recommended practices include doubling angles, measuring distances backward and forward, and comparing turned angles with compass bearings. Care in measuring and reducing slope distance is advised. Where significant points cannot be occupied, side shots are used.

Traverses are oriented to reference meridians by compass, celestial observations, or by tie-ins with geodetic monuments.

Traversing is the most widely used plane surveying control method. Triangulation is a common control method in geodetic surveying.

## ACTIVITIES

1. Run a transit line down the approximate center of one block of an urban street. Locate all the man-made features. Include fronts of buildings, poles, curbs, manholes, driveways, sidewalks, fences, etc. Use all the methods described in this chapter. Where residents do not want the surveyors to enter the property, use a method of measurement that makes trespassing unnecessary.
2. With a protractor, triangles, and scales (1″ = 20′), make a plat showing all the features located in activity 1.
3. With a protractor, triangles, and scale (1″ = 20′), make a plat of the notes illustrated in figure 6-2.

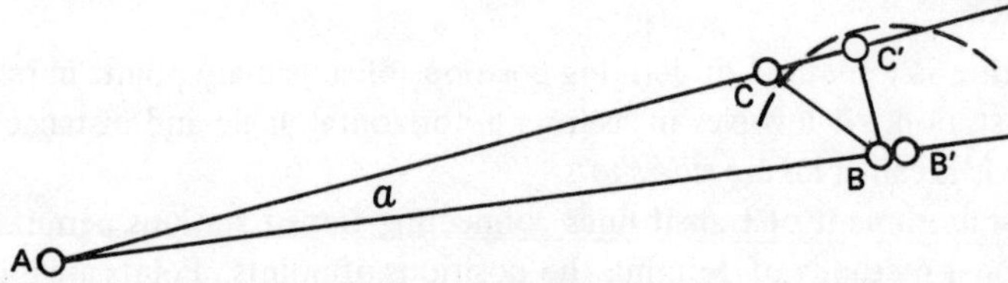


**Fig. 6-24 Refinement of the random line method**

4. Set hubs at vertices of triangle ABC, figure 6-24, with an estimated distance of 200 feet between A and B. Pace some 15 or 20 feet for side BC. Set up the transit at A. Assume it is impossible to run a direct line to B and that line AC is a suitable random line. Prolong line AC to a point C′ where a perpendicular through B apparently strikes line AC prolonged, using procedure B. Measure distance BC′. Calculate angle $a$ and distance AB. Set up the transit at C′, backsight on A, and turn 90°00′. Mark point B′ where the foresight strikes line AB prolonged. Adjust the calculated length of distance AB by adding or subtracting distance BB′. Measure distance AB and compare.
5. Set stations 0, 1, and 2 at 100.00-foot intervals along a straight line. With the transit at station 2, assume the line is obstructed by a building between stations 2 and 3. Detour around it by the parallel line method, and set stations 3, 4, and 5 in line with stations 1 and 2. Check the work by setting up the transit again on station 2. Backsight on station 0, plunge, and measure any discrepancies at stations 3, 4, and 5.
6. In the same situation described in activity 3, detour the building by the isosceles triangle method. Set stations 3, 4, and 5. Again check the work by setting up the instrument at station 2 and observing any discrepancies at stations 3, 4, and 5.
7. Set points at the vertices of a triangle with unequal sides of estimated (paced) lengths of 90 to 110 feet. Lines bisecting the three interior angles of any triangle intersect at its center. Set a hub at the center by finding where two bisecting lines meet. Check its location by bisecting the third angle.
8. Reference a transit station by two or more ties. Reference another transit station by a pair of reference hubs along each of two radiating lines.
9. Set hubs to mark corners of a polygon having five sides of unequal length. Make each side approximately 100 to 200 feet long. Run a traverse using the deflection angle method. Rerun the same traverse using the traverse by azimuth method.

## REVIEW QUESTIONS

### A. Multiple Choice

1. The intersection method of locating subsidiary points becomes inaccurate when the intersecting lines of sight form

a. large angles.
b. small angles.
c. angles of approximately 45 degrees.
d. angles of approximately 90 degrees.

2. When two ties locate a subsidiary point, it is preferable that the angle at which they meet be
a. large.
b. small.
c. approximately 45 degrees.
d. approximately 90 degrees.

**B. Short Answer**

3. List four methods by which horizontal angles at traverse stations are turned.
4. List five methods by which subsidiary points are located.
5. List two methods by which obstacles encountered in prolonging a line are circumvented.
6. List two methods by which transit stations are referenced.
7. What is the geometric figure traced by a closed traverse?
8. What trigonometric law is used to calculate the distances to subsidiary points located by intersection?

**C. Problems**

9. The interior angles for an entire polygon total (n - 2) 180 degrees. How many degrees should the algebraic sum of deflection angles for an entire polygon total?
10. The forward azimuth of a transit line is 207 degrees. At station 9+25, the azimuth of subsidiary point 3 is 157 degrees. At station 10+75, the azimuth of subsidiary point 3 is 67 degrees. Calculate the distance to point 3 from station 9+25 and from station 10+75.
11. Point 6 is located by two ties, one from station 3+50, and the other from station 4+50. Each tie to point 6 is 100.00 feet. What are the interior angles from transit line to point 6 at stations 3+50 and 4+50.
12. Angles to the right totaling 1260 degrees are turned during a clockwise survey around a field. How many sides does the field have?
13. Interior angles totaling 540 degrees are turned during a survey of a field. How many sides does the field have?
14. An isosceles triangle offset is made around an obstacle. Using figure 6-14, angle $a$ is 30 degrees and side b is 50.00 feet long. What is the distance AB?

# CHAPTER 7

# Traverse Computations

## OBJECTIVES

After studying this chapter, the student will be able to:

- orient and balance a traverse.
- calculate rectangular coordinates for traverse stations and related points.
- find a line's bearing and length from its known departure and latitude.

## ERROR OF CLOSURE

Relative positions of all stations on a traverse are calculated from measured angles and distances. By definition, a closed traverse terminates at its beginning point. Yet calculated positions of terminal points never exactly coincide with positions of beginning points. Discrepancies occur because of small, unavoidable errors in measurement. The amount by which angles alone fail to close is called the *angular error of closure.* The linear discrepancy is called the *linear error of closure,* figure 7-1. The linear error of closure divided by the traverse perimeter is commonly called the error of closure. A preferred alternate term now in use is *precision of closure.* It is expressed as a fraction with a numerator of 1. An error of closure or precision of closure of $\frac{1}{3870}$ is read one part in 3870.

If errors are excessive, survey results are rejected. Where significant errors or mistakes in measurement cannot be traced, new field measurements are made.

Traverse computations include a process for distributing allowable errors in a logical manner. The positions of all traverse stations are adjusted. This process is called balancing the traverse.

## ADJUSTMENT OF ANGLES

The first step in balancing a traverse is to determine the angular error of closure. That error is the amount by which sums of traverse angles fail to meet known geometric conditions. Those conditions are summarized as follows:

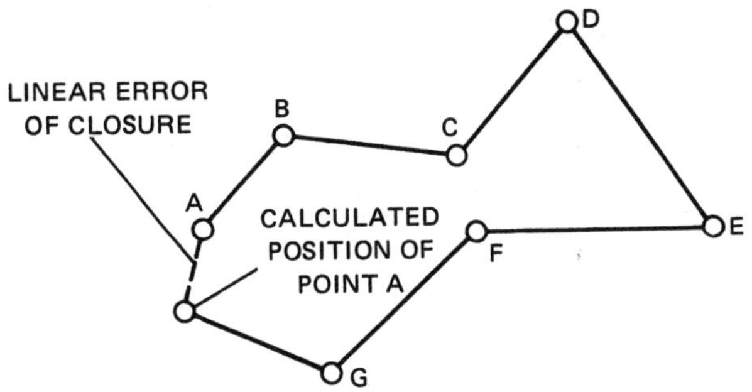


**Fig. 7-1 Linear error of closure**

- Deflection angles. The sums of clockwise deflection angles differ from the sums of counterclockwise deflection angles by 360 degrees when measured in one direction around from start to finish.
- Traverse by azimuth. The final angle turned returns the forward azimuth to the beginning traverse line azimuth.
- Angle to the right. Angles to the right outside of a polygon total (n + 2) 180 degrees.
- Interior angles. Interior angles of a polygon total (n - 2) 180 degrees.

Ordinary surveys close within a few minutes. The tolerated angular error of closure depends upon the precision required and the number of traverse angles. Government agencies publish specifications for surveys of various classes. Specifications of the agency having jurisdiction where the survey is made should be observed. One rule-of-thumb for finding the angular error of closure is expressed by the formula:

$$M < I\sqrt{n}$$

M is the maximum allowable angular error of closure in minutes. The symbol $<$ means less than. I is the closest interval to which angles are measured and n is the number of angles turned. If I is one minute and n is 16, the angular error of closure should not exceed four minutes.

In the traverse by azimuth method, azimuths of successive traverse lines are listed in field notes. Even if angles are measured by another method, the whole adjustment process is simplified by compiling a similar listing. The list begins with a known or assumed azimuth for one course. Azimuths of other courses are obtained by adding the deflection angles algebraically in sequence. A course with an azimuth determined by reference to monuments or by celestial observation makes an ideal beginning. Where orientation is by compass only, the listing begins with any course having an acceptable bearing. The bearing is then converted to an azimuth and begins the list. If no field observations of directions are made, the listing process begins by assuming one course has zero azimuth. When azimuths of traverse lines are arrayed in sequence, the angular closure is revealed. It is the amount by which the final measured angle fails to return the forward azimuth to the beginning azimuth.

The next step is to absorb the angular error of closure by adjusting the measured angles. No measured angle is adjusted by an amount smaller than the closest interval to which angles are measured. For example, adjustments are no smaller than one-half minute when angles are being read only to the closest half-minute. Ordinarily, the number of angles needing adjustment are fewer than the number measured. Angles chosen for adjustment are those most likely to be in error. Those most likely to have sighting error are adjacent to the shorter traverse lines. Other angles are chosen if peculiarities of sighting conditions are believed to have made them less reliable.

For the traverse shown in figure 7-2, course AB is known to have a true bearing of N 40°03′ E. Measured deflection angles are as listed in column 2 of figure 7-3. Angle A is the last angle measured. The known bearing is converted to an azimuth of 220°03′. Azimuths of other traverse lines appear in column 4 as calculated by adding the measured angles algebraically. The last measured angle fails to return course AB's azimuth to its beginning azimuth by 3 minutes. Since angles are read to the nearest minute, three angles should be adjusted by one minute each. The shorter traverse lines are adjacent to points A, D, and E. To offset the error, each angle needs clockwise adjustment. Angles A and E, measured to the right are therefore increased by one minute. Angle D, measured to the left, is then decreased by one minute. Adjusted azimuths are shown in column 5. They are converted back to bearings as shown in column 6 as a basis for further calculations.

## ADJUSTMENT OF DIRECTION BASED ON COMPASS OBSERVATION

Angle adjustments and consequent changes in directions of traverse lines neither improve nor impair their orientation to a reference meridian. A one-degree clockwise error in course AB's observed bearing in figure 7-2 makes all adjusted azimuths one degree too great. Neither the calculated angular error of closure nor the angular adjustments are affected by such an error. The whole traverse simply appears to be rotated one degree clockwise from its proper position.

Some rotation occurs whenever compass bearings are accepted as a basis for computing directions. Local magnetic disturbances are apparent when significant differences occur between forward and backward compass observations of

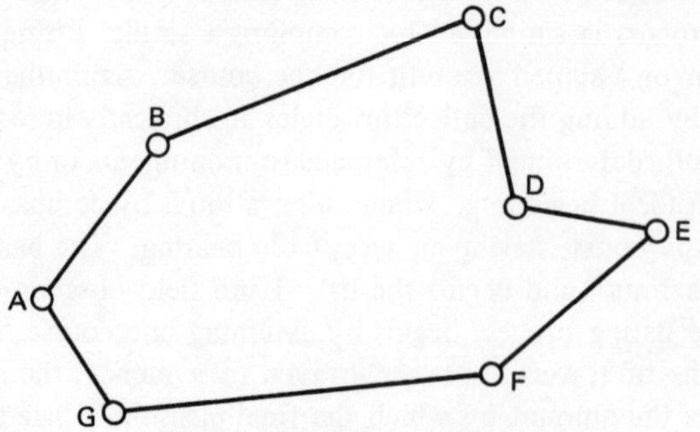


**Fig. 7-2 Illustrative traverse**

| 1<br>STA. | 2<br>DEFLECTION ANGLE | 3<br>KNOWN BEARING | 4<br>UNBALANCED CALCULATED AZIMUTH | 5<br>ADJUSTED AZIMUTH | 6<br>ADJUSTED BEARING |
|---|---|---|---|---|---|
| A | | | | | |
| | | N 40°03′E | 220°03′ | 220°03′ | N 40°03′E |
| B | 20°10′R | | | | |
| | | | 240°13′ | 240°13′ | N 60°13′E |
| C | 102°28′R | | | | |
| | | | 342°41′ | 342°41′ | S 17°19′E |
| D | 30′<br>67°~~31~~′L | | | | |
| | | | 275°10′ | 275°11′ | S 84°49′E |
| E | 24′<br>129°~~23~~′R | | | | |
| | | | 44°33′ | 44°35′ | S 44°35′W |
| F | 39°54′R | | | | |
| | | | 84°27′ | 84°29′ | S 84°29′W |
| G | 58°22′R | | | | |
| | | | 142°49′ | 142°51′ | N 37°09′W |
| A | 12′<br>77°~~11~~′R | | | | |
| | | | 220°00′ | 220°03′ | N 40°03′E |
| B | | | | | |

**Fig. 7-3 Angle and bearing adjustment based on true north**

the same course. Only courses without such discrepancies are acceptable as a beginning course for listing azimuths. The amount of rotation depends partially upon which one of the acceptable courses is adopted. Regardless of which course is adopted, azimuths derived from it are in a trial stage. A refinement is needed to place equal dependence upon every acceptable compass bearing observed around the entire traverse.

Columns one through five of figure 7-4, page 94, illustrate what occurs in a trial stage. Compass bearings observed for each course are listed in column 3. True bearings of all courses are unknown. Assume that course AB, with a compass bearing of N 39°30′E, is chosen at random as the beginning course. Calculated azimuths of all courses would then be as in column 4. Since the deflection angles are the same as in figure 7-3, an angular error of closure of three minutes is again found. Also, the same adjustment of angles A, D, and E

| 1<br>Sta. | 2<br>Deflection Angle | 3<br>Observed Compass Bearing | 4<br>Unbalanced Calculated Azimuth | 5<br>Adjusted Azimuth Based on Trial Bearing | 6<br>Compass Bearing Converted To Azimuth | 7<br>Deviation Column 6 From Column 5 | 8<br>Adjusted Azimuth Depending on All Observed Bearings | 9<br>Adjusted Bearings Depending on All Observed Bearings |
|---|---|---|---|---|---|---|---|---|
| A | | | | | | | | |
| | | N 39 1/2°E | 219°30′ | 219°30′ | 219°30′ | 0 | 220°08′ | N 40°08′E |
| B | 20°10′R | | | | | | | |
| | | N 60°E | 239°40′ | 239°40′ | 240°00′ | +20′ | 240°18′ | N 60°18′E |
| C | 102°28′R | | | | | | | |
| | | S 17 1/2°E | 342°08′ | 342°08′ | 342°30′ | +22′ | 342°46′ | S 17°14′E |
| D | 30′<br>67°~~31~~′L | | | | | | | |
| | | S 84 1/2°E | 274°37′ | 274°38′ | 275°30′ | +52′ | 275°16′ | S 84°44′E |
| E | 24′<br>129°~~23~~′R | | | | | | | |
| | | S 47 1/2°W | 44°00′ | 44°02′ | 47°30′ | ~~+208~~ | 44°40′ | S 44°40′W |
| F | 39°54′R | | | | | | | |
| | | S 85°W | 83°54′ | 83°56′ | 85°00′ | +64′ | 84°34′ | S 84°34′W |
| G | 58°22′R | | | | | | | |
| | | N 36 1/2°W | 142°16′ | 142°18′ | 143°30′ | +72′ | 142°56′ | N 37°04′W |
| A | 12′<br>77°~~11~~′R | | | | | | | |
| | | N 39 1/2°E | 219°27′ | 219°30′ | 219°30′ | 0 | 220°08′ | N 40°08′E |
| B | | | | | | | | |

**Fig. 7-4** Angle and bearing adjustment based on compass observation

are made. Adjusted azimuths based on the trial bearings are as listed in column 5. All are 33 minutes less than those derived in figure 7-3. This is the difference between the true bearing and the observed compass bearing of course AB. Unfortunately, many surveyors are content with such differences resulting from random selection of a beginning course, and second stage refinement is not attempted.

Second stage procedure is illustrated in columns six through nine of figure 7-4. The recommended practice is to convert the compass readings for each traverse line to azimuths and list these azimuths in column 6. Enter in column 7 their deviations from the trial bearing azimuths shown in column 5. Eliminate any large deviation indicating an erratic compass reading. Find the average deviation of the remaining acceptable compass azimuths. Adjust the azimuths of all the traverse lines by that average amount.

In column 7, the erratic value of 208 is eliminated. The average deviation is 38 minutes clockwise. This indicates that course AB's observed bearing has an azimuth 38 minutes less than the average for all the acceptable compass bearings. Column 8 lists the azimuths adjusted by adding 38 minutes to values in column 5. The adjusted bearing for course AB is N 40°08′E, or only 5 minutes clockwise of its true bearing. Azimuths of all other courses are also just 5 minutes clockwise of their values derived in figure 7-3. The second stage adjustment is well worth the effort. In this example the rotation error is reduced from 33 minutes counterclockwise to five minutes clockwise. Furthermore, regardless of which acceptable course is randomly chosen for trial purposes, the stage-two procedure produces the same 5 minute rotation error.

Column 7 indicates how much variation in rotation occurs if stage-two procedure is not followed. Adoption of course GA instead of AB as the beginning course rotates the traverse 72 minutes. Bearings observed by another crew with another transit around the same traverse might produce even greater variation.

The importance of refinements in procedures for orienting traverses can be seen by noting the effects of rotation. An angular difference of one degree makes a 1.75-foot perpendicular offset at a distance of 100 feet. Offsets are proportional to angular differences and distances. If point E is 600 feet from point A, a 72-minute rotation indicates the position of point E to be displaced by more than 12 feet.

In the above examples, adjusted azimuths and bearings based on compass readings are in close agreement with true bearings. Such near coincidence occurs only when the compass's zero index reading is set to compensate for magnetic declination. Unadjusted values would differ from those shown in figure 7-4 by the amount of the magnetic declination. Second-stage procedure, with the reliance on all acceptable compass readings, has the same importance in either case.

## DEPARTURES AND LATITUDES

Calculation of departures and latitudes is a process of resolving lines into east-west and north-south components. The east-west component is called its

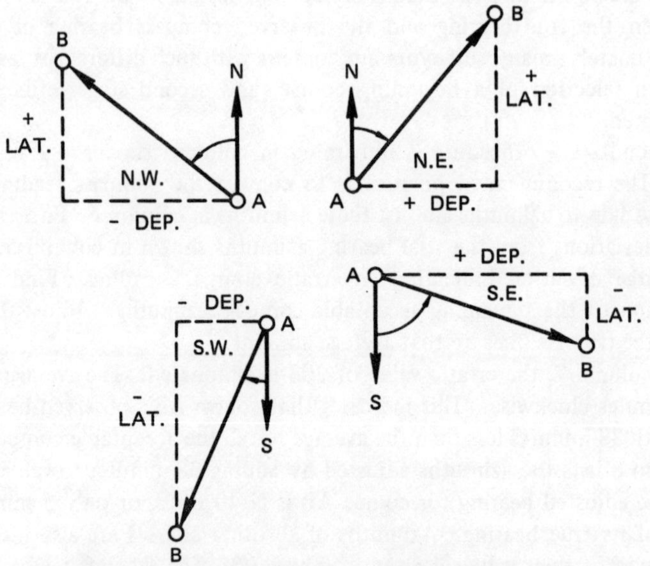


Fig. 7-5 Departures and latitudes by quadrants

*departure.* It has a plus value if it extends eastward and a minus value if it extends westward. The north-south component is called its *latitude.* It has a plus value if it extends northward, and a minus value if it extends southward. Figure 7-5 shows how departures and latitudes of line AB appear when moved from one quadrant to another.

By eliminating all diagonal directions, the process makes it easier to define the relative positions of points. Point E, figure 7-6, can simply be defined as 618.78 feet east and 62.01 feet north of point A. Otherwise the definition is:

> N 40°04′E for a distance of 221.85 feet, thence N 60°14′E for a distance of 349.74 feet, thence S 17°20′E for a distance of 286.51 feet, thence S 84°49′E for a distance of 87.35 feet.

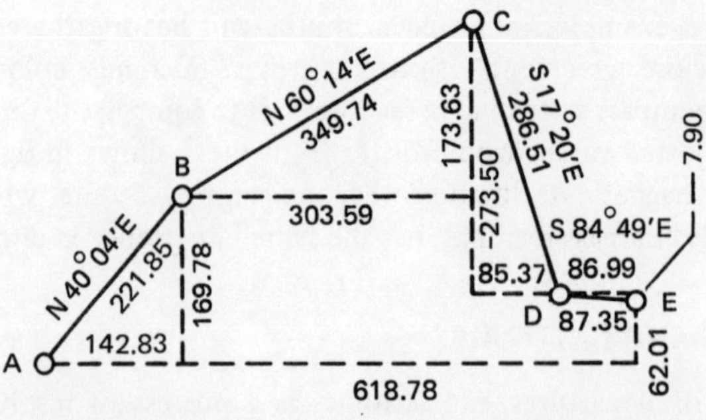


Fig. 7-6 Departures and latitudes as components of courses

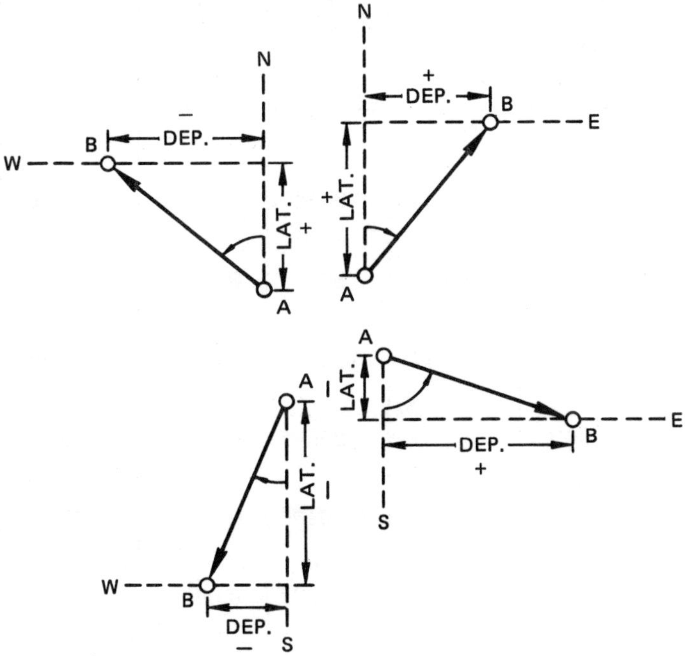


**Fig. 7-7 Departures and latitudes as related to bearing angles**

Departures and latitudes illustrated in figure 7-5 can also be represented as shown in figure 7-7. There it is seen that any traverse line's departure is its length multiplied by the sine of its bearing. Its latitude is its length multiplied by the cosine of its bearing. Computations for the seven-sided traverse shown in figure 7-2 are begun in the form shown in figure 7-8, page 98.

In a closed traverse, algebraic sums of departures and latitudes each approach zero. Amounts by which they fail to reach zero reflect the linear error of closure. In figure 7-8, the total departure error is -0.62 feet and total latitude error is +0.23 feet. This means that the calculated position of the terminal point is 0.62 feet west and 0.23 feet north of the beginning point, figure 7-9, page 98.

When a line's departure and latitude is known, its bearing is calculated from the formula:

$$\text{Tan bearing} = \frac{\text{Dep}}{\text{Lat}}$$

Its length is calculated from the formula:

$$\text{Distance} = \frac{\text{Dep}}{\text{Sine bearing}} = \frac{\text{Lat}}{\text{Cos. bearing}}$$

$$\text{or Distance} = \sqrt{\text{Dep}^2 + \text{Lat}^2}$$

Calculation of a line's bearing and length from its known departure and latitude is called *inversing.*

| 1<br>Sta. | 2<br>Adjusted Bearing | 3<br>Distance | 4<br>Sine | 5<br>Cosine | 6<br>Departure | 7<br>Latitude |
|---|---|---|---|---|---|---|
| A | | | | | | |
| | N 40°03′E | 221.84 | 0.64346 | 0.76548 | +142.75 | +169.81 |
| B | | | | | | |
| | N 60°13′E | 349.65 | 0.86791 | 0.49672 | +303.46 | +173.68 |
| C | | | | | | |
| | S 17°19′E | 286.44 | 0.29765 | 0.95467 | + 85.26 | -273.46 |
| D | | | | | | |
| | S 84°49′E | 87.32 | 0.99591 | 0.09034 | + 86.96 | - 7.89 |
| E | | | | | | |
| | S 44°35′W | 171.47 | 0.70195 | 0.71223 | -120.36 | -122.13 |
| F | | | | | | |
| | S 84°29′W | 424.13 | 0.99537 | 0.09614 | -422.17 | - 40.78 |
| G | | | | | | |
| | N 37°09′W | 126.71 | 0.60390 | 0.79706 | - 76.52 | +101.00 |
| A | | 1667.56 | | | +618.43 | +444.49 |
| | | | | | -619.05 | -444.26 |
| | | | | | - 0.62 | + 0.23 |

Fig. 7-8

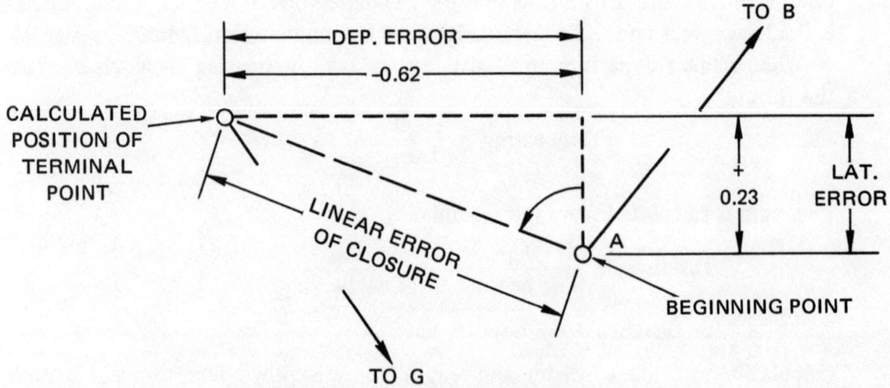


Fig. 7-9 Departure and latitude errors

Calculations of length and bearing of the linear error of closure are shown below:

$$\text{Distance} = \sqrt{(0.62)^2 + (0.23)^2}$$
$$= 0.66 \text{ feet}$$
$$\text{Tan Bearing} = \frac{-0.62}{0.23}$$
$$= -2.69565$$
$$\text{Bearing} = \text{N } 69^\circ 39' \text{W}$$

The precision of closure is the length of the linear error of closure, 0.66 feet, divided by the perimeter, 1667.56 feet. It is expressed as the ratio $\frac{1}{2530}$.

## FINAL ADJUSTMENT OF TRAVERSE

A major step in balancing a traverse is to eliminate the linear error of closure. It is done by distributing the errors in departures and latitudes. Most surveyors distribute errors by either the compass rule or the transit rule. When angles and distances are measured with the same precision, the compass rule is better. When angles are measured with greater precision than the distances, the transit rule is better.

Under either rule, each departure receives a share of the total departure error. Each latitude receives a share of the total latitude error. Under the compass rule, the share that each receives is proportionate to the length of the course. Under the transit rule, the share that each departure or latitude receives is proportionate to its own length. Adjustment of each departure and latitude is made to offset the amount of error distributed to it. Corrections are subtracted algebraically when the total error in departures or latitudes is plus. Corrections are added algebraically when the total error in departure or latitudes is minus. By doing this, numerical values that are too large are decreased, while numerical values that are too small are increased.

Figure 7-10, page 100, is an extension of figure 7-8. It shows corrections for each departure and latitude as computed by each rule. Corrections under the compass rule are computed as follows:

$$\text{Departure correction per 100 ft. of course length} = \frac{+0.62}{1667.56} \times 100 = +.037$$

$$\text{Latitude correction per 100 ft. of course length} = \frac{-0.23}{1667.56} \times 100 = -.014$$

| COURSE | COURSE LENGTH | DEP CORRECTION | LAT CORRECTION |
|---|---|---|---|
| AB | 221.84 | +0.08 | -0.03 |
| BC | 349.65 | +0.13 | -0.05 |
| CD | 286.44 | +0.11 | -0.04 |
| DE | 87.32 | +0.03 | -0.01 |
| EF | 171.47 | +0.06 | -0.02 |
| FG | 424.13 | +0.16 | -0.06 |
| GA | 126.71 | +0.05 | -0.02 |
| | 1667.56 | +0.62 | -0.23 |

| 1 | 6 | 7 | 8 | 9 | 10 | 11 | 12 | 13 | 14 | 15 |
|---|---|---|---|---|---|---|---|---|---|---|
| | | | Compass Rule | | | | Transit Rule | | | |
| | | | Departure | | Latitude | | Departure | | Latitude | |
| Sta. | Departure | Latitude | Cor. + | Adjusted Dep. | Cor. - | Adjusted Lat. | Cor. + | Adjusted Dep. | Cor. - | Adjusted Latitude |
| A | | | | | | | | | | |
| | +142.75 | +169.81 | 0.08 | +142.83 | 0.03 | +169.78 | 0.07 | +142.82 | 0.04 | +169.77 |
| B | | | | | | | | | | |
| | +303.46 | +173.68 | 0.13 | +303.59 | 0.05 | +173.63 | 0.15 | +303.61 | 0.05 | +173.63 |
| C | | | | | | | | | | |
| | + 85.26 | -273.46 | 0.11 | + 85.37 | 0.04 | -273.50 | 0.04 | + 85.30 | 0.07 | -273.53 |
| D | | | | | | | | | | |
| | + 86.96 | - 7.89 | 0.03 | + 86.99 | 0.01 | - 7.90 | 0.04 | + 87.00 | 0.00 | - 7.89 |
| E | | | | | | | | | | |
| | -120.36 | -122.13 | 0.06 | -120.30 | 0.02 | -122.15 | 0.06 | -120.30 | 0.03 | -122.16 |
| F | | | | | | | | | | |
| | -422.17 | - 40.78 | 0.16 | -422.01 | 0.06 | - 40.84 | 0.22 | -421.95 | 0.01 | - 40.79 |
| G | | | | | | | | | | |
| | - 76.52 | +101.00 | 0.05 | - 76.47 | 0.02 | +100.98 | 0.04 | - 76.48 | 0.03 | +100.97 |
| A | +618.43 | +444.49 | | +618.78 | | +444.39 | | +618.73 | | +444.37 |
| | -619.05 | -444.26 | | -618.78 | | -444.39 | | -618.73 | | -444.37 |
| | - 0.62 | + 0.23 | +0.62 | - - - | -0.23 | - - - | +0.62 | - - - | -0.23 | - - - |

Fig. 7-10 Traverse computation – Part 2

Corrections under the transit rule are computed as follows:

$$\text{Departure correction per 100 ft of length} = \frac{+0.62}{1237.48} \times 100 = +.050$$

$$\text{Latitude correction per 100 ft of length} = \frac{-0.23}{888.75} \times 100 = -0.26$$

| COURSE | DEP | DEP CORRECTION | LAT | LAT CORRECTION |
|---|---|---|---|---|
| AB | 142.75 | +0.07 | 169.81 | -0.04 |
| BC | 303.46 | +0.15 | 173.68 | -0.05 |
| CD | 85.26 | +0.04 | 273.46 | -0.07 |
| DE | 86.96 | +0.04 | 7.89 | -0.00 |
| EF | 120.36 | +0.06 | 122.13 | -0.03 |
| FG | 422.17 | +0.22 | 40.78 | -0.01 |
| GA | 76.52 | +0.04 | 101.00 | -0.03 |
| | 1237.48 | +0.62 | 888.75 | -0.23 |

Adjustment of departures and latitudes is completed when the algebraic sums of departures and latitudes each equal zero.

A final step is recomputing the lengths and bearings of each course as determined by the adjusted departures and latitudes. Results for the foregoing example adjusted by the compass rule are shown in figure 7-11, page 102. In practice, the essential columns of figure 7-8, 7-10, and 7-11 are arranged in a single table.

## RECTANGULAR COORDINATES

The relative positions of points are defined by adding and subtracting departures and latitudes. Numerous additions and subtractions are avoided by relating all survey points to common north-south and east-west axes. A north-south axis is called a y-axis, and an east-west axis is called an x-axis. Perpendicular distances to points east or west of a y-axis are x-coordinates. Those north or south of an x-axis are y-coordinates. X- and y-coordinates of a point are seldom of equal length. They normally form a rectangular pattern with x- and y-axes and are therefore called rectangular coordinates, figure 7-12, page 102.

The point of intersection of the x- and y-axes is called the *origin of coordinates*. Although it can be anywhere, it is best located west and south of all points of interest. By putting it there, the algebraic signs of all x- and y-coordinates are plus. The y-axis extends through or is west of the most western point of interest. The x-axis extends through or is south of the most southern point, figure 7-13, page 103.

| 1<br>Sta. | 9<br>Adjusted Departure | 11<br>Adjusted Latitude | 16<br>Tan. | 17<br>Final Adjusted Bearing | 18<br>Sin. | 19<br>Cos. | 20<br>Adjusted Distance |
|---|---|---|---|---|---|---|---|
| A | | | | | | | |
| | +142.83 | +169.78 | 0.84127 | N40°04′E | | 0.76530 | 221.85 |
| B | | | | | | | |
| | +303.59 | +173.63 | 1.74879 | N60°14′E | 0.86805 | | 349.74 |
| C | | | | | | | |
| | + 85.37 | -273.50 | 0.31214 | S17°20′E | | 0.98459 | 286.51 |
| D | | | | | | | |
| | + 86.99 | - 7.90 | 11.01139 | S84°49′E | 0.99591 | | 87.35 |
| E | | | | | | | |
| | -120.30 | -122.15 | 0.98485 | S44°34′W | | 0.71243 | 171.46 |
| F | | | | | | | |
| | -422.01 | - 40.84 | 10.33325 | S84°28′W | 0.99534 | | 423.99 |
| G | | | | | | | |
| | - 76.47 | +100.98 | 0.75728 | N37°08′E | | 0.79723 | 126.66 |
| A | | | | | | | 1667.56 |

Fig. 7-11 Traverse computation – Part 3

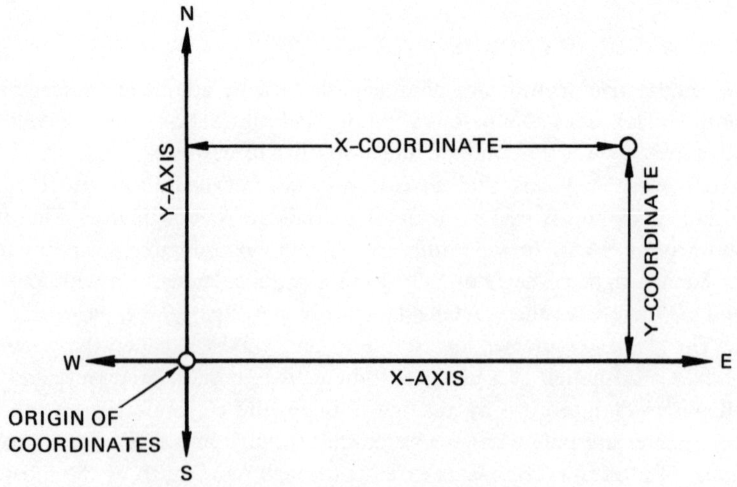


Fig. 7-12 System of rectangular coordinates

Any point's x- and y-coordinates are also the departure and latitude of a line between the origin and the point. The difference between x-coordinates for any two points is the departure of a line between them. The difference between y-coordinates is the latitude of a line between them, figure 7-14. With coordinates available, two subtractions provide the departures and latitudes for a direct line between any two points. Without coordinates, addition and subtraction of departures and latitudes for a series of courses is necessary.

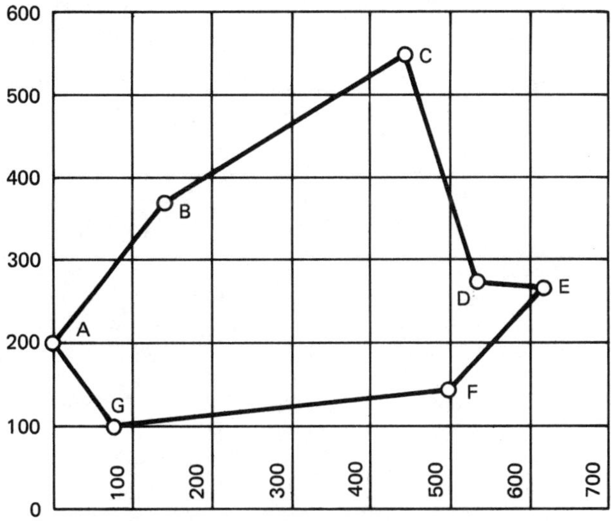


**Fig. 7-13 Fitting rectangular coordinates to a traverse**

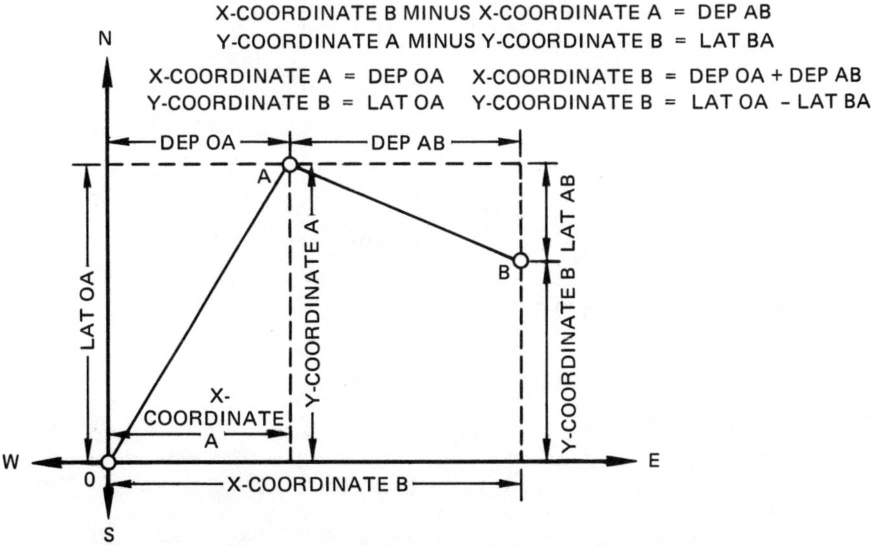


**Fig. 7-14 Correlation of coordinates with departures and latitudes**

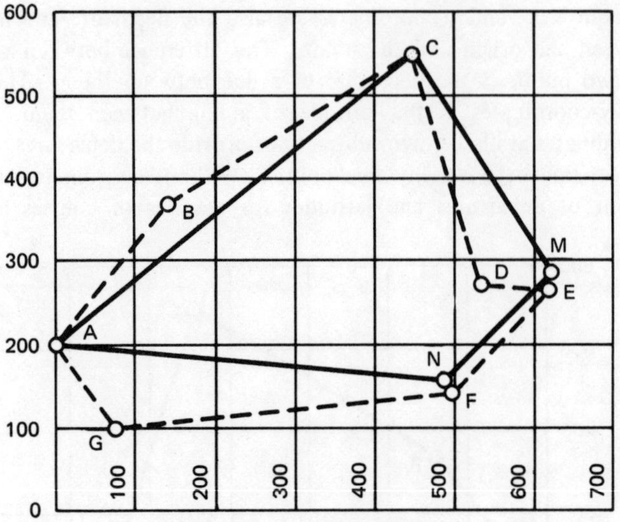


Fig. 7-15 A control traverse

| 1 | 9 | 11 | 21 | 22 |
|---|---|---|---|---|
| | | | Coordinates | |
| Sta. | Adjusted Departure | Adjusted Latitude | X | Y |
| A | | | 0.00 | 200.00 |
| | +142.83 | +169.78 | | |
| B | | | 142.83 | 369.78 |
| | +303.59 | +173.63 | | |
| C | | | 446.42 | 543.41 |
| | + 85.37 | -273.50 | | |
| D | | | 531.79 | 269.91 |
| | + 86.99 | - 7.90 | | |
| E | | | 618.78 | 262.01 |
| | -120.30 | -122.15 | | |
| F | | | 498.48 | 139.86 |
| | -422.01 | - 40.84 | | |
| G | | | 76.47 | 99.02 |
| | - 76.47 | +100.98 | | |
| A | | | 0.00 | 200.00 |

Fig. 7-16 Control traverse coordinates

In figure 7-15, the polygon ABCDEFG represents a control traverse for a four-sided field ACMN. An obstacle between A and C makes it necessary to establish point B. Corners M and N are large trees located by side shots from E and F. Obstacles between C and E makes point D necessary. The traverse is balanced and rectangular coordinates for all traverse points are computed, figure 7-16. Departures and latitudes for side shots EM and FN, and rectangular coordinates for points M and N are determined, figure 7-17, page 106.

The bottom half of figure 7-17 shows the calculated bearing and length of each side of the field ACMN. These calculated results start by listing the coordinates of A, C, M, N, and again A in sequence. Enter the departures and latitudes of AC, CM, MN, and NA as differences between coordinates of their terminal points. Determine the bearings and lengths of each side.

Coordinates simplify the calculation of distances and directions for setting points. Running and adjusting a random line between terminals that are not visible to each other is explained in chapter 6. Another method is to run a random traverse with any convenient intermediate points between such terminals, figure 7-18, page 107. Distance and direction between terminals are easily calculated from their rectangular coordinates. The traverse method is more practical than the random-line method in rough or wooded terrain.

Appreciation of rectangular coordinates grows as their usefulness in solving a wider variety of problems is experienced. The form used for figure 7-17 has sufficient flexibility for any of the examples of figures 7-8, 7-10, 7-11, and 7-16.

## CALCULATIONS

The following are suggested to assist in calculating traverses and coordinates:

- the departure divided by the sine occasionally gives values slightly different from the latitude divided by the cosine. The sine or cosine having the greater value gives the more reliable answer.
- placing parentheses around spaces in figures where calculated values are to be derived, makes checking easier. This method is used in figure 7-17.

For balancing traverses and calculating coordinates, familiarity with alternative calculating methods and calculating equipment is helpful. Students should be alert at all times to identify and apply rough checks mentally.

Slide rules are quick and effective for calculating distribution of errors by either the compass or transit rule. They aid in making rough checks on traverse calculations, and in interpolating values of trigonometric functions.

Calculators are essential for efficient calculation of traverse, rectangular coordinates, adjusted distances, and bearings. Calculators equipped with trigonometric functions should be used only after first mastering the use of trigonometric tables.

| 1 Sta. | 2 Bearing | 3 Distance | 4 Sin | 5 Cos. | 6 Tan | 7 Departure | 8 Cor. | 9 Latitude | 10 Cor. | 11 Coordinates X | 12 Coordinates Y | 13 Computed Bearing | 14 Computed Distance | 15 Remarks |
|---|---|---|---|---|---|---|---|---|---|---|---|---|---|---|
| E | | | | | | | | | | 618.78 | 262.01 | | | From Fig. 7-16 |
| | N5°00′E | 20.4 | 0.08716 | 0.99619 | | (+1.78) | | (+20.32) | | | | | | |
| M | | | | | | | | | | (620.56) | (282.33) | | | |
| F | | | | | | | | | | 498.48 | 139.86 | | | From Fig. 7-16 |
| | N38°30′W | 18.6 | 0.62251 | 0.78261 | | (−11.58) | | (+14.56) | | | | | | |
| N | | | | | | | | | | (486.90) | (154.42) | | | |
| A | | | | | | | | | | 0.00 | 200.00 | | | From Fig. 7-16 |
| | | | 0.79264 | | (1.1996) | (+446.42) | | (+343.41) | | | | (N52°26′E) | (563.21) | |
| C | | | | | | | | | | 446.42 | 543.41 | | | From Fig. 7-16 |
| | | | | 0.83192 | (0.66700) | (+174.14) | | (−261.08) | | | | (S33°42′E) | (313.83) | |
| M | | | | | | | | | | 620.56 | 282.33 | | | From upper half this table |
| | | | 0.72256 | | (1.04493) | (−133.66) | | (−127.91) | | | | (S46°16′W) | (184.98) | |
| N | | | | | | | | | | 486.90 | 154.42 | | | From upper half this table |
| | | | 0.99565 | | (10.6823) | (−486.90) | | (+45.58) | | | | (N84°39′W) | (489.03) | |
| A | | | | | | | | | | 0.00 | 200.00 | | | From Fig. 7-16 |

Fig. 7-17 Side shot and boundary coordinates

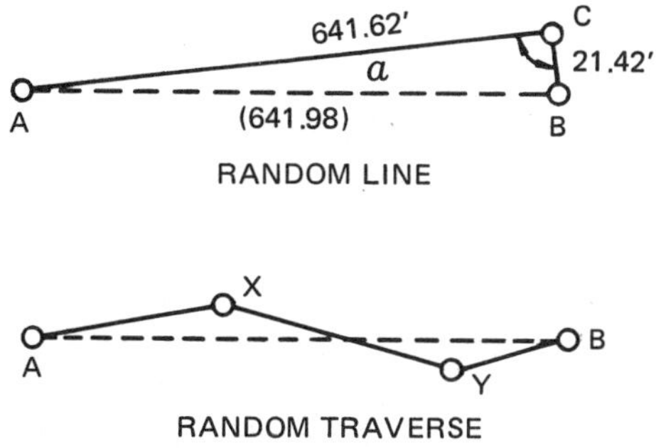


**Fig. 7-18 Random traverse vs. random line method**

## CALCULATORS

Sophisticated electronic circuitry and modern technology has brought the price of small battery-operated and desk-top calculators low enough to make them available to almost everyone, including the surveyor. Portable calculators with full trigonometric capability are used often to check data in the field. In the office, they eliminate many hours of laborious computations previously done by hand. They are essential for efficient calculation of traverse, rectangular coordinates, adjusting distances, and bearings.

Some calculators, such as those in figure 7-19, are programmable, performing many operations automatically. They calculate traverses, inverses, horizontal

**Fig. 7-19 AC/DC powered programmable display calculator**

**Fig. 7-20 This calculator stores information and programs on magnetic cards**

and vertical curve data, coordinates, areas, and perimeters. They convert azimuths to bearings and bearings to azimuths.

Desk-top programmable calculators, figure 7-20, use magnetic tape or magnetic card systems to store programs and information. The operator punches in data from the field notes. The calculator can balance a traverse, print out coordinates for all traverse points and side shots, calculate the area, and list direction and length of the property boundaries. It corrects tape measurements for temperature; reduces slope distances to horizontal distances and differences in elevation; translates coordinates to the state plane coordinate system; reduces stadia notes; and calculates curve data, profile grades, and earthwork volumes.

Wide printer attachments, figure 7-21, are available that print out plots of lines and curves using dots between tick marks at coordinate points. It prints the number of each point, the bearings and lengths of lines in tabular form in a corner of the plot, and lists curve data. It attaches to a desk-top programmable calculator.

## SUMMARY

Small unavoidable errors in measurement of angles and distances prevent the exact closure of traverses. The angular discrepancy is called the angular error of closure. The linear discrepancy is called the linear error of closure.

**Fig. 7-21 Page printer system and working plot attached to desk-top programmable calculator**

The terms error of closure and precision of closure are interchangeable. Both refer to the ratio of the linear error of closure to the length of the traverse perimeter. They are expressed as a fraction with a numerator of 1.

Balancing the traverse is a process of adjusting angles and distances by logically distributing errors. During the process, the traverse is oriented with respect to a reference meridian.

Distribution of angular error of closure is accomplished before the linear error of closure is calculated.

Traverse orientation needs adjustment when it is based on a random selection of compass observations for a single course. A second stage procedure places reliance upon the average of all acceptable compass observations for all courses.

Distances are resolved into east-west and north-south components called departures and latitudes. Any traverse line's departure is its length multiplied by the sine of its bearing. Its latitude is its length multiplied by the cosine of its bearing.

The algebraic sum of departures for all courses of a closed traverse approach zero. The algebraic sum of latitudes do the same thing. Amounts by which they fail to reach zero are the departure and latitude of the linear error of closure.

The linear error of closure's length equals the square root of the sum of the squares of its departure and latitude.

Disposition of linear errors of closure is made by adjusting departures and latitudes of a traverse's courses. Errors are distributed by either the compass rule or the transit rule. Distribution by the compass rule is in proportion to the length of the courses. Distribution by the transit rule is in proportion to lengths of the departures and latitudes themselves. Corrections have algebraic signs opposite to errors. The adjustment of departures and latitudes is complete when their algebraic sums each equal zero.

The final step in balancing a traverse is the recomputation of all bearings and courses on the basis of adjusted departures and latitudes.

After departures and latitudes are adjusted, positions of points are defined by rectangular coordinates.

A system of rectangular coordinates requires establishment of an east-west x-axis, and a north-south y-axis. An x-coordinate of a point is its departure measured from the y-axis. A y-coordinate of a point is its latitude measured from the x-axis.

The difference in x-coordinates of terminal points of any line is the line's departure. The difference in y-coordinates of any line is the line's latitude. The departure and latitude of any line is calculated by subtracting one set of terminal point coordinates from the other. The departure and latitude of any line is then easily translated to the line's bearing and length.

Plane coordinates make it easier to set points, run lines around obstacles, and do calculations for a wide variety of problems.

It is important to develop skill with slide rules, calculators, and tables of trigonometric functions.

## ACTIVITIES

1. Balance the traverse run by deflection angles in chapter 6, activity 9 using the compass rule, and the transit rule.
2. Select x- and y-axes and compute rectangular coordinates of the traverse points as balanced by the compass rule in activity 1.
3. Run a random traverse between two points that are not visible to each other. Calculate the coordinates of both points. From those coordinates, calculate the bearing and length of a direct line between them.
4. Choose a bearing for a line 500.00 feet long that causes it to penetrate a thickly wooded area. Run a random traverse using several intermediate points so positioned as to require little, if any, brush cutting. When occupying a point near the estimated position of the far terminal, calculate the bearing and distance of the remaining course. Set a point at the far terminal.
5. On an available plot of land, run a traverse and take any necessary side-shots to locate corners. Balance the traverse and calculate the bearings and lengths of all boundary lines.

## REVIEW QUESTIONS

### A. Multiple Choice

1. The exact closure of a traverse is possible
   a. when care is taken in measuring distances.
   b. when care is taken in measuring angles.
   c. when traverses are properly oriented.
   d. never.

2. The departure of a course can be defined as
   a. the course's length times the cosine of its bearing.
   b. the east-west component of the course.
   c. the difference in x-coordinates of a course's terminal points.
   d. the difference in y-coordinates of a course's terminal points.

## B. Short Answer

3. What is another term for error of closure?
4. In balancing a traverse, which kind of error is identified and distributed first?
5. What is the difference in x-coordinates of terminal points of any line called? What is the difference in y-coordinates of terminal points of any line called?
6. For a closed traverse, how many degrees do the sums of clockwise deflection angles differ from the sums of counter-clockwise deflection angles?
7. A traverse line is located in the NE quadrant when algebraic signs of both departure and latitude are plus. In what quadrant is a traverse line located when:
   a. the departure is plus and latitude minus?
   b. the departure is minus and latitude plus?
   c. both departure and latitude are minus?

## C. Problems

8. A deed calls for a boundary course CD to run N 60°00′E for a distance of 1200.00 feet. Lines run by two surveyors from point C while searching for point D are 21 feet apart after measuring 1200.00 feet. What is the difference in bearings of the two lines?
9. What is the departure and latitude of the course CD in question 8?
10. For a closed traverse the summation of departures is -0.43 ft and the summation of latitudes is -0.36 ft. The perimeter is 1400.00 ft. Under the compass rule, what is the departure correction for the 161.50 ft Northeast course A–B? What is the latitude correction for the 112.40 ft Northwest course B–C?
11. Coordinates of points B and F are:

| Point | X-coordinate | Y-coordinate |
|---|---|---|
| B | 531.42 | 101.27 |
| F | 938.53 | 306.55 |

What is the bearing and length of a line from B to F?

12. For a closed traverse, the summation of departures is -0.43 feet and the summation of latitudes is -0.36 feet. The perimeter is 1400.00 feet. Under the compass rule, what is the departure correction for the 161.50-foot northeast of course AB? What is the departure correction for the 112.40-foot northwest of course BC?

13. The departure of course MN is 300.00 feet. The latitude of course MN is -150.00 feet. What is the bearing and length of course MN?

14. In figure 7-22, point C is to be set 278.43 feet from point B. Bearing of BC is N44°00′E. Line of sight along BC is obstructed by thick woods. A random traverse is run using points X and Y where lines of sight are clear. Course BX has a bearing of N32°41′E and a length of 117.72 feet. Course X-Y has a bearing of N71°49′E and a length of 125.04 feet. Assume the origin of coordinate at B. From calculated coordinates for Y and C, derive the departure and latitude of course YC. What is the bearing and length of course YC? What is the deflection angle to be turned at Y to set point C?

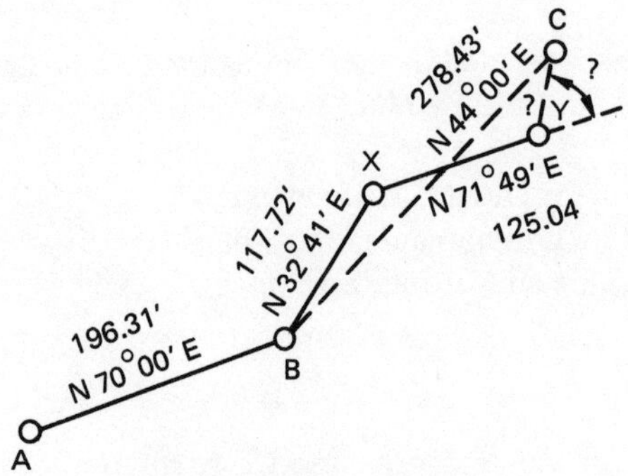


**Fig. 7-22 Random traverse**

15. Forward and backward compass bearings for seven courses are as follows:

| Point | Backward | Forward |
|---|---|---|
| A | | |
| | N14°W | S14 1/2°E |
| B | | |
| | S89 1/2°W | S87 1/2°E |
| C | | |
| | S57°W | N54°E |
| D | | |
| | S9 1/2°E | N9°W |
| E | | |
| | S76 1/2°E | N77 1/2°W |
| F | | |
| | N60°E | S60 1/2°W |
| G | | |
| | N17 1/2°E | S17°W |
| A | | |

Which courses are unacceptable as a basis for orienting the traverse?

16. Deflection angles turned at stations listed in question 15 are: A, 31°42′L; B, 72°08′L; C, 36°17′L; D, 67°05′L; E, 67°19′L; F, 43°29′L; G, 42°02′L. Courses adjacent to stations C and G have the shortest lengths. Give equal weight to the numerical values of backward and forward bearing of course AB as a basis for orientation. Calculate the adjusted azimuths for all courses on a trial basis as illustrated in figure 7-4, columns one through five.

17. Using the data for questions 15 and 16, follow the second stage procedure as illustrated in columns 6 through 9 of figure 7-4. Determine the adjusted bearings depending on all acceptable observed compass bearings. Give equal weight to the numerical values of backward and forward bearings for each course.

18. Copy and complete the following table, calculating corrections by the transit rule:

| **Unbalanced** | | **Corrections** | | **Balanced** | |
|---|---|---|---|---|---|
| **Dep** | **Lat** | **Dep** | **Lat** | **Dep** | **Lat** |
| +113.42 | - 80.61 | | | | |
| +142.53 | +121.17 | | | | |
| +216.09 | + 46.53 | | | | |
| -384.73 | +286.21 | | | | |
| - 87.01 | -373.58 | | | | |

# CHAPTER 8

# Basic Map Plotting and Drafting

## OBJECTIVES

After studying this chapter, the student will be able to:

- plot positions by rectangular coordinates; or protractor, drafting compass, and engineer's scale.
- make a land survey map conforming to specifications of governing state boards of registration.
- use basic map-making equipment properly and demonstrate basic map-making skills.

## BASIC TOOLS AND SKILLS

A surveyor makes rough sketches in his notebook to show the approximate relative positions of points. Maps are made in the office showing relative positions of points to exact scale.

Basic map-making equipment consists of:

- drawing board
- T square
- pair of triangles, one 45° and one 30° -60°
- protractor
- engineer's scale
- set of drawing instruments
- hard pencils, 4H to 6H
- pen
- eraser
- erasing shield
- black drawing ink

Basic drafting skills consist of:

- manipulating T square and triangles to draw parallel lines, perpendiculars, and intersecting lines at angles in multiples of 15 degrees
- hand lettering
- scaling distances with engineer's scale
- making angles with a protractor
- inking with ruling pens
- making circles with a compass

Unless a surveyor produces neat maps of good quality, the whole survey is usually judged as poor. This is true even though the field work and computations are performed skillfully.

## MAP PLOTTING METHODS

Mapping begins by plotting positions of traverse points or other horizontal control points. Traverses are balanced and traverse computations completed before mapping begins. In some instances, preliminary plotting of control points directly from the field notes is done. This aids in tracking down major errors. Preliminary plotting helps in the study of unfinished traverses for the best survey routes to traverse terminal points or other features. Before modern calculators became available, plotting of traverses directly from field notes was more frequently practiced. In the past, graphical balancing of a survey was sometimes accepted as an alternative to plotting from computed survey results. This practice is hard to justify today, considering new mathematical aids for the surveyor.

Plotting positions by rectangular coordinates is the most reliable and acceptable method. Good quality cross-section paper, preferably ten lines to the inch each way, is used for ordinary work. It is available with non-photographic blue grid lines which are screened out during map reproduction. Only lines or data entered boldly with pencil or ink on this cross-section paper appear on reproductions. This eliminates the need for retracing or inking on overlays of tracing cloth or other transparent material. A map scale is selected which permits the largest layout of the surveyed area within the desired marginal limits. Values of x- and y-coordinates are entered around the perimeter nearly every inch. They are arranged so that the surveyed area is centered between the margins, figure 8-1, page 116.

Positions of traverse points and other control points are plotted by their x- and y-coordinates. Each position is marked by making a small diameter circle around each point with a bow compass. Lines connecting traverse points are drawn but not allowed to extend inside of those small circles, figure 8-2, page 116. Lines are ordinarily dotted or dashed to avoid domination over more important features. Scaled lengths from center-to-center of the small circles are checked against computed course lengths. Bearings of lines scaled with a pro-

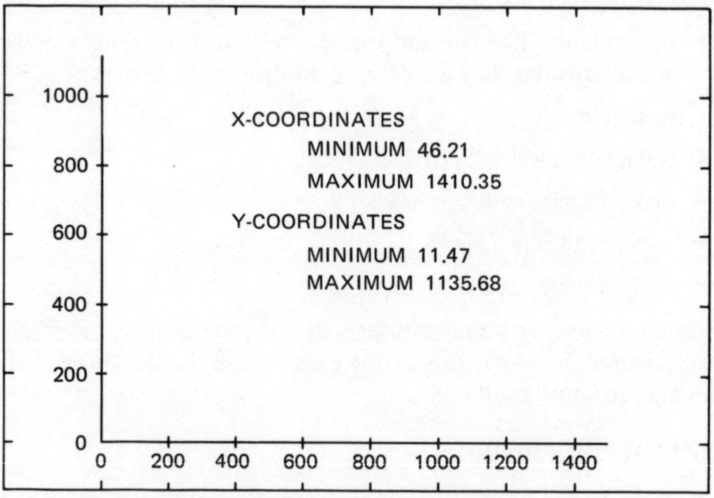


**Fig. 8-1 Arrangement of coordinate values**

tractor are checked against computed bearings. Property corners not occupied as traverse points are also plotted by their x- and y-coordinates. They are likewise marked by circles, connected, and checked. If property boundaries are the dominant feature of the map, their connecting lines are solid.

Many subsidiary points measured by side shots or other field methods are plotted with a protractor, drafting compass, and scale. For side shots, the same angles and distances as measured in the field are plotted. Positions of points located by two ties are plotted at the intersection of arcs swung with a compass. Lines are not ordinarily drawn between control points and subsidiary points.

A tangent method of plotting angles is sometimes used instead of a protractor, figure 8-3. The tangent method is more accurate than a protractor when the ten-unit distance is longer than the radius of the protractor.

For example, in figure 8-3, angle BAC of 25°00′ is to be drawn. Base line AB is extended 10 inches, or any other convenient length. The tangent of 25°00′ is 0.46631. Therefore, vertical line BC is marked off 4.6631 inches long, which is found by multiplying length AB times the tangent of the angle. From point C a line is drawn back to the vertex A to define the angle.

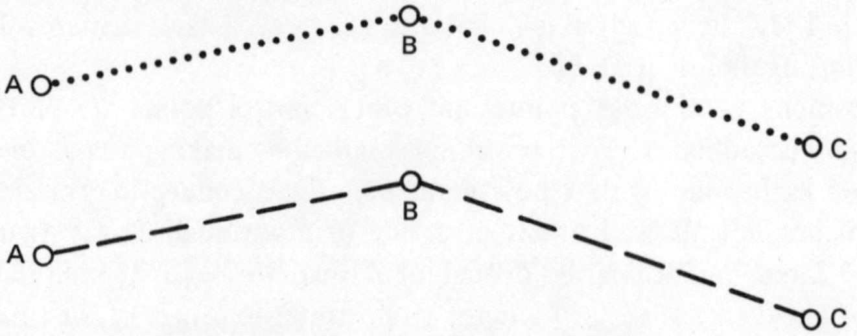


**Fig. 8-2 Symbols for traverse points and lines**

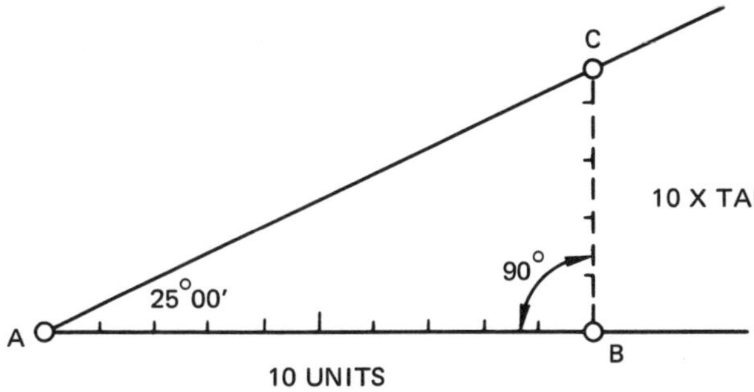


**Fig. 8-3 Tangent method of plotting angles**

Traverses may also be plotted on cross-section paper without referring to x- and y-coordinates. Directions of courses are first listed as azimuths or bearings. The first angle's north-south base is drawn parallel to a vertical printed line. The azimuth or bearing is laid off from the vertical base. A line is then drawn in the forward direction. The length of the first course is laid off along that line. The same process is repeated at each point. North-south bases of all angles defining the azimuth or bearing coincide with or are parallel to the vertical printed lines. This orientation prevents drifting of the figure by accumulation of unavoidable plotting errors common to protractor or tangent methods.

## ELEMENTS OF A FINISHED MAP

Common features of most maps are:

- complete title
- map scale and scale bar
- reference meridian (north arrow)
- legend
- margin

For maps of land surveys, specifications of governing state authorities are followed. In addition to the common features, the following information is usually included:

- length and bearing of each boundary
- names of owners of the tract that is mapped, and the names of owners of adjoining tracts
- area of tract mapped
- description of monuments (existing or set) together with reference ties
- locations and names of roads, streams, and other landmarks
- precision of closure
- surveyor's certification

Figure 8-4 is an example of a finished map. The title states:

- what the map represents
- where the survey is located
- who it was performed for
- when it was made
- the surveyor's name

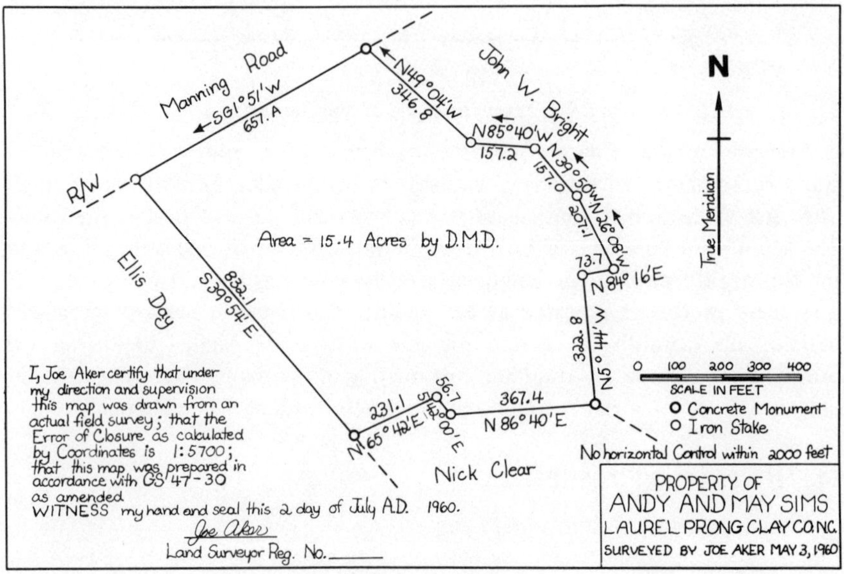


**Fig. 8-4 A finished map**

The wording is well arranged, neat, and symmetrical. A graphical scale is preferred. Stating the number of feet to an inch is worthless if reproductions to larger or smaller sizes are made. Reference meridians show whether they are true or magnetic north. A legend explains symbols for all map features shown. A margin gives a map a finished look and makes binding and filing easier.

Horizontal control points established by traverse or other methods are usually plotted on preliminary drafts of maps. They do not necessarily appear on the finished maps. Each map is made for a specific purpose and is not cluttered with unnecessary information.

## SUMMARY

A surveyor makes maps showing the relative positions of points to scale. Basic work is performed with simple, inexpensive equipment.

A surveyor's work is judged largely on the neatness and quality of the maps, regardless of the accuracy of the field work.

Mapping begins by plotting the positions of the traverse or other horizontal control points.

Plotting positions by rectangular coordinates is the most reliable and acceptable method. Many subsidiary points are plotted by protractor, drafting compass, and scale.

A tangent method of plotting angles is more accurate than a protractor when distances to be plotted are larger than the radius of the protractor.

Cross-section paper is usually used for plotting control points.

Common features of most maps include a complete title, a map scale, re--erence meridian, legend, and margin.

Maps of land surveys must meet the specifications of the governing state authorities.

Horizontal control points established by traverse or other methods do not have to appear on finished maps.

## ACTIVITIES

1. With protractor and engineer's scale, plot the traverse run as in chapter 6, activity 9, directly from the field notes. Scale the linear error of closure and compare it with the calculated error of closure.
2. By the tangent method, plot the traverse run in chapter 6, activity 9, directly from the field notes. Compare the scaled and calculated errors of closure.
3. Plot the adjusted traverse by rectangular coordinates using the coordinates as calculated in chapter 7, activity 2. Scale the lengths of courses and compare them with the adjusted lengths of courses as computed.
4. After studying the deed descriptions for a tract of land and its adjoining tracts, make a field survey to locate the corners and all man-made improvements. Balance the traverse. Compute the coordinates of all corners, and the lengths of its boundaries. Plot the tract boundaries and the locations of all man-made improvements. Use a scale and a form acceptable for recording plots in the county where the tract is situated. Include the title block, map scale, reference meridian, legend, appropriate margin, length and bearing of each boundary, names of owners of adjoining tracts, and precision of closure.

## REVIEW QUESTIONS

### A. Multiple Choice

1. The first points plotted on a map identify
   a. prominent features such as property corners.
   b. horizontal control points.
   c. subsidiary points.
   d. departures and latitudes.
2. The most reliable and acceptable method of plotting is by
   a. a good quality cross-section paper.
   b. protractor, drafting compass, and scale.
   c. the tangent method of plotting angles.
   d. rectangular coordinates.

3. When plotting with a protractor, the best results are obtained when the base line of the protractor with the 0-degree reading is along
   a. a datum
   b. a traverse line.
   c. a magnetic north line.
   d. or parallel to a vertical printed line on cross-section paper.
4. When plotting by rectangular coordinates, the x- and y-axes are laid out
   a. along marginal lines.
   b. so as to be oriented to true north.
   c. so that the surveyed area is centered between margins.
   d. parallel to marginal lines.
5. Directions of lines are more accurate when plotted by
   a. the tangent method of plotting deflection angles with a short base line.
   b. the tangent method of plotting deflection angles with a long base line.
   c. rectangular coordinates.
   d. skillful handling of a protractor with a large diameter graduated circle.
6. A graphical scale is preferred to a scale that only states the number of feet to an inch because
   a. a map user finds them easier.
   b. they give a professional appearance.
   c. reproductions may be larger or smaller.
   d. graphical scales are traditional.

**B. Short Answer**

7. How often is plotting of traverses directly from field notes practiced?
8. Should horizontal control points show on a finished map?
9. List five common features of most maps.
10. List five or more requirements for information usually specified by state authorities that must appear on maps of land surveys in addition to common features.
11. What is a worthwhile reason for plotting a traverse directly from field notes?
12. Where does the date and location of a survey appear on a map?

# CHAPTER 9

# Areas

## OBJECTIVES

After studying this chapter, a student will be able to:

- determine the approximate areas of tracts of land from maps.
- calculate the precise areas of measured tracts directly from field notes without any mapping, except for those bordering on circular curves.
- recognize the value of viewing all areas bounded by straight lines as networks of triangles.

## AREA BY PLANIMETER

When precise determinations of land area are needed, they are calculated from field measurements. Where maps are available, it is possible to determine the approximate areas by calculations based on scaled values, or by planimeter.

A *planimeter,* figure 9-1, page 122, is an instrument with two pivoting arms which measures areas plotted on paper. The paper is laid out smoothly on a drawing board. A pole arm pivots around an anchor point at one of its ends. Its opposite end is a pivotal point for one end of a tracing arm. At the tracing arm's opposite end is a tracing point. As the tracing point is made to follow around the area's perimeter, a graduated roller revolves. Upon returning the tracing point to the starting point, the roller's revolutions are proportional to the circumscribed area.

The number of revolutions is indicated by three scales: (1) a disc geared to the roller indicates the number of complete revolutions, (2) a drum on the roller indicates the number of tenths and hundredths of a revolution, (3) a vernier for the drum indicates the number of thousandths of a revolution. The difference between the initial and final readings is the number of revolutions. One revolution on a fixed arm instrument is usually 10.00 square inches. It may be made equal to some other value on adjustable arm instruments. Complete instructions are furnished by the manufacturers.

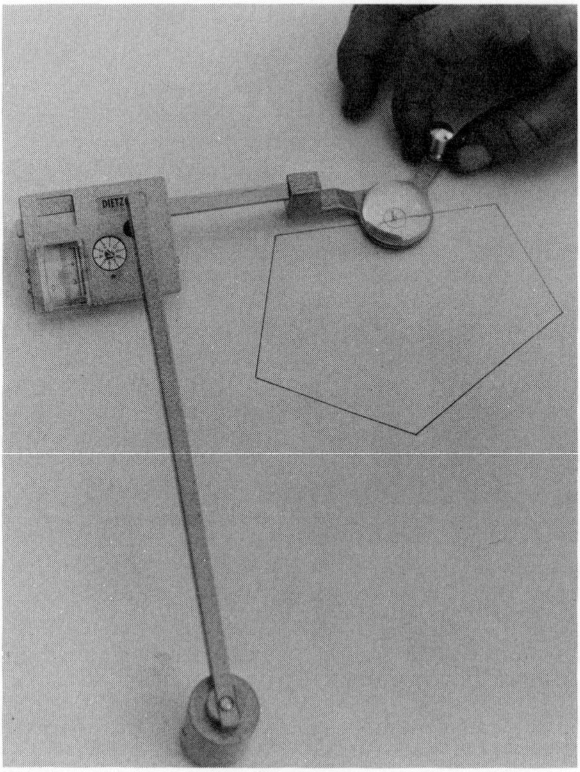


**Fig. 9-1 Planimeter**

To operate the planimeter, a starting point is chosen. This point is marked near the middle of a straight line on the perimeter of the area being measured. The tracing arm is arranged parallel to that straight line with the pole arm perpendicular to it, figure 9-1. This causes the roller to revolve slowly at the beginning and the end of its travel, producing more consistent results.

When converting a map area to ground units, the linear scale of the map is squared. Where a linear scale is one inch = 100 feet, one square inch represents 10,000 square feet.

## AREAS BOUNDED BY STRAIGHT LINES

Areas bounded by straight lines are viewed as a network of triangles for calculating purposes. Areas of oblique triangles are calculated by trigonometry when:

- three sides are known
- two sides and an included angle are known
- one side and three angles are known.

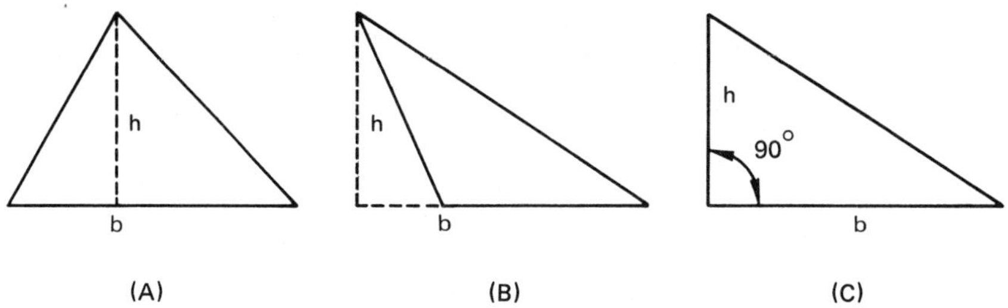


**Fig. 9-2 Area = 1/2 bh in all three triangles**

A common method is to calculate the area as one-half the base times the altitude. The altitude may strike the base directly, figure 9-2A, or its extension, figure 9-2B. If it coincides with a side, the triangle is a right triangle, figure 9-2C. In all three cases the formula below is true:

$$\text{Area} = 1/2\ bh$$

An irregular tract of land ABCDEF, figure 9-3, is visualized as consisting of four triangles. The area of polygon ABCDEF is calculated as:

(A) $$A = \frac{(BC)(h_1) + (AC)(h_2) + (AF)(h_3) + (DE)(h_4)}{2}$$

The length of each base and altitude is scaled from the drawing if not already known. Lengths of bases BC, AF, and DE are likely known from field measurements. Altitudes can also be measured in the field.

Formulas and procedures have been developed for computing the areas of triangles grouped in patterns, figure 9-4, page 124. Perpendiculars are drawn from each point of an irregular boundary ABCDEFG to line AG. Diagonals Bc, Dc, De, and Fe are drawn. Five pairs of triangles are formed, each pair having one perpendicular as a common base. One-half of base Bb times the distance Ac is the area of the first pair of triangles. Viewing all pairs of triangles similarly, only five multiplications give the combined area of the ten triangles. More specifically, the area of polygon ABCDEFG is:

(B) $$A = 1/2\ [(Bb)(Ac) + (Cc)(bd) + (Dd)(ce) + (Ee)(df) + (Ff)(eG)]$$

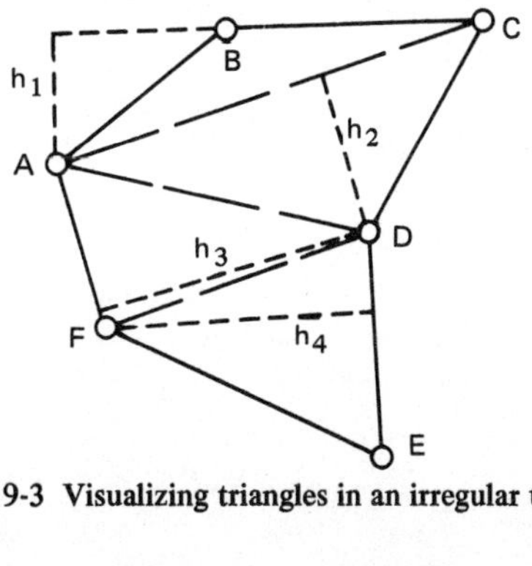


**Fig. 9-3 Visualizing triangles in an irregular tract**

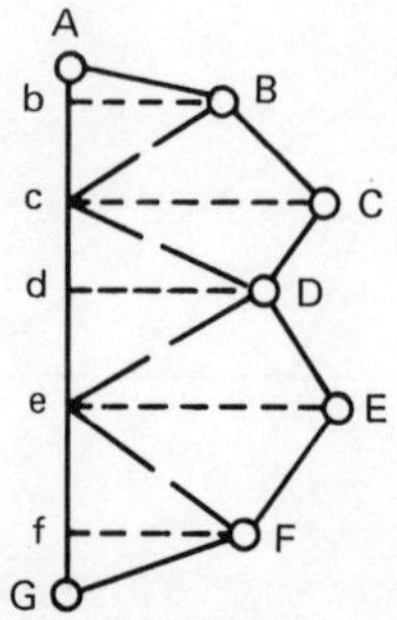


**Fig. 9-4 Pattern of triangles**

Perpendiculars may also be drawn to any external line, such as MN, figures 9-5 and 9-6. Area ABCDEFG is seen to be the difference between shaded area bBCDEFGg and shaded area bBAGg. Triangles are formed by drawing perpendiculars from each point of the polygon to line MN. Diagonals are drawn from the alternate perpendiculars as shown. Each end perpendicular serves as a base for one triangle. Each intermediate perpendicular serves as a common base for two triangles.

(C) Area bBCDEFGg = 1/2 [(Bb)(bc) + (Cc)(bd) + (Dd)(ce) + (Ee)(df) + (Ff)(eg) + (Gg)(fg)]

Area bBAGg = 1/2 [(Bb)(ba) + (Aa)(gb) + (Gg)(ga)]

Area ABCDEFG = Area bBCDEFGg–Area bBAGg

(D) Area ABCDEFG = 1/2 [-(Aa)(gb) + (Bb)(ac) + (Cc)(bd) + (Dd)(ce) + (Ee)(df) + (Ff)(eg) - (Gg)(fa)]

This pattern is the basis for two widely used methods for calculating areas bounded by traverses or property lines. They are the coordinate method and the double meridian distance method. Most state boards of registration require surveyors to be familiar with both of these methods.

## COORDINATE METHOD

A tract of land ABCDE, figure 9-7, is plotted by coordinates. It has an area equal to the shaded area bBCDd minus the shaded area bBAEDd. Area bBCDd, figure 9-8, page 126, when viewed as a group of four triangles equals 1/2 [(Bb)(bc) + (Cc)(bd) + (Dd)(cd)]. Area bBAEDd, figure 9-9, page 126, when viewed as a group of six triangles equals 1/2 [(Bb)(ba) + (Aa)(be) + (Ee)(ad) + (Dd)(ed)]. All perpendiculars Aa, Bb, Cc, Dd, and Ee are x-coordinates of points A, B, C, D, and E. All distances with end points identified by small letters only are differences in y-coordinates. Substituting x- and y-coordinate designations for the letters, the large shaded area is:

$$\text{Area bBCDd} = 1/2\ [X_B(Y_B - Y_C) + X_C(Y_B - Y_D) + X_D(Y_C - Y_D)]$$

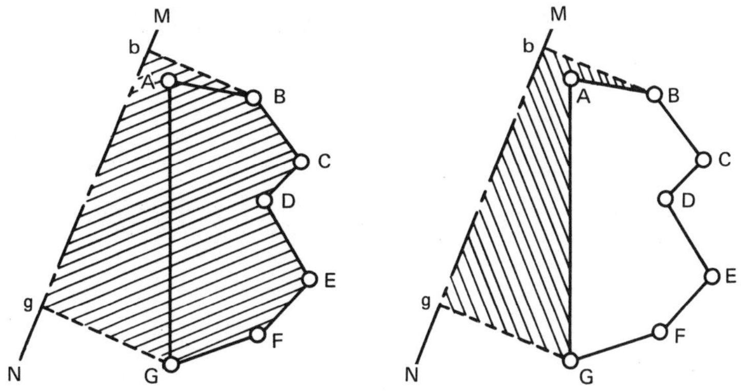


**Fig. 9-5 Polygon related to external line MN**

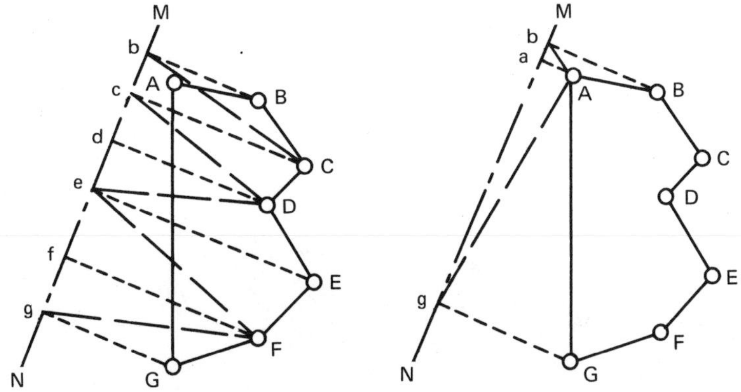


**Fig. 9-6 Pattern of triangles related to external line MN**

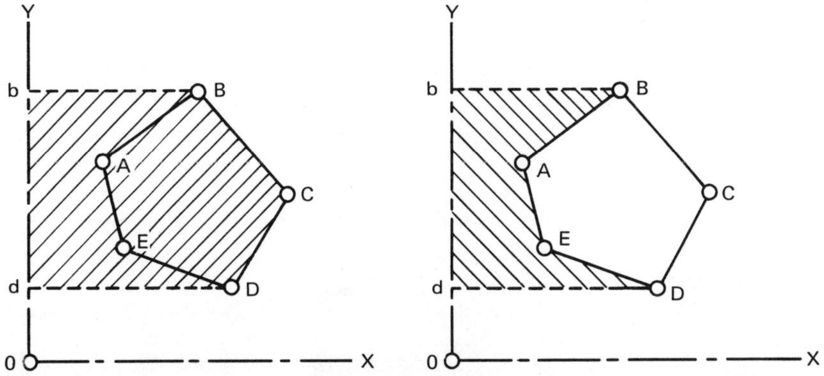


**Fig. 9-7 Tract plotted by coordinates**

The small shaded area is:

$$\text{Area bBAEDd} = 1/2\ [X_B(Y_B-Y_A)+X_A(Y_B-Y_E)+X_E(Y_A-Y_D) +X_D(Y_E-Y_D)]$$

The difference in shaded areas is:

$$\text{Area ABCDE} = 1/2\ [X_BY_B-X_BY_C+X_CY_B-X_CY_D+X_DY_C-X_DY_D] - 1/2\ [X_BY_B-X_BY_A+X_AY_B-X_AY_E+X_EY_A-X_EY_D +X_DY_E-X_DY_D]$$

(E) or $$\text{Area ABCDE} = 1/2\ [X_A(Y_E-Y_B)+X_B(Y_A-Y_C)+X_C(Y_B-Y_D) +X_D(Y_C-Y_E)+X_E(Y_D-Y_A)]$$

(F) or $$\text{Area ABCDE} = 1/2\ [X_AY_E+X_BY_A+X_CY_B+X_DY_C+X_EY_D -X_AY_B-X_BY_C-X_CY_D-X_DY_E-X_EY_A]$$

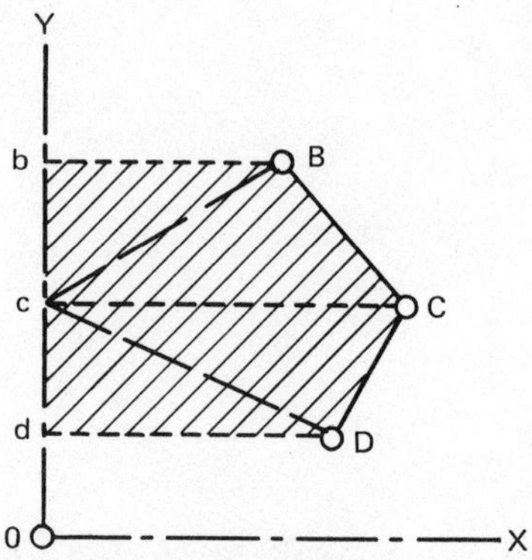


**Fig. 9-8 Positive pattern of triangles for area calculation**

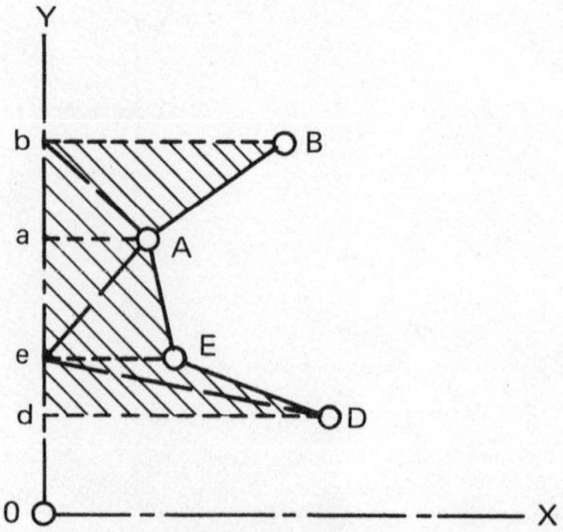


**Fig. 9-9 Negative pattern of triangles for area calculation**

Equations (E) and (F) are the bases for two alternate rules for determining areas by coordinates:

Rule 1: The area of any polygon is one-half the sum of each points' x-coordinate multiplied by the difference in y-coordinates of adjacent points, always subtracting the following y-coordinate from the preceding y-coordinate.

Rule 2: The area of any polygon is one-half the sum of each points' x-coordinate multiplied by the y-coordinate of its preceding point, less one-half the sum of each point's x-coordinate multiplied by the y-coordinate of its following point.

Calculation of the area ABCDE shown in figure 9-7 by Rule 1 is as follows:

| STA | X | Y | | RULE 1 PRODUCTS + | RULE 1 PRODUCTS - |
|---|---|---|---|---|---|
| A | 150 | 480 | (150)(-400) = | | 60,000 |
| B | 370 | 600 | (370)(+130) = | 48,100 | |
| C | 600 | 350 | (600)(+500) = | 300,000 | |
| D | 500 | 100 | (500)(+150) = | 75,000 | |
| E | 210 | 200 | (210)(-380) = | | 79,800 |

423,100   139,800
139,800
2) 283,300
141,650

By Rule 2, it is as follows:

| STA | X | Y | RULE 2 PRODUCTS + | | RULE 2 PRODUCTS - | |
|---|---|---|---|---|---|---|
| A | 150 | 480 | (150)(200) = | 30,000 | (150)(600) = | 90,000 |
| B | 370 | 600 | (370)(480) = | 177,600 | (370)(350) = | 129,500 |
| C | 600 | 350 | (600)(600) = | 360,000 | (600)(100) = | 60,000 |
| D | 500 | 100 | (500)(350) = | 175,000 | (500)(200) = | 100,000 |
| E | 210 | 200 | (210)(100) = | 21,000 | (210)(480) = | 100,800 |

763,600   480,300
480,300
2) 283,300
141,650

Algebraic signs in these calculations are reversed if courses are listed in a counterclockwise direction. The net area always carries a minus sign for counterclockwise listings. The negative is regarded as irrelevant since only the numerical value is important.

## DOUBLE MERIDIAN DISTANCE METHOD

Another traditional method is the double meridian distance method. It is used when latitudes and departures have not been algebraically accumulated for a listing of x- and y-coordinates.

*Meridian distances* are similar to x-coordinates. They are defined as distances east or west from a north-south reference meridian. To avoid difficulties with algebraic signs, reference meridians pass through the most western points of traverses. Using this method, the area of the polygon ABCDE from figure 9-7 is shaded area bBCDd minus shaded area bBAEDd, figure 9-10.

$$\text{Area bBCDd} = \frac{(Bb)(bc)}{2} + \frac{(Cc)(bc)}{2} + \frac{(Cc)(cd)}{2} + \frac{(Dd)(cd)}{2}$$

$$\text{Area bBAEDd} = \frac{(Bb)(Ab)}{2} + \frac{(Ee)(eA)}{2} + \frac{(Ee)(de)}{2} + \frac{(Dd)(de)}{2}$$

The difference in these areas is:

$$\text{(G)} \quad \text{Area ABCDE} = -\frac{(Bb)}{(2)}(Ab) + \frac{(Bb+Cc)}{(2)}(bc) + \frac{(Cc+Dd)}{(2)}(cd) - \frac{(Dd+Ee)}{(2)}(de) - \frac{(Ee)}{(2)}(eA)$$

$$\text{(H)} \quad \text{or Area ABCDE} = 1/2\,[\,-(Bb)(Ab) + (Bb + Cc)(bc) + (Cc + Dd)(cd) - (Dd + Ee)(de) - (Ee)(eA)]$$

In equation (G), terms $\frac{Bb}{2}$, $\frac{Bb+Cc}{2}$, $\frac{Cc+Dd}{2}$, $\frac{Dd+Ee}{2}$ and $\frac{Ee}{2}$ are the average of meridian distances to the terminal points of each of the five courses. Terms Ab, bc, cd, de, and eA are the latitudes of each of the five courses. The meridian distance to a course's midpoint is the average of meridian distances to its terminal points. Therefore, equation (G) provides the basis for rule 3.

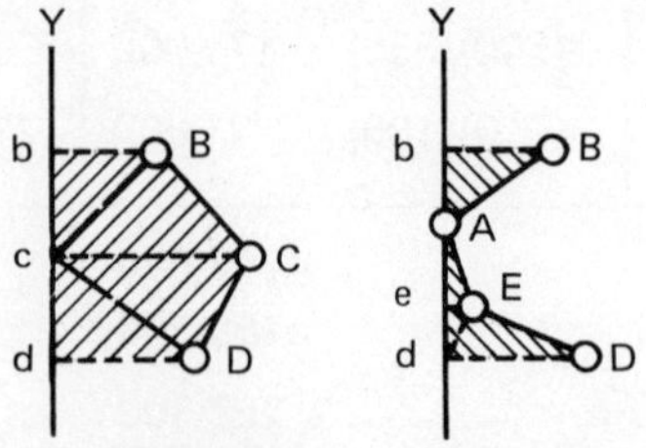


**Fig. 9-10 Tract shifted to reference meridian**

Rule 3: The area of a polygon is the algebraic sum of the products of midpoint meridian distances and the latitudes for each course.

In equation (H), terms Bb, Bb + Cc, Cc + Dd, Dd + Ee, and Ee are double the values of meridian distances to the midpoints of each course. When multiplied by the latitudes for each course, they produce double areas. From this comes the term double meridian distance, and rule 4:

Rule 4: The area of a polygon is one-half the algebraic sum of products of midpoint double meridian distances and the latitudes for each course.

The special feature of the double meridian distance method is a set of three convenient rules. By these rules, double meridian distances are derived from a listing of departures for a traverse. Any other means of deriving the combined length of meridian distances to each courses' terminal points also serve as well. The three rules are:

Rule 5(a): The double meridian distance of the first course (beginning at the point through which the reference meridian passes) is equal to the departure of that course.

Rule 5(b): The double meridian distance of any course is equal to the double meridian distance of the preceding course, plus the departure of the preceding course, plus the departure of the course itself.

Rule 5(c): The double meridian distance of the last course is numerically equal to its departure, but with an opposite algebraic sign.

Calculations of area ABCDE by rule 4 and rules 5(a), 5(b), and 5(c) is as follows:

| STA | DEP. | LAT. | DMD | | AREA (SQ. FT.) | |
|---|---|---|---|---|---|---|
| | | | | | + | - |
| A | | | | | | |
| | 220 | 120 | 220 | (220)(120) | 26,400 | |
| B | | | | | | |
| | 230 | -250 | 670 | (670)(-250) | | 167,500 |
| C | | | | | | |
| | -100 | -250 | 800 | (800)(-250) | | 200,000 |
| D | | | | | | |
| | -290 | 100 | 410 | (410)(100) | 41,000 | |
| E | | | | | | |
| | - 60 | 280 | 60 | (60)(280) | 16,800 | |
| A | | | | | | |
| | | | | | 84,200 | 367,500 |

84,200

2)283,300

-141,650

The only reason to use double meridian distances is to provide a basis for the traditional double meridian distance rules. It is simpler to use basic rule 3. Undoubled midpoint meridian distances used in rule 3 are easily derived from listings of departures or x-coordinates.

By rule 3, area calculations are as follows:

| STA | DEP. | LAT. | M.D. TO MID POINT | AREA + | (SQ. FT.) - |
|---|---|---|---|---|---|
| A | | | | | |
| | 220 | 120 | 110 | 13,200 | |
| B | | | | | |
| | 230 | -250 | 335 | | 83,750 |
| C | | | | | |
| | -100 | -250 | 400 | | 100,000 |
| D | | | | | |
| | -290 | 100 | 205 | 20,500 | |
| E | | | | | |
| | - 60 | 280 | 30 | 8,400 | |
| A | | | | | |
| | | | | 42,100 | 183,750 |
| | | | | | 42,100 |
| | | | | | -141,650 |

The midpoint meridian distance of any course is seen to be any one of the following:

- the sum of departures for preceding courses plus one-half the departure of the course itself
- the preceding midpoint meridian distance plus one-half the sum of the departures for the preceding course and the course itself
- the average of the x-coordinates for the two ends of a course

Courses are listed in a clockwise order in the above illustration of rules 3 and 4. The area computed by either the midpoint meridian method or double meridian distance method came out with a negative sign. Had courses been listed in a counterclockwise order, the area would have emerged with a positive sign. The algebraic sign is not important.

## IRREGULAR BOUNDARIES

Boundaries frequently follow irregular courses, such as meandering roads or streams. When this happens, traverse lines are run along convenient courses near the boundaries. Points of significant change in boundary directions are located by offsets perpendicular to the traverse line. The area between the traverse line and those points is calculated by equation (C).

Sometimes perpendicular offsets at regular intervals, every 50 or 100 feet, define sufficiently accurate areas, figure 9-11. A rule simpler to apply is derived as follows:

From equation (C),

$$A = 1/2 \; [(Aa)(ab) + (Bb)(ac) + (Cc)(bd) + (Dd)(ce) + (Ee)(df) + (Ff)(eg) + (Gg)(fg)]$$

When intervals ab = bc = cd = de = ef = fg, each interval is represented by symbol "d", and:

$$A = 1/2\ [(Aa)(d) + (Bb)(2d) + (Cc)(2d) + (Dd)(2d) + (Ee)(2d) + (Ff)(2d) + (Gg)(d)]$$

or:

$$\text{(I)} \quad A = d\,(Bb + Cc + Dd + Ee + Ff) + \frac{d}{2}\,(Aa + Gg)$$

Equation (I) is the basis for:

Rule 6: The area of a polygon bounded by a straight line and courses between points perpendicularly offset at regular intervals from the straight line, is the interval times the sum of intermediate offsets plus one-half the interval times the sum of end offsets.

When end offsets are zero, points A and G in figure 9-12, the area is simply the interval times the sum of the offsets, or:

$$\text{(J)} \quad A = d\,(Bb + Cc + Dd + Ee + Ff)$$

## SIMPSON'S ONE-THIRD RULE

Rules have been devised for refining calculations where irregular boundaries follow curved paths. Simpson's one-third rule is effective for adjoining concave portions, or for adjoining convex portions. For meandering lines with reverses in curvature, it is of little use. It can even produce less accurate results than to ignore the curvature altogether. It assumes that a curved boundary through three successive points measured at regular intervals is parabolic. Offsets must be at regularly spaced intervals along the traverse line. There must be an odd number of offsets. If an even number have been measured, exclude the last end area and calculate it separately, ignoring the curvature.

Simpson's rule is to weigh the end offsets by a factor of 1, weigh all other odd offsets by a factor of 2, and weigh all even offsets by a factor of 4. Add the weighed values for all offsets. Multiply the sum by the common interval between the offsets, and divide by 3.

The recommended practice is to control the field measurements of meandering lines so as to make their refinement by rules unnecessary. The skillful choice of intervals to locate offsets at all significant changes in direction produces the best results.

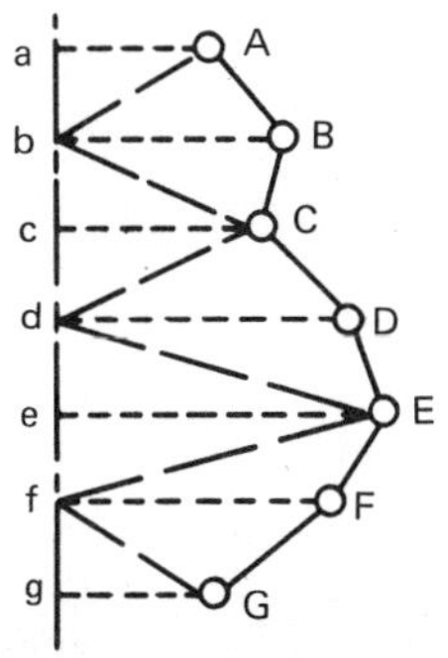


**Fig. 9-11 Perpendicular offsets for irregular boundary**

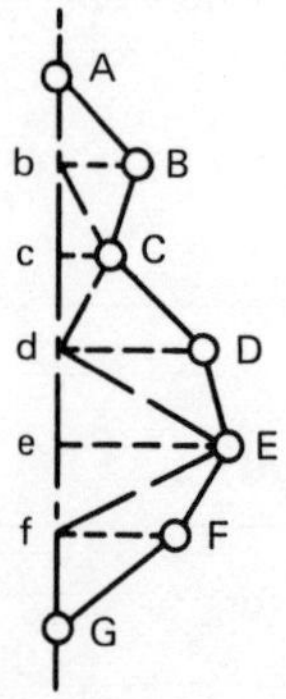


**Fig. 9-12 Alternate arrangement of perpendicular offsets**

## SUMMARY

Approximate areas can be calculated from values scaled from maps, or by a planimeter. For more precise determinations, calculations are based on field measurements.

A planimeter measures a map area by moving a tracing point along the area's boundary. The factor for converting the map area to the land area is the linear map scale squared.

Areas bounded by straight lines are viewed as networks of triangles for calculating purposes. Triangles are grouped in certain patterns for devising labor-saving methods of calculation.

Two methods for calculating areas are the coordinate method and the double meridian distance method. By either method, algebraic signs of calculated areas depend upon the order in which the courses are listed. Numerical values are the same and equally acceptable regardless of the algebraic signs.

Areas between traverse lines and meandering lines are visualized as triangular areas. These triangles are also grouped for calculations. Simpson's one-third rule helps when curving lines along meandering boundaries cannot be disregarded. The best practice is to control the field measurement of meandering lines sufficiently to make refinement by such rules unnecessary.

## ACTIVITIES

1. With a planimeter, measure the area bounded by the traverse plotted in chapter 8, activity 3, and the area of the tract of land plotted in chapter 8, activity 4.

2. Divide the areas of chapter 8, activities 3 and 4 into triangles. From scaled dimensions of the triangles, compute the areas bounded by the traverse and the tract boundaries.

3. Calculate the traverse and tract boundaries of chapter 8, activities 3 and 4 by the coordinate method, and the double meridian distance method.

4. Run a straight transit line to one side of meandering road or stream. Measure the offsets to the center of the road or stream. Calculate the area between the base line and the meandering line by formula (C) or (I).

## REVIEW QUESTIONS

### A. Multiple Choice

1. For calculating purposes, areas bounded by straight lines are viewed as networks of
   a. squares. c. triangles.
   b. rectangles. d. polygons.

2. Meridian distances are the same as
   a. latitudes. c. x-coordinates.
   b. departures. d. y-coordinates.

3. The double meridian distance of any course is not the sum of the
   a. meridian distances of its terminals.
   b. departures of its terminals.
   c. x-coordinates of its terminals.
   d. DMD of the preceding course, departure of the preceding course, and the departure of the course itself.

### B. Short Answer

4. State boards of registration usually require surveyors to be thoroughly familiar with two methods of calculating areas. What are they?

5. In what order are courses listed to ensure getting results with positive algebraic signs for the coordinate method? For the DMD method?

### C. Problems

6. A map area measured by planimeter is 36.42 square inches. The map scale is one inch = 200 feet. What is the land area?

7. A square field 400 feet on each side has coordinates as follows:

| Corner | X | Y |
|---|---|---|
| A | 100 | 200 |
| B | 100 | 600 |
| C | 500 | 600 |
| D | 500 | 200 |

   Show the calculations for this area by both rule 1 and rule 2 of the coordinate method.

8. Repeat review question 7, but with corners listed in the order ADCB.

9. Calculate the departures and latitudes for all courses in review question 7. Show the area calculation by DMD.

10. Repeat review question 9, but with corners listed in the order ADCB.

11. Refer to figure 9-4. Consider ABCDEFG as lines connecting points along a meandering stream. Line AG is a traverse line. Dashed lines Bb, Cc, Dd, Ee, and Ff are perpendicular offsets. Field measurements are assumed to have the following values:

| | | | |
|---|---|---|---|
| Ab | 12 | Bb | 53 |
| bc | 35 | Cc | 87 |
| cd | 28 | Dd | 67 |
| de | 38 | Ee | 92 |
| ef | 41 | Ff | 60 |
| fG | 18 | | |

Calculate the area between the traverse line and the meandering stream using equation (B).

12. Refer to figures 9-11 and 9-12. Assign a value of 50 feet to the regular intervals between perpendicular offsets. Field measurements are assumed to have the following values:

| | **Fig. 9-11** | **Fig. 9-12** |
|---|---|---|
| Aa | 47 | 0 |
| Bb | 71 | 24 |
| Cc | 63 | 16 |
| Dd | 91 | 44 |
| Ee | 101 | 54 |
| Ff | 84 | 37 |
| Gg | 47 | 0 |

By equation (I), what is the area included within the perimeter of figure 9-11? By equation (J), what is the area included within the perimeter of figure 9-12?

# CHAPTER 10

# Finding True Meridian

## OBJECTIVES

After studying this chapter, the student will be able to:

- measure the horizontal angles between lines on the earth and a star or the sun.
- make all measurements necessary to calculate the true bearing of a star or the sun from any point.
- calculate from star or solar measurements the true bearing of lines on earth.

## POLARIS AND THE SUN

The true direction of any line is found by observing Polaris (the North Star), or the sun. With careful work, results within ± 0.5 minutes for Polaris and within ± 2 minutes for the sun are achieved. Closer results are possible by averaging repeated observations.

A transit set up at a transit station A is backsighted on an adjoining transit station B, figure 10-1. With the foresight on Polaris or the sun, the horizontal

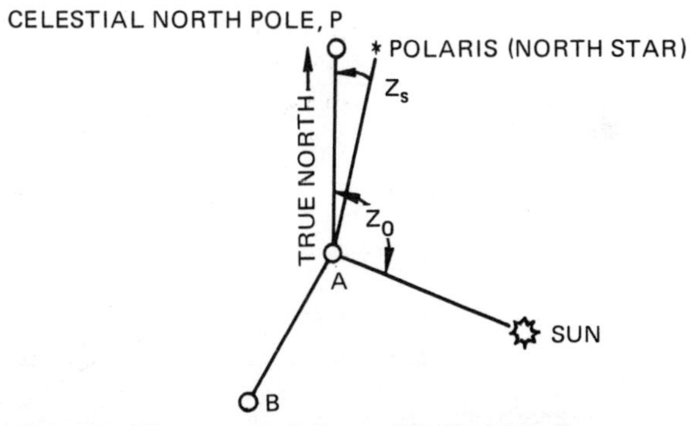


**Fig. 10-1 Line AB on earth relative to the celestial bodies**

angle and time of observation are noted. When observing the sun, the vertical angle, approximate ground elevation, and atmospheric temperature are also recorded. The horizontal angle between true north and Polaris $Z_s$ or the sun $Z_o$ is calculated. Angles in a clockwise direction are considered plus and those in a counterclockwise direction minus. The algebraic difference between the turned angle and $Z_s$ or $Z_o$ is horizontal angle BAP. That angle gives the direction of line AB with respect to true north. In the continental United States, $Z_s$ is less than 1° 17′ and lies either clockwise or counterclockwise from true North. $Z_o$ varies from less than 90° to about 150°, clockwise in the morning and counterclockwise in the afternoon.

## THE LOCATION OF POLARIS IN THE SKY

True direction can be found by observing any one of a number of stars. In the northern hemisphere, Polaris is the most easily identified. It is so close to true north that it is referred to as the North Star. To observers looking northward, stars appear to revolve counterclockwise around the earth's axis. Their center of rotation is a point in the sky in the northern extension of the earth's axis. The sky is conceived as a celestial sphere of infinite radius. The center of rotation of stars in the northern sky is called the north pole of the celestial sphere. The radius of a star's circular path around this center is called its polar distance, p. For Polaris, p is approximately 1°.

Out about 30° to one side of Polaris are seven stars defining the Big Dipper. Out about 30° to the opposite side of Polaris are five stars defining the "W" shaped constellation called Cassiopeia. Every surveyor should learn these two constellations because they help to locate Polaris. Polaris lies almost midway along a straight line between stars at the clockwise ends of both constellations, figure 10-2. One is the end star of the Dipper handle. The other is the upper left-hand star in the "W" of the constellation Cassiopeia. No other stars of the same magnitude are near it. At the counterclockwise end of the Dipper, two stars form the outer part of the bowl. They are called pointers because a line through them points almost directly to Polaris. Polaris is always found on the same side of the celestial north pole as Cassiopeia.

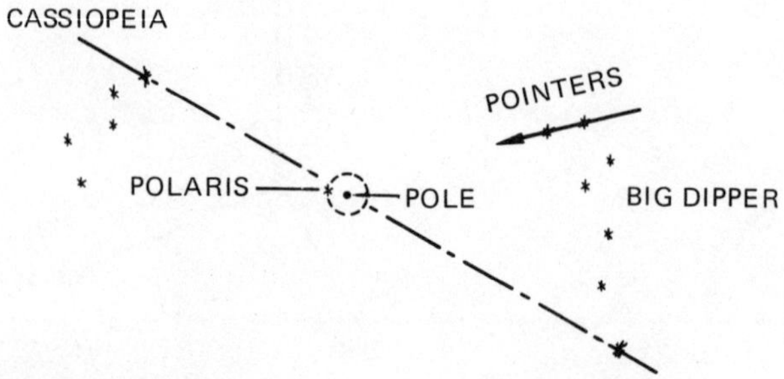


**Fig. 10-2 Polaris relative to two conspicuous constellations**

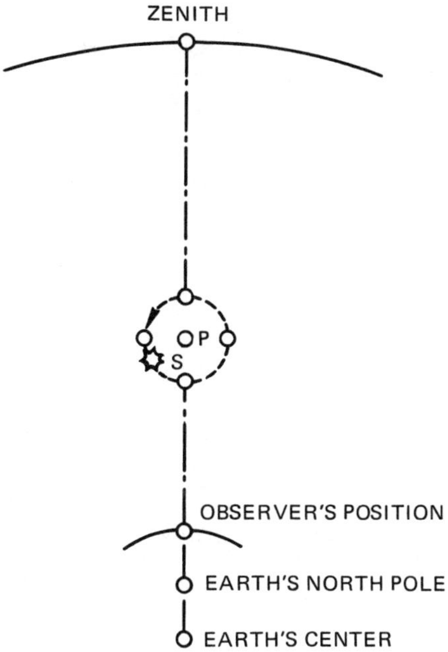


**Fig. 10-3 Significant points in a plane of the observer's meridian**

The central vertical line in figure 10-3 connects five significant points in a plane through an observer's north-south meridian:

- the zenith, which is a point on the celestial sphere directly overhead,
- the celestial north pole,
- the earth's north pole well over the horizon,
- the earth's center,
- the observer's position.

Looking up vertically, an observer looks directly at the zenith, Z, figure 10-4. Looking northward at a vertical angle equal to the observer's latitude, is the celestial north pole, P. This line of sight to the celestial pole is parallel to the

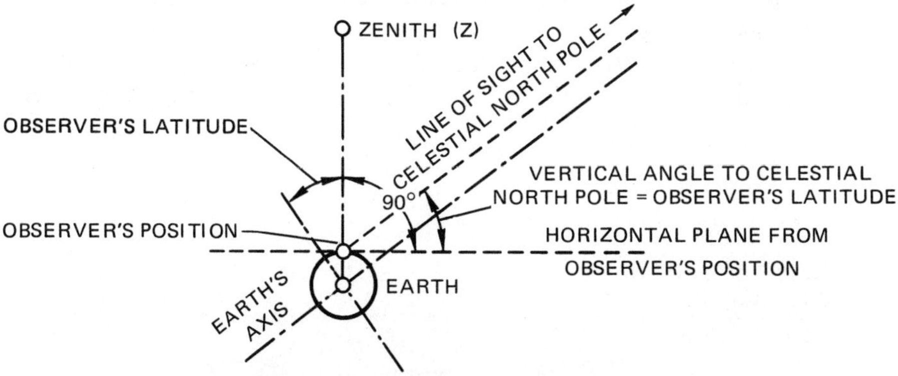


**Fig. 10-4 Another view of significant points**

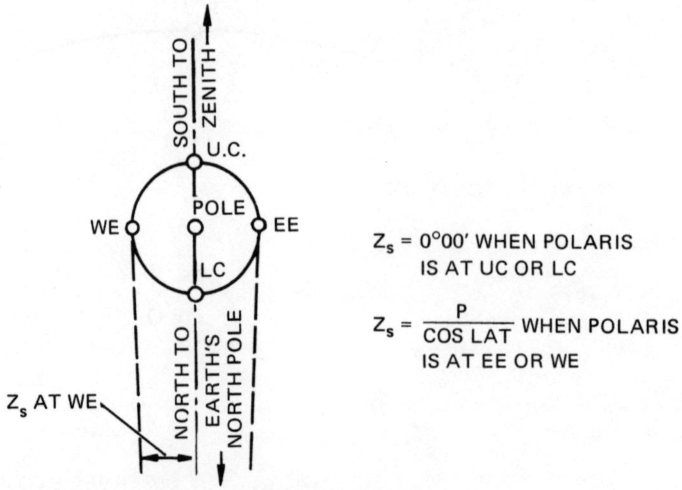


**Fig. 10-5 $Z_S$ relative to Polaris' circular path**

earth's axis of rotation. The distance to the pole is so infinite that the offset between the parallel lines is considered zero.

Polaris is said to be at upper culmination (UC), figure 10-5, when it is observed directly above the celestial pole. In that position it is south of the celestial pole and toward the zenith. About twelve hours later, it is directly below and north of the celestial pole, toward the earth's north pole. There it is at its lower culmination (LC). At both upper and lower culmination, it is in the plane of the observer's north-south meridian. In that plane $Z_S$ is $0^\circ$. About halfway between these positions, Polaris reaches its greatest east or west bearing. At these points, it is said to be at eastern or western elongation (EE or WE). There $Z_S$ becomes Polaris' polar distance divided by the cosine of the observer's latitude. Observation at upper culmination, lower culmination, eastern elongation, or western elongation simplifies calculations. However, the occurrence may be at an inconvenient time or when clouds prevent observation. A disadvantage of observations at the upper or lower culmination is that they must be most exacting with regard to time keeping. These observations occur when Polaris appears to be traveling horizontally, figure 10-6. The value of $Z_S$ is then changing at its fastest rate, about 20-angular seconds per minute of time. An advantage of observations at the east or west elongation is that they are least exacting. They

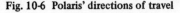

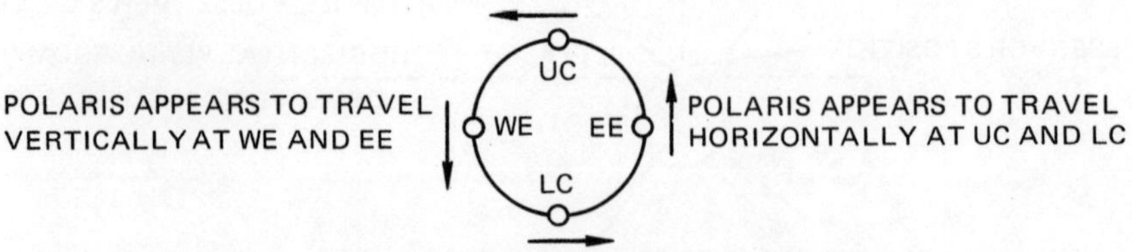


**Fig. 10-6 Polaris' directions of travel**

occur when Polaris appears to be traveling vertically. The value of $Z_s$ is then changing at its slowest rate, about 1- or 2-angular seconds per minute of time.

Surveying equipment manufacturers publish ephemerises annually giving theory, field procedures, sample calculations, and tables of values needed. An ephemeris is also published annually by The United States Bureau of Land Management. An ephemeris must be obtained for the year in which work is performed.

Stars do not return to exactly the same position at the end of a year. One result is that the 365-day year is corrected every leap year by an additional day. Stars do not return to exactly their same positions after 24 hours, figure 10-7. Observations made at the same time each night show their positions to be advancing counterclockwise daily. They gain an angle of 0° 59.1′ during a 24-hour period, or about 180° every 6 months. Positions of stars with respect to true north are determined from ephemeris tables for any moment. The process involves finding where a star is on its circular track at the time of observation. Its bearing east or west of the true north-south meridian at that time is computed.

## POLARIS' BEARING

Polaris' bearing, $Z_s$, depends upon an observer's latitude, longitude, and time of observation. Latitude and longitude are obtained easily from a United States Geological Survey quadrangle map. The only field measurement that affects $Z_s$ is the time of observation.

For observations at upper culmination, lower culmination, eastern elongation, or western elongation, an observer predetermines their time of occurrence. At any other time an observer proceeds with the observation and then makes the calculations. Steps involved in calculating $Z_s$ are, figure 10-8:

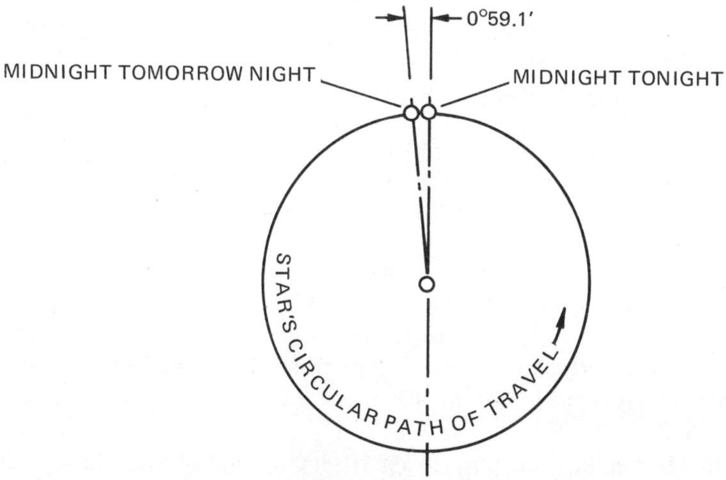


**Fig. 10-7 Daily angular travel of the stars**

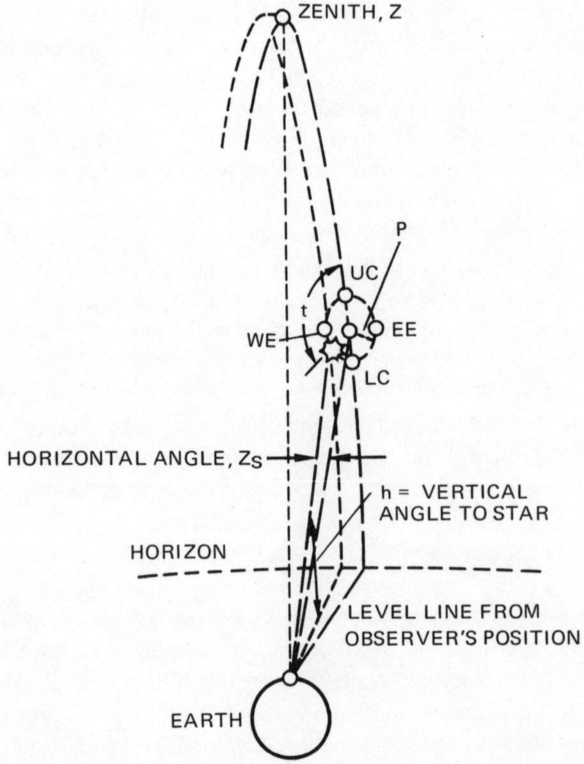


**Fig. 10-8 Angles t, p, and h**

1. determining the counterclockwise angle, t, to which Polaris has rotated at the moment of observation, measured from the upper culmination.
2. determining Polaris' polar distance, p, on the day of observation.
3. obtaining Polaris' true altitude, h, by applying a correction to the latitude of the place of observation.
4. computing Polaris' bearing, $Z_S$, by the formula:
   (A)

$$Z_S = \left(\frac{\sin t}{\cos h}\right)(p)$$

**Angle t**

Lines PZ and PS, figure 10-9, are segments of great circles where the celestial sphere is intersected by planes through north-south meridians.

Segment PZ is in a plane through an observer's north-south meridian when Polaris is at its upper culmination. It runs from the celestial north pole, P, southward through Polaris' position at its upper culmination. It extends south to the zenith, Z.

Segment PS is in the plane through a north-south meridian over which Polaris is passing at any moment. It runs from P to Polaris' position, S, at any moment.

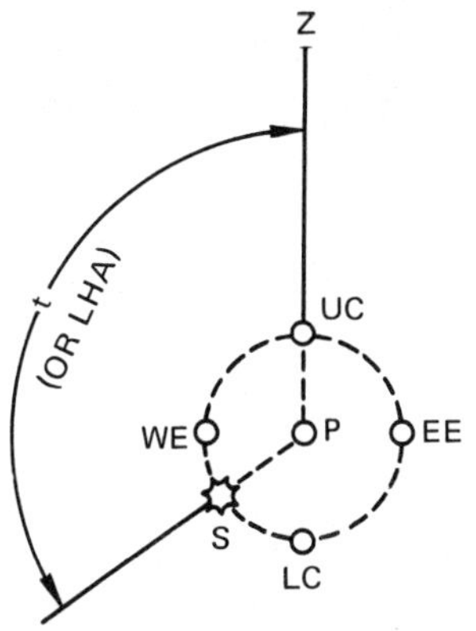


**Fig. 10-9 Angle t (LHA)**

The polar angle at P between lines PZ and PS is the angle t. It is also called the local hour angle (LHA).

The north-south zero reference meridian for reckoning longitude, time, and hour angles of stars passes through Greenwich, England. The Greenwich hour angle (GHA) is a polar angle measured westward from the Greenwich meridian to a meridian over which a star or the sun is passing, figure 10-10.

For observations at longitude 0°, the local hour angle and Greenwich hour angle are the same. At any point in the western hemisphere, the Greenwich hour angle exceeds the local hour angle by the observer's longitude, or:

(B) GHA = LHA + Longitude

For example, an observer at Greenwich when Polaris nears the lower culmination finds the Greenwich hour angle, local hour angle, and angle t values to be 180°±. To an observer at longitude 75° W at the same moment, Polaris also has a Greenwich hour angle of 180°±. However, the local hour angle, and angle t have values of 105°±. There is a wait of approximately 5 hours for Polaris to travel 75° to view it at its lower culmination. Then angle t becomes 180° at this point of observation.

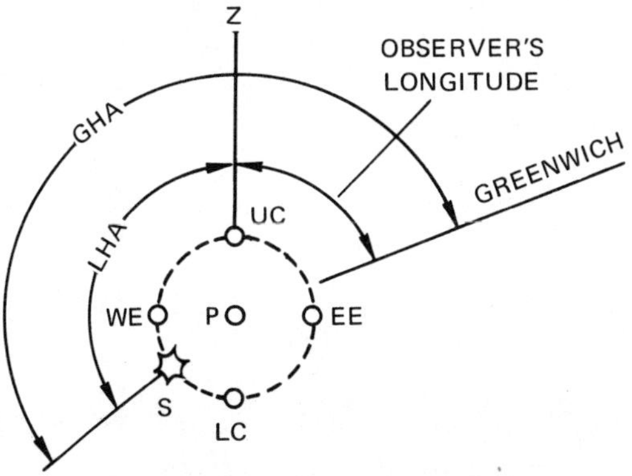


**Fig. 10-10 GHA, LHA, and longitude**

Note that the Greenwich hour angle at any moment of time is the same at any point of observation. On the other hand, the local hour angle (or t) depends upon the observer's longitude. Ephemeris tables therefore give the stars' positions in terms of Greenwich hour angle at any moment in a year. They leave it to the observer to select values from the tables for the moment of observation. The observer converts those values to terms of local hour angle for the longitude. This procedure is followed whether predetermining the time of Polaris' positions or calculating angle t after observations.

## Time

Conversion of values between Greenwich hour angle, local hour angle, angle t, and longitude requires recognition of differences in the time resulting from:

- time interval between different longitudes
- time interval between time zones
- differences between rate of angular travel of sun and stars
- differences in latitude

Greenwich civil time (GCT) is the elapsed time since the moment of midnight, $0^h$, at the Greenwich meridian. Hours are expressed cumulatively for the entire day; that is, 3:15 PM is expressed as 15h 15′.

Local civil time (LCT) is the elapsed time since the moment of midnight at a precise meridian of longitude. Moments of midnight occur at any western longitude later than at Greenwich. Local civil time lags behind Greenwich civil time by amounts equal to the longitude divided by 15°. Four minutes of time are subtracted from Greenwich civil time for each degree of longitude. Hours for local civil time are also expressed cumulatively for the entire day.

Standard time zones do not recognize civil time changes for small differences in longitude. Central meridians for standard time zones are at 15-degree intervals of longitude. All time pieces in a standard time zone are set to read the same time. They are a whole number of hours slower than Greenwich civil time because the sun travels 15° of longitude each hour. Hours are expressed cumulatively for 12 hours before noon (AM) and 12 hours after noon (PM). Central meridians for standard time zones are as follows:

| Time Zone | Central Meridian | Hours Behind GCT |
|---|---|---|
| EST | 75° | 5 |
| CST | 90° | 6 |
| MST | 105° | 7 |
| PST | 120° | 8 |

At the instant of 22h 22.7′ Greenwich civil time, the local civil time at longitude 84° W is:

$$\text{LCT} = 22\text{h}\ 22.7' - \frac{84^\circ}{15^\circ}$$

$$= 22\text{h}\ 22.7' - 5\text{h}\ 36.0' = 16\text{h}\ 46.7'$$

At that same moment, the eastern standard time at longitude 84° W, or anywhere else in the eastern standard time zone, is 5 hours slower than Greenwich civil time, or:

EST = 22h 22.7′ – 5h 00.0′ = 17h 22.7′, or 5:22.7 PM,

and anywhere in the central standard time zone:

CST = 22h 22.7′ – 6h 00.0′ = 16h 22.7′, or 4:22.7 PM

Thus, local civil time at longitude 84° W is:

- 5 hours 36 minutes slower than Greenwich civil time
- 0 hours 36 minutes slower than eastern standard time
- 0 hours 24 minutes faster than central standard time.

Stars travel 9.8 seconds per hour faster than the sun, completing their circuit in 23h 56.1′. Polaris gains .011 minutes for every 1° of longitude, making corrections in time intervals necessary. At longitudes 75° W and 84° W, Polaris has gained 0.82′ and 0.92′ respectively. Such a gain necessitates corrections of –0.8′ and –0.9′, making times of observations:

- at longitude 75° W, 4h 59.2′ after longitude 0° observation.
- at longitude 84° W, 5h 35.1′ after longitude 0° observation.

No further correction is necessary if Polaris is at its upper or lower culmination. At the east or west elongation, times of observation vary slightly with the latitude of the point of observation. The exact amount is furnished in the ephemeris tables.

## Predetermination of Time of Observation

Tables of a solar ephemeris furnish the time of Polaris' culminations and elongations. Values are for observations at longitude 0°, latitude 40° N. Tables also furnish small corrections of time for western and eastern elongation for latitudes other than 40° N.

Suppose an observer wishes to determine the time of culminations and elongations at longitude 84° W, latitude 35° N, on May 11, 1979. With values from the tables the following calculations and corrections are made, figure 10-11.

| Item | Long. | Lat. | EE | UC | WE | LC |
|---|---|---|---|---|---|---|
| GCT (from table) | 0° | 40° N | 4h 59.5′ | 10h 55.7′ | 16h 52.0′ | 22h 53.8′ |
| Time Interval Between Longitudes | | | +5h 36.0′ | +5h 36.0′ | +5h 36.0′ | 5h 36.0′ |
| Time Interval Between Time Zones | | | -5h 00.0′ | -5h 00.0′ | -5h 00.0′ | -5h 00.0′ |
| Uncorrected EST | 84° W | 40° N | 5:35.5′AM | 11:31.7AM | 5:28.0PM | 11:29.8PM |

**Fig. 10-11 (continued)**

| Item | Long. | Lat. | EE | UC | WE | LC |
|---|---|---|---|---|---|---|
| Correction For Polaris' Travel | 84° W | 40° N | -0.9′ | -0.9′ | -0.9′ | -0.9′ |
| Correction For Latitude (from table) | 84° W | 35° N | -0.5′ | - | +0.5′ | - |
| Corrected EST | 84° W | 35° N | 5:34.1AM | 11:30.8AM | 5:27.6PM | 11:28.9PM |

Fig. 10-11

**Calculations of Angle t at Any Time of Observation**

At upper culmination, angle t is 0° and at lower culmination it is 180°. At western elongation it is 89°± and at eastern elongation it is 271 °±, the exact amount depending upon the observer's latitude.

At any other time of observation, angle t is calculated by converting the standard time of observation to Greenwich civil time. Find the Greenwich hour angle for Greenwich civil time at the time of observation. Subtract the longitude from the Greenwich hour angle to obtain the local hour angle (equal to angle t) for the time of observation. Example: From point A, Longitude 83°55.4′ W (from map) Latitude 35°32.3′ N (from map), an observation is made. On May 11, 1979 at 9:34:48 PM EST, the clockwise angle from point B to Polaris is 174°24′. Angle t is found as follows:

Time of Observation = 9:34:48 PM EST May 11
= 21:34:48 EST (24 hour basis)
= 26h 34.8 GCT May 11
= 2h 34.8 GCT May 12

An ephemeris table furnishes for $O^h$ May 12, 1977 Polaris' GHA of 196° 36.1′. Polaris travels 360°59.1′ every 24 hours. That means that angle t measuring Polaris' progress from meridian to meridian increases at the rate of 15.0411° per hour. In 2h 34.8′ from $O^h$ to time of observation, Polaris' angular travel is:

$$2h\ 34.8' \times 15.0411° = 38.806° \text{ or } 38°48.4'$$

That value is also given in the ephemeris table.

| | |
|---|---|
| GHA for $O^h$ May 12, 1979 | 196°36.1′ |
| Increase in GHA for 2h 33.6′ | 38°48.4′ |
| GHA | 235°24.5′ |
| Less West longitude | 83°55.4′ |
| LHA = t | 151°29.1′ |

**Polar Distance**

An ephemeris table shows Polaris' polar distance, p, to increase from 0° 49.66′ on January 1, 1979 to 0°50.16′ in early July. It then decreases to

$0°49.42'$ at the year's end. Fifty years earlier it had been $1°05.11'$, indicating an average annual decrease of about 18 seconds. From the same ephemeris table, the appropriate value is selected for May 12, 1979, which is:

$$p = 0°50.00'$$

### Altitude

The altitude, h, of Polaris is simply the vertical angle at the time of observation. The observed vertical angle is acceptable only when the transit is in perfect adjustment and the greatest care is exercised. A more frequently used method is to apply a correction to the observer's latitude. Corrections depending upon the value of angle t are given in ephemeris tables. For latitude $35°32.3'$ N and angle t of $151°29.1'$, the altitude, h, is:

$$35°32.3' - 43.8' = 34°48.5'$$

### Calculation of $Z_s$

In the above example, where

$$\begin{aligned} t &= 151°29.1' \\ p &= 0°50.00' \\ h &= 34°48.5' \end{aligned}$$

$$Z_s = \left(\frac{\sin t}{\cos h}\right) p = \left(\frac{.47739}{.82107}\right)\left(50.00\right) = 0°29' \text{ counterclockwise}$$

$Z_s$ is counterclockwise when t is less than 180° and clockwise when it exceeds 180°.

### True Meridian

In the above example, it is found that $Z_s$ is $0°29'$ counterclockwise, giving line AS, figure 10-12, a bearing of N $0°29'$ W. With the transit at A, backsight on B, and foresight on S, a $174°24'$ angle is turned clockwise. Angle BAP gives the direction of the true north-south meridian with respect to line AB

$$\begin{aligned} BAP &= 174°24' + 0°29' \\ &= 174°53' \end{aligned}$$

and the true bearing of line AB is S $5°07'$ W.

## FIELD PROCEDURE FOR POLARIS OBSERVATION

In preparation for a Polaris observation the observer's longitude and latitude is scaled from a United States Geological Survey or other reliable map. They are read to the nearest tenths of minutes. A reliable watch is obtained. It is compared with radio time, which is furnished continuously at 5, 10, and 15 megahertz. The local hour angle is roughly estimated by observing the Big Dipper and Cassiopeia. Estimates are also made by consulting star charts or an ephemeris. Two good flashlights are needed.

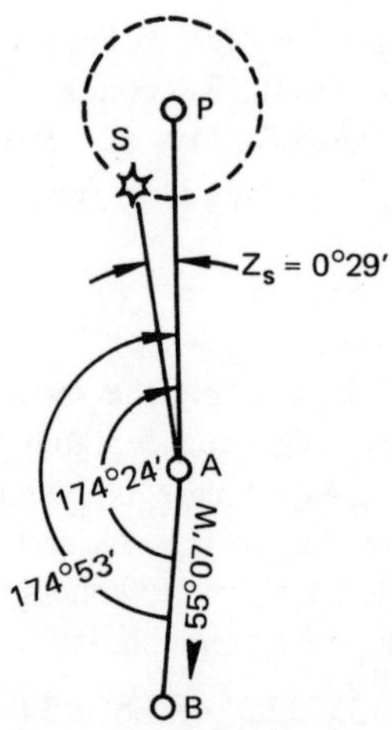


**Fig. 10-12 Turned horizontal angle and $Z_s$**

Evening operations can begin at dusk. Early morning operations begin early enough to ensure completion one-half hour before sunrise. Observations are, however, possible even when there is too much daylight to see Polaris by the naked eye. A transit is set up at one end of a line whose true bearing is desired. Horizontal angles to Polaris are repeated, and the times of observation recorded at each turning. All operations are completed promptly.

When it is dark, a flashlight illuminates the backsight, and another flashlight illuminates the telescope's cross hairs. At the backsight, a flashlight is held behind a plumb bob string or a pencil held on a tack. At the telescope, a flashlight is held to shine into the objective end at a slight angle. Some prefer to shine it on a piece of paper extending beyond the sun shade. The paper is shaped cylindrically and held by a rubber band encircling the telescope.

After a backsight is taken, the telescope is focused on a distant point. It is then pointed north at a vertical angle equal to the observer's latitude. Polaris is usually found within a degree of the line of sight. Search in the direction of the upper left-hand star of the "W" of Cassiopeia. Manipulation of the horizontal and vertical tangent screws brings Polaris into view. Should considerable manipulation occur, the first angle readings are regarded only as aids in locating Polaris. A new beginning is made by taking a fresh backsight before accepting any readings for record purposes.

Notes include a sketch showing the:

- line for which a true bearing is desired.
- point occupied.
- direction of the angle turned to Polaris.

## LOCATION OF THE SUN IN THE SKY

The sun's position at any moment is defined by its Greenwich Hour Angle and its declination, figure 10-13. The Greenwich hour angle indicates which

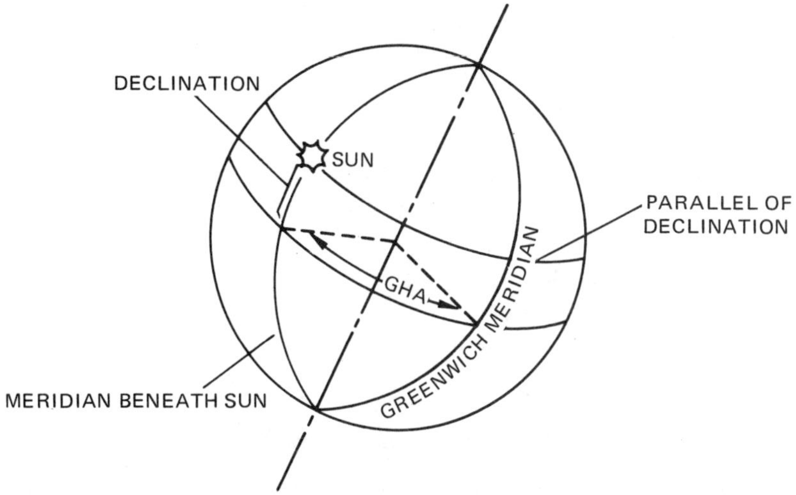


**Fig. 10-13 GHA and declination**

longitudinal meridian it is passing over. The declination indicates its angular distance reckoned from the celestial equator toward the celestial pole. A parallel of declination on the celestial sphere is comparable to a parralel of latitude on earth. On June 22 the sun's path is at a declination of about N 23 1/2°. On December 22 its path is at a declination of about S 23 1/2°. During the six intervening months, its path veers gradually from one of these extreme declinations to the other. It crosses the celestial equator during the last weeks of March and September.

The sun's rate of movement is not constant. The time of its mid-day passage over an observer's meridian varies from 16 minutes before, to 14 minutes past its average.

## THE SUN'S BEARING

The sun's bearing, $Z_o$, depends upon:

- an observer's latitude, 1,
- the measured altitude, h, corrected for refraction and parallax.
- the sun's declination at the time of observation, d.

An observer's latitude is obtained from a United States Geological Survey quadrangle for the local area. The measured altitude is an observer's measured vertical angle. Corrections are based upon the observer's elevation above sea level and atmospheric temperature at time of observation. The sun's declination is furnished in ephemeris tables for the moment of observation. The bearing of $Z_o$ is computed by the formula:

(C) $$\cos Z_o = \frac{\sin d - \sin h \sin 1}{\cos h \cos 1}$$

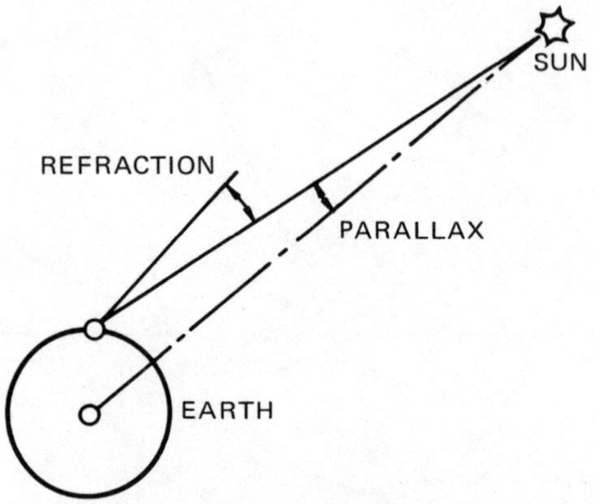


**Fig. 10-14 Refraction and parallax**

Measured vertical angles to the sun are affected by refraction and parallax, figure 10-14. *Refraction* is a downward bending of light rays as they enter the earth's atmosphere. It causes a telescope to point higher than the sun when apparently pointed at it. Refraction corrections are always negative and are only a few minutes for vertical angles of 20°. They increase as much as half a degree for low angles close to the horizon. Refraction corrections also vary slightly with changes in temperature and air density. Correlations between air density and elevation permit adjustment for air density to be based on the observer's elevation.

*Parallax* is a distortion resulting from measurements of vertical angles from the earth's surface rather than its center. Parallax is offset by small angular corrections which are always positive. A solar ephemeris furnishes corrections for refraction and parallax under standard conditions. Tables give adjustment factors for refraction corrections based on the elevation and temperature of the observation point.

For a measured altitude 36° 14′, temperature 64° F, and elevation 1900 feet, the refraction correction is:

$$-(1.31)(.95)(.97) \text{ or } -1.21$$

The parallax correction is +0.12′, giving a net correction to the measured altitude of -1.1′.

For example, from point A, latitude 35°32.3′ N (from map), temperature 64° F, elevation 1900 feet (from map), an observation is made. On May 11, 1979 the average of six vertical angles turned to the sun is 36°00.2′. With a backsight on B, the average of six horizontal angles is 92° 01′ counterclockwise. The average time of observations is 8:31 AM eastern standard time.

| | | |
|---|---|---|
| Eastern Standard Time | | 8:31 AM |
| Correction for Time Zone | | 5:00 |
| GCT | | 13:31 |
| Sun's declination $0^h$ | | |
| May 11 | = | N 17°40.0′ (from table) |

| | | |
|---|---|---|
| Change since $0^h$ 13:52 x 0.65 = | | + 8.8′ (from table) |
| Sun's d | = | N 17°48.8′ |
| Measured Altitude | | 36°00.2′ |
| Refractions & Parallax | | 1.1′ |
| True Altitude, h | | 35°59.1′ |

$$\text{Cos } Z_o = \frac{(\sin 17°48.8') - (\sin 35°59.1')(\sin 35°32.3')}{(\cos 35°59.1')(\cos 35°32.3')}$$

$$= \frac{(.30592) - (.58757)(.58125)}{(.80917)(.81373)} = -0.05408$$

$$Z_o = 93°06' \text{ clockwise}$$

**True Bearing Line AB**

In the above example, it is found that $Z_o$ is 93°06′ clockwise, giving line AO a bearing of S 86°54′ E. Angle BAP gives the direction of the true north-south meridian with respect to line AB, figure 10-15,

$$\text{BAP} = -92°01' - 93°06'$$
$$= -185°07'$$

and the bearing of line AB is S 5°07′ W.

## FIELD PROCEDURE FOR SOLAR OBSERVATION

In preparation for a solar observation the observer's latitude and ground elevation is obtained from a United States Geological Survey or other reliable map. The latitude is read to the nearest tenth of a minute. The elevation is read to the nearest 100 feet above sea level. The timepiece should be correct within a few minutes. New observers should practice sighting and tracking techniques before engaging in actual observations.

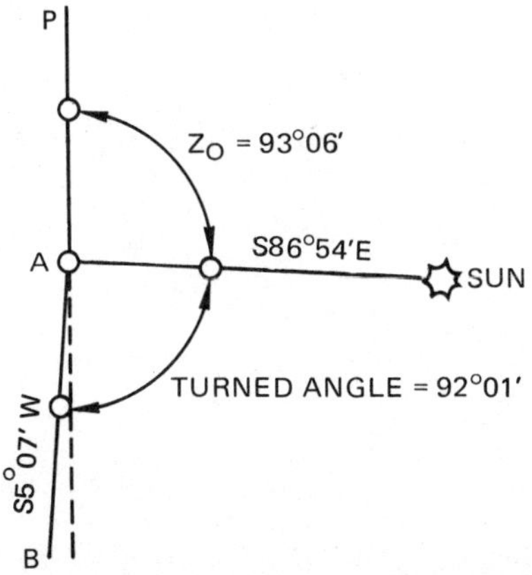


**Fig. 10-15 Turned horizontal angle and $Z_o$**

CAUTION: Looking at the sun through a telescope can cause serious eye injury.

The recommended sighting technique begins with focusing the sun's image on a white piece of paper. The paper is held several inches behind the telescope's eyepiece, figure 10-16. The telescope is aimed in the general direction of the sun. Clamps are tightened when the telescope's gray shadow on the white paper becomes circular. Manipulation of the tangent screws brings the sun's dazzling round image to the approximate center of the gray circular shadow. The objective lens and the eyepiece are then focused, and the white paper's position adjusted until the images of the sun and the cross hairs are sharp. Practice in locating the sun's image tangent to both the vertical cross hair and the central horizontal cross hair then begins. This is a tracking exercise which can be practiced for each one of the four quadrants. Care is taken in identifying the central horizontal cross hair, because only two horizontal cross hairs are seen. The sun's image moves diagonally downward and to the right in the morning. It moves upward and to the right in the afternoon. For observations in left-hand quadrants, manipulation of the horizontal tangent screw keeps the image's vertical leading edge tangent to the vertical cross hair. The sun's motion brings the image's trailing horizontal edge tangent to the central horizontal cross hair. At the moment both are tangent, tracking stops, time is called, and readings are taken. For observations in right-hand quadrants, manipulation of the vertical tangent screw keeps the image's horizontal leading edge tangent to the central horizontal cross hair. The sun's motion brings the image's vertical trailing edge tangent to the vertical cross hair. Again, tracking stops, time is called, and readings are taken.

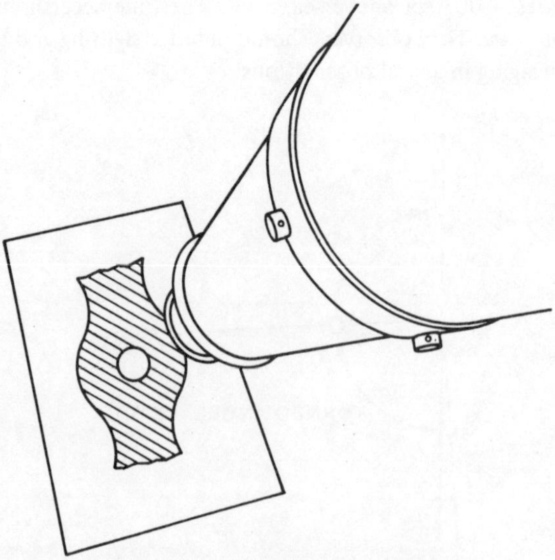

**Fig. 10-16 Catching the sun's image**

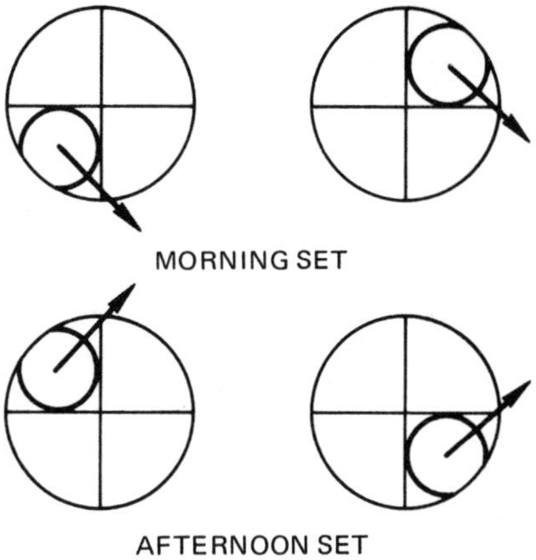


**Fig. 10-17 Quadrants for AM and PM solar observations**

Observations are not made in early morning or late afternoon hours when the sun is at a low altitude. At those times, refraction corrections are large and less reliable. Observations are not made within two hours of noon. Small field errors at that time become magnified during calculations of azimuths. Periods of 0800 to 1000 and 1400 to 1600 local civil time are regarded as most favorable.

A transit is set up at one end of a line where a true bearing is desired. Normal backsights are taken on a hub at the other end. An even number of observations are made with the foresights alternating between diagonally opposite quadrants. The telescope is also alternated between direct and inverted positions as is regularly done when doubling angles. Morning observations are in lower-left and upper-right quadrants, figure 10-17.

Observations are completed promptly, preferably within a span of ten minutes. Readings for vertical angles, horizontal angles, and times of observation are each averaged. Air temperature is also recorded.

Notes include a sketch showing:

- the line for which true bearing is desired.
- the point occupied.
- the direction of the horizontal angle turned to the sun.

## SUMMARY

The true bearing of a star from any point can be calculated for any moment. Calculation of the sun's true bearing is calculated when the vertical angle, ground elevation, and atmospheric temperature are known. Surveyors measure horizontal angles between lines on the earth and a star or the sun. They also record the necessary information for calculating the true bearing of the star or sun. After making that calculation, they determine the line's true bearing. It differs from the true bearing of the star or sun by the measured horizontal angle.

All stars appear to make slightly more than one counterclockwise revolution around the earth's axis daily. Their center of rotation in the northern sky is called the north pole of the celestial sphere. The radius of a star's circular path around this center is its polar distance, p. Polaris, the North Star, is the most easily identified star in the northern sky. For Polaris, p is approximately 1°. When in a north-south alignment with the celestial north pole, Polaris is at its lower or upper culmination. When in east-west alignment, it is at its eastern or western elongation.

Observation of Polaris' position on its circular track is made between dusk and dawn. Its exact position for the moment of time of observation is calculated. From this information its true bearing is calculated. Surveying equipment manufacturers publish ephemerises annually giving theory, field procedure, sample calculations, and tables of values needed.

The position of Polaris is defined by t, the counterclockwise angle to which Polaris has rotated from its upper culmination. The vertical angle to Polaris' position is its altitude, h.

Polaris' true bearing is $\left(\frac{\sin t}{\cos h}\right)(p)$.

Polaris calculations require knowledge of differences in time designation. Greenwich civil time is a whole number of hours faster than standard time. That number equals one-fifteenth of the longitude of a time zone's central meridian. Local civil time lags behind Greenwich civil time by amounts equal to the longitude of the observer divided by 15°.

Stars travel 9.8 seconds per hour faster than the sun. They complete a 360-degree circuit in 23h 56.1′ and in 24 hours traverse 360°59.1′. This faster rate of travel is considered in calculating star positions.

Observations on Polaris include:

- noting from maps the longitude and latitude of the point of observation.
- setting a timepiece accurately by radio time signals.
- measuring the horizontal angle between a line on the earth and Polaris and noting the exact time of observation.
- sketching in a notebook the line for which the true bearing is desired, the point occupied, and the measured angle.

The sun's position at any moment is defined by its Greenwich hour angle and its declination.

The Greenwich hour angle indicates which longitudinal meridian it is passing over.

The declination varies from 23 1/2° north of the equator to 23 1/2° south of it.

The sun's bearing depends upon:

- the observer's latitude, l.
- the measured altitude, h, corrected for refraction and parallax.
- the sun's declination when observed, d.

The horizontal angle between true north and the sun's position when observed is calculated by the formula:

$$\cos Z_o = \frac{\sin d - (\sin h)(\sin l)}{(\cos h)(\cos l)}$$

Values for the sun's declination and for refraction and parallax corrections are given in solar ephemerises. Corrections depend upon measured altitude, air temperature, and the approximate ground elevation.

Practice of sighting and tracking techniques is recommended before making a solar observation.

Most favorable times of day for a solar observation are 0800 to 1000 and 1400 to 1600 local civil time.

Solar observations include:

- noting from maps the latitude and ground elevation of the point of observation.
- setting a timepiece within a few minutes of the correct time.
- measuring the horizontal angle between a line on earth and the vertical angle to the sun.
- noting the time and air temperature when angles are measured.
- sketching in a notebook the line for which a true bearing is desired, the point occupied, and the measured horizontal angle.

## ACTIVITIES

1. Practice sighting and tracking techniques for solar observations with a transit.
2. Practice locating Polaris with a transit beginning one hour before dark.
3. Establish a straight line several hundred feet long marked by terminal hubs at Points A and B. With the transit at A and a backsight on B, make a Polaris observation. Calculate the true bearing of line AB.
4. At the same location as activity 3, make a solar observation. Again calculate the true bearing of line AB and compare the results.

## REVIEW QUESTIONS

### A. Multiple Choice

1. Polaris is traveling horizontally from left to right at its
   a. upper culmination. c. eastern elongation.
   b. lower culmination. d. western elongation.
2. As seen on paper held behind the telescope's eyepiece, morning solar observations are in the lower-left and
   a. upper-left quadrant. c. upper-right quadrant.
   b. lower-left quadrant. d. lower-right quadrant.

### B. Short Answer

3. True direction of a line is determined by observing Polaris with what type of precision as compared to solar observation?

4. Compared to solar observations, how complex are the field operations for Polaris observations?
5. To observers looking north, stars appear to revolve around the earth's axis in what direction?
6. As seen on paper held behind a telescope's eyepiece, the sun's image travels diagonally downward and in what direction in the AM?
7. Polaris is very close to the center of rotation of all stars. What is the center called?
8. What are two conspicuous constellations that enable an observer to locate Polaris?
9. How many degrees do stars appear to revolve every 24 hours? Stars revolve 360°00′ in how many hours and minutes?
10. What two factors make correction of the sun's measured altitude necessary?
11. What danger must be avoided in making a solar observation?
12. What are the most favorable times of day for solar observations?

## C. Problems

13. A horizontal angle of 86°14′ is turned counterclockwise from a line on earth to Polaris at its lower culmination. If a 0-degree azimuth is due north, what is the azimuth of the line? Under the same conditions except that Polaris is at its upper culmination, what is the azimuth of the line?
14. Polaris observations are made May 12, 1980 to determine the bearings of four different lines. Horizontal angles from the four lines to Polaris and the calculated $Z_s$ angles are listed in figure 10-18. What are the true bearings of the four lines?

| Line | Horizontal Angle | | $Z_s$ | |
|---|---|---|---|---|
| | Counter Clockwise | Clockwise | Counter Clockwise | Clockwise |
| AB | 46° 33′ | | 0° 14′ | |
| GH | | 18° 11′ | 0° 05′ | |
| MN | 219° 41′ | | | 0° 04′ |
| RS | | 211° 52′ | | 0° 11′ |

Fig. 10-18

15. The pointer stars of the Big Dipper are observed at 9 PM one night to be in horizontal alignment. At what time that night are they observed in vertical alignment? How many months later are they observed in vertical alignment at 9 PM?
16. A surveyor at latitude 44°05′ observes Polaris at its eastern elongation. At the time of observation, the value of p given by an ephemeris is 0°50.28′.

What is the value of $Z_s$? Six hours later Polaris is observed at its upper culmination. What then is the value of $Z_s$?

17. Find the missing values in figure 10-19.

| Longitude | EST | GCT | LCT |
|---|---|---|---|
| 70° | 5:00 PM | 2200 | |
| 83° | | 1900 | |
| 75° | | | 1712 |
| 77°30′ | | | 1712 |

**Fig. 10-19**

18. An observer finds from a solar ephemeris that the Greenwich hour angle at the time of the Polaris observation is 230°45.2′. What is done to convert this value to local hour angle?

19. A horizontal angle is turned to Polaris with the telescope direct at 9:22.06 PM, pacific standard time, July 23, 1980. A second horizontal angle is turned with telescope inverted at 9:24:18 PM, the same date. What Greenwich civil time is used to find the Greenwich hour angle in a solar ephemeris?

20. At $0^h$ Greenwich civil time, June 22, 1980, the sun's declination is N 23° 26.3′. At what latitude is a person if the sun passes directly overhead at that time?

21. What is the Greenwich hour angle of the sun as it crosses the meridian at longitude 104°21.0′?

22. Bearing $Z_o$ is calculated to be 98°03′ for an afternoon solar observation. What is the sun's bearing?

# CHAPTER 11

# State Plane Coordinate Systems

## OBJECTIVES

After studying this chapter, the student will be able to:

- tie-in a local plane survey to nearby geodetic horizontal control monuments or other reference points for which state plane coordinates are known.
- calculate grid bearings and distances for a plane survey tied-in to state plane coordinates.
- locate on the ground any point with known state plane coordinates from any other nearby point with known state plane coordinates.

## LAMBERT AND MERCATOR PROJECTIONS

Whenever possible, local surveys made by plane surveying methods are tied to horizontal control monuments. These monuments are established by government agencies employing geodetic surveying methods. These methods compensate for the effects of the earth's curvature. High order triangulation is used for geodetic surveying in the United States. Control monuments are located by triangulation and primary traversing with a precision far greater than practiced in plane surveying. One system of identifying, relating, and referencing control monuments and other points is the state plane coordinate system.

Surveyors engaged in plane surveying are confronted with the same limitations as map makers. For example, a postage stamp can be pasted on a beach ball without wrinkling, but not on a golf ball. Wall paper cannot be pasted on even a beach ball without wrinkling. Similar distortions are unavoidable when spherical reference lines are projected onto plane surfaces for ordinary mapping purposes. The state plane coordinate system begins with a plane surface and achieves projections with acceptable degrees of distortion. It permits calculating the relationship of widely separate control points and plane surveys without resorting to geodetic surveying methods. Tie-in of property corners, local topographic control points, and construction base lines preserves knowledge of their

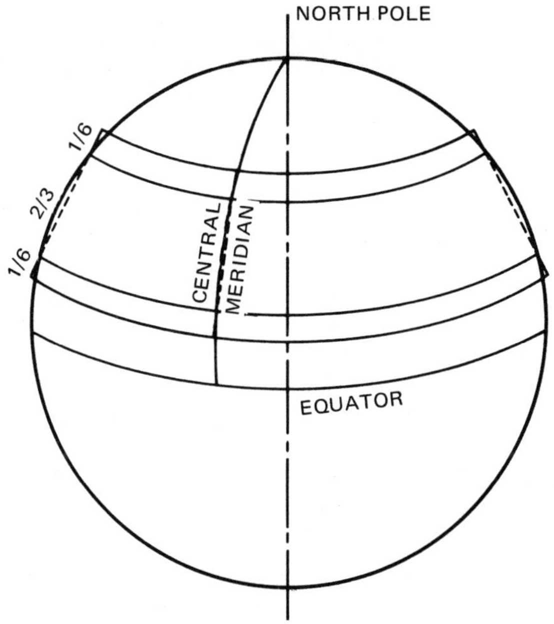


**Fig. 11-1 Lambert conformal conic projection**

locations permanently. Even if removed, these points remain in known relationship to the entire national control network. At any time in the future, they can be reestablished.

The state plane coordinate system was conceived in 1932 by two state highway engineers from North Carolina, O. B. Bestor and George F. Syme. This system consists of rectangular coordinates laid out on a plane surface. They chose a *lambert conformal conic projection* because North Carolina's longest dimension is in an east-west direction. In this type of projection, horizontal angles are undistorted, figure 11-1.

The system was adopted by The National Geodetic Survey, which extended it to other states. The transverse Mercator projection is used where a state's longest dimension is in a north-south direction. In the larger states, areas are zoned to keep errors due to projection distortions down to 1/10,000. Within that tolerance, zones are any length, but widths are limited to 158 miles. State coordinates systems in the continental United States are based on 67 lambert and 44 traverse Mercator zones.

## STATE PLANE COORDINATES BASED ON LAMBERT PROJECTION

A lambert conformal conic projection is a section of a cone centered on the earth's axis. It intersects the earth's surface at sea level along two standard parallels of latitude. For a state plane coordinate zone, a conical section at a desired latitude is selected. It has standard parallels *1/6 of the way* in from each edge. If stretched out on a plane surface, the conical section appears as in

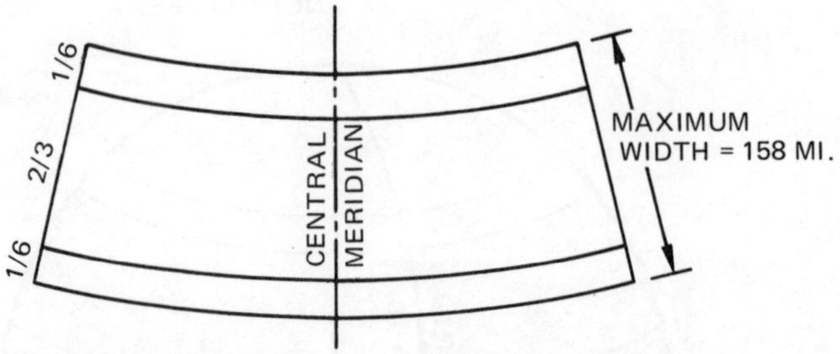


**Fig. 11-2 Conic section portrayed on a plane surface**

figure 11-2. A grid of rectangular coordinates is laid out on this surface. It is oriented with the y-axis along a central meridian approximately midway between the east-west limits of the zone. Its x-axis is perpendicular to the y-axis where y = 0.

The North Carolina state plane coordinate system's standard parallels are at latitudes 34°20′ and 36°10′. The central meridian is at longitude 79°00′ W, figure 11-3. An x-coordinate value of 2,000,000 is assigned to all points on the 79th meridian. A y-coordinate value of 0 is on the 79th meridian at latitude 33°45′00″. Zero coordinates are outside of the state to the west and south. All values are plus for any point in the state. Grid north coincides with true north at the 79th meridian, but departs clockwise from more eastern meridians. It departs counterclockwise from the more western meridians. The angle of rotation between meridians and grid north is called theta ($\theta$). Values of $\theta$ at any longitude are published by The National Geodetic Survey and the North Carolina Division of Geodetic Survey. It is 2°06′ clockwise at longitude 75°20′ W near the zone's eastern end. It is 3°10′ counterclockwise at longitude 84°30′ W near the zone's western end. True bearings are rotated to grid bearings by values of

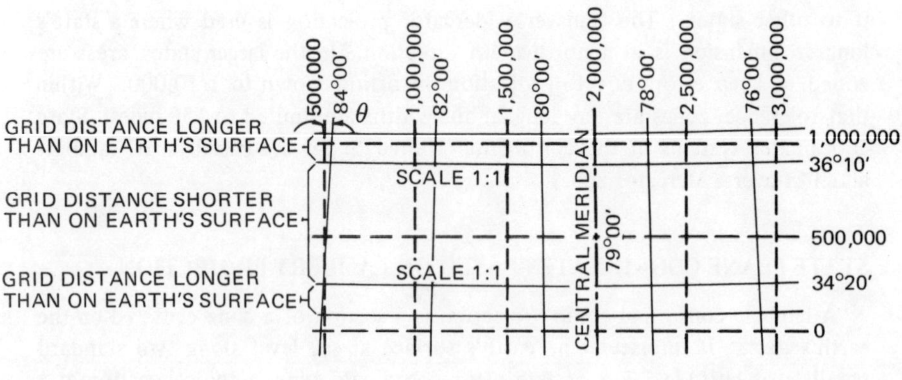


**Fig. 11-3 Orientation of the North Carolina state plane coordinates**

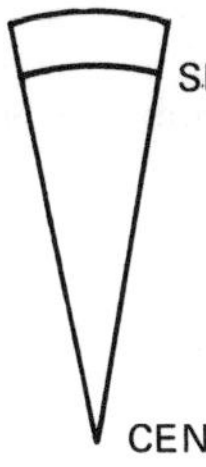


**Fig. 11-4 Surveyed distance reduced to sea level**

$\theta$ when calculating state plane coordinates. Surveyed distances are also converted to grid distances by a sea level factor and a scale factor.

## Sea Level Factor

A sea level factor is necessary because grid distances are at sea level. Horizontal surveyed distances above sea level are reduced in length. This is done by comparing a points' elevation above sea level to the earth's radius, figure 11-4. The mean radius of the earth in North Carolina is 20,902,600 feet. A negative correction of .048 feet per 1000 feet of elevation per 1000 feet of surveyed distance is required. Therefore, a surveyed distance of 1342.65 feet at an elevation of 4000 feet, corrected to sea level becomes:

$$\begin{aligned}\text{Sea level distance} &= 1342.65 - (.048)\left(\frac{4000}{1000}\right)\left(\frac{1342.65}{1000}\right)\\ &= 1342.65 - 0.26\\ &= 1342.39 \text{ feet}\end{aligned}$$

Sea level factors expressed as ratios are also published by the North Carolina Division of Geodetic Survey.

## Scale Factor

Surveyed distances change slightly when projected radially from the earth's curving surface to the plane surface. Projected lines become shorter anywhere between the two standard parallels of a lambert projection. Projected lines become longer than the earth's curving surface anywhere outside of those parallels, figure 11-5. Only at the standard parallels where the projection touches the earth's surface is the scale of the projection exact. The lengths of lines are adjusted by applying a scale factor. Values of scale factors at any latitude, expressed as a ratio, are published by The National Geodetic Survey.

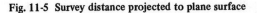

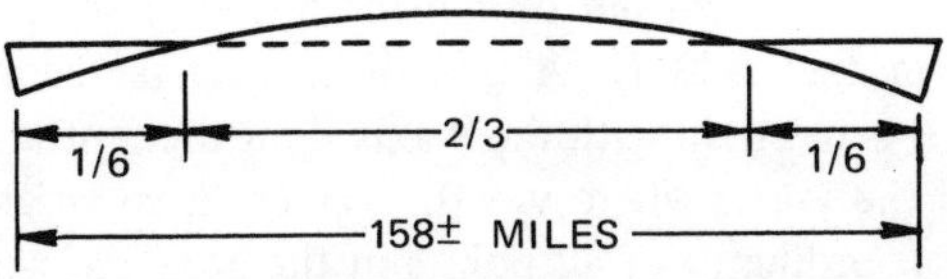


**Fig. 11-5 Survey distance projected to plane surface**

Assume the state plane coordinates are known for point A, a control monument. State plane coordinates of another point B, are calculated as follows:

Point A at latitude 35°50′43″ longitude 78°30′48″ has state plane coordinates: X = 2,144,201.16 and Y = 762,888.48.

Line AB at an elevation of 300 feet has a true bearing of N 25°47′ E and a surveyed horizontal length of 600.06 feet.

| | | |
|---|---|---|
| Grid bearing AB | = | true N 25°47′ E - $\theta$ |
| | = | N 25°47′ E -0°16′ |
| | = | N 25°31′ E |
| Grid length | = | Surveyed horizontal length multiplied by both sea level factor and scale factor |
| | = | 600.06 x 0.9999856 x 0.999926 |
| | = | 600.01 feet |
| Y-coordinate of point B | = | 762,888.48 + (cos of the grid bearing of AB) (grid distance) |
| | = | 762,888.48 + (.90246)(600.01) |
| | = | 762,888.48 + 541.48 |
| | = | 763,429.96 |
| X-coordinate of point B | = | 2,144,201.16 + (sin of the grid bearing AB) (grid distance) |
| | = | 2,144,201.16 + (.43077)(600.01) |
| | = | 2,144,201.16 + 258.47 |
| | = | 2,144,459. 63 |

Tie-in of a plane survey to a single control monument is possible only if the survey is oriented to true north. Where two control points visible to each other are available with known state plane coordinates, celestial observations are unnecessary except as a check.

## STATE PLANE COORDINATES BASED ON A TRANSVERSE MERCATOR PROJECTION

A *transverse Mercator projection,* figure 11-6, is a section of a cylindrical ring. It intersects the earth's surface at sea level along two standard meridional lines, parallel to a central meridian. For a state plane coordinate zone, a cylindrical section centered at a desired longitude is selected. The projection is less than 158 miles wide, with standard meridional lines 1/6 of the way in from each edge. If stretched out on a plane surface, a section of the cylindrical ring appears as in figure 11-7. A grid of rectangular coordinates is laid out on this surface. It is oriented with the y-axis along the central meridian and x-axis perpendicular to the y-axis where y = 0. Values are so assigned to the x- and y-coordinates that coordinates of all points in the zone are positive. The central meridian is usually assigned a value of 500,000, figure 11-8.

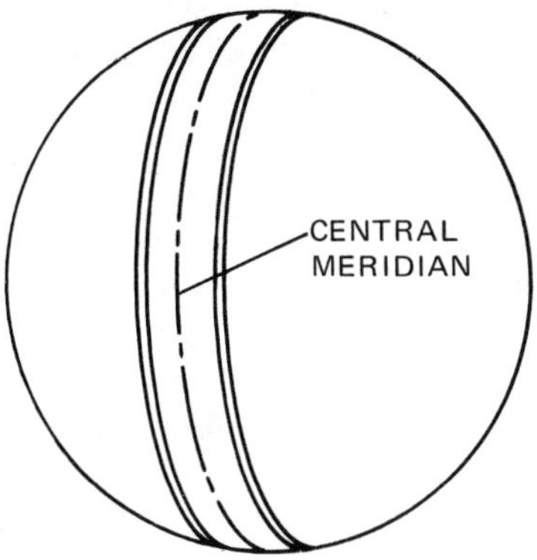


**Fig. 11-6 Transverse Mercator projection**

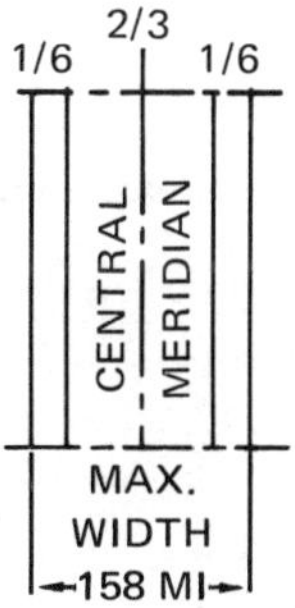


**Fig. 11-7 Cylindrical section portrayed on a plane surface**

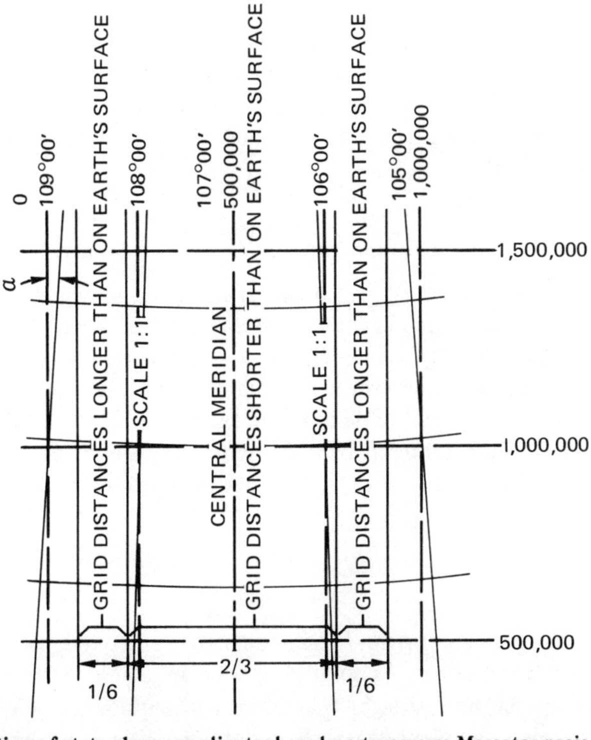


**Fig. 11-8 Orientation of state plane coordinates based on transverse Mercator projection**

Grid north coincides with true north at the central meridian. Grid north departs clockwise from more eastern meridians and counterclockwise from more western meridians. The angle of rotation between the meridians and grid north is called alpha ($\alpha$). Its value depends upon the perpendicular distance of the point from the central meridian, and the latitude of the point. Values needed for calculating alpha for any point are published by The National Geodetic Survey.

Surveyed distances are adjusted to grid distances in about the same manner whether based on the lambert or the transverse Mercator projections. The sea level factor is identical, and the scale factor is similar in theory. Values of scale factors at any x-coordinate value, expressed as a ratio, are published by The National Geodetic Survey.

## Precision Measuring Equipment

Throughout this book, angle and distance measurements have been made with the three basic surveyor's tools: the tape, transit, and level rod. This equipment is sufficient for land surveying. However, precision geodetic work requires greater precision. For example, the work performed by the National Geodetic Survey requires setting bench marks to very precise elevations and locations since other surveys use them for reference points. More precise optical equipment, such as the theodolite, was created to meet this demand. When it was discovered that radio waves, light waves, and lasers could be used for measuring distance, surveyors adopted these systems to speed up field work and make it easier. As the prices and weight of the equipment reduce because of technological progress, even those surveyors not engaged in geodetic surveying are using electronic equipment for ordinary land surveys, construction, highway, and sewer work.

## Theodolites

Theodolites were developed many years ago to measure horizontal and vertical angles to meet the precision requirements of geodetic surveying. Some models reading directly to one minute, with estimation to one-tenth of a minute, cost no more than some transits. Some models, such as the one shown in figure 11-9, read to one second. Still more precise models read to 0.2 seconds.

Theodolites are centered over a point by means of an optical plummet instead of a plumb-bob. Angles expressed in degrees, minutes, and seconds are read digitally.

## Levels

Automatic levels, figure 11-10, maintain horizontal lines of sight by self-leveling compensators. As soon as a circular bubble is centered, the compensator brings the line of sight horizontal and automatically keeps it there. On some models, preliminary leveling is done with three leveling screws. Others

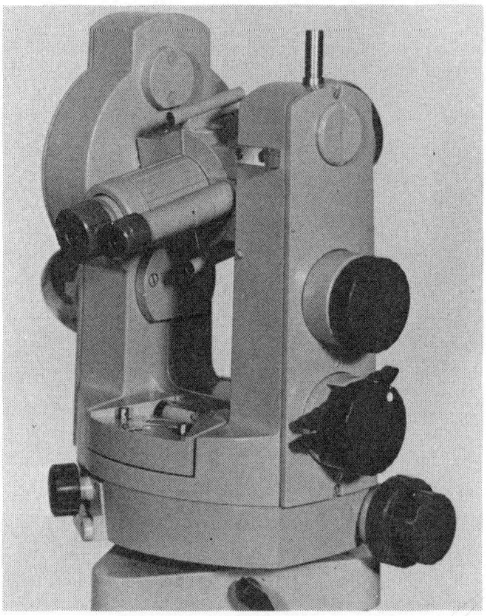

**Fig. 11-9 Theodolite**

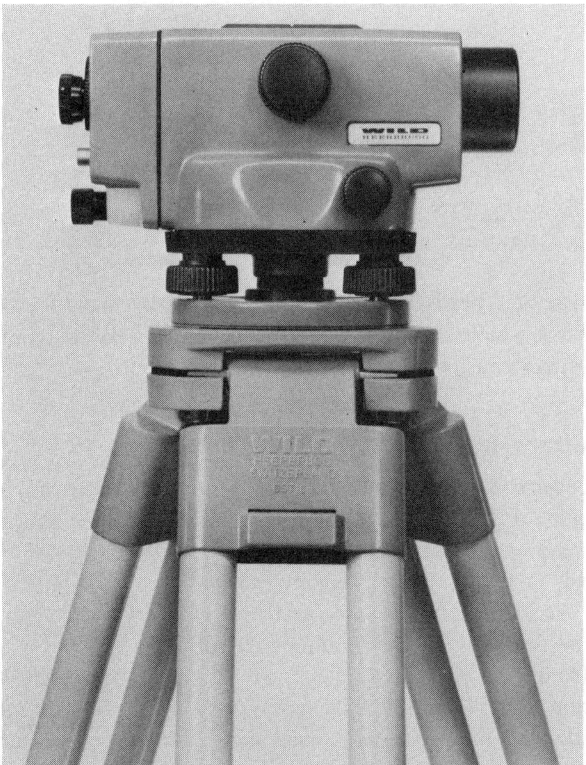


**Fig. 11-10 Engineer's automatic level**

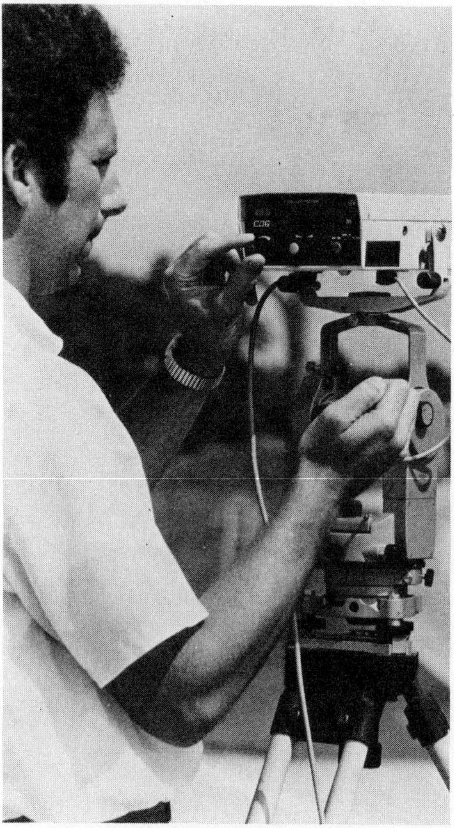

Fig. 11-11 Microwave distance meter

have ball and socket heads. Tracking levels, especially made for construction, have laser beams which shine a red dot wherever the level is pointed within a 600-foot working radius.

**Electronic Distance Measurement**

With microwave equipment, figure 11-11, slope distance can be measured with a precision of 1/330,000 ± 0.04 feet. Distances can be as short as a few hundred feet and as long as 50 miles. The equipment consists of two identical units mounted on tripods, each weighing 35 pounds. Each unit has a built-in transmitter, receiver, antenna, measurement circuitry, and communication system. For best results, the line should be free of obstacles.

With electro-optical equipment, figure 11-12, slope distance can be measured with a precision of 1/500,000 ± 0.03 feet. There are short-range models weighing 24 pounds, and long-distance heavier models than can measure up to 40 miles. They consist of two different units. One is a light pulse transmitter, the other a prism reflector. Infra-red models used in combination with theodolites are called tachometers. They display horizontal distance, difference in elevation, and coordinate differences.

**Complete Electronic Surveying Station**

Equipment is now being produced combining high precision theodolites, electro-optical distance meters, data collectors, and calculators, figure 11-13.

**Fig. 11-12 Electro-optical distance meter**

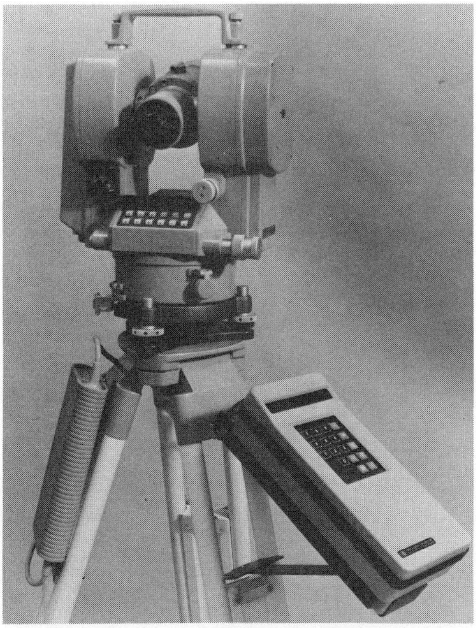

**Fig. 11-13 Data collector interfaced to electronic total station**

Maximum measurable distance is three miles. Such equipment is pointed quickly and easily to a prism set over a point. Horizontal and vertical angles are measured to one second. Angles are in degrees, minutes and seconds. Distances are automatically corrected for earth curvature, refraction, and environmental conditions. Horizontal and vertical components of slope distances are displayed in response to pressing a button. Data is transmitted directly to a computer or to a data collecting device for later processing.

**Other Special Equipment**

Self-reducing alidades for plane tables, figure 11-14, feature constant viewing angle for telescope and circle reading eyepieces. They give direct reading of horizontal distance and difference in elevation.

Laser eyepieces are available as attachments for theodolites, plummets, and levels. They provide visible lines of sight for precise alignment in construction, tunneling, and mining.

Zenith and nadir optical plummets are available to establish vertical control lines for major construction projects. The model shown in figure 11-15, projects vertical lines of sight with a precision of 1/30,000.

North-seeking gyroscope attachments for theodolites, figure 11-16, find north anywhere without celestial observation. They are useful in mines and tunnels, and where bad weather conditions prevent celestial observations for sustained periods.

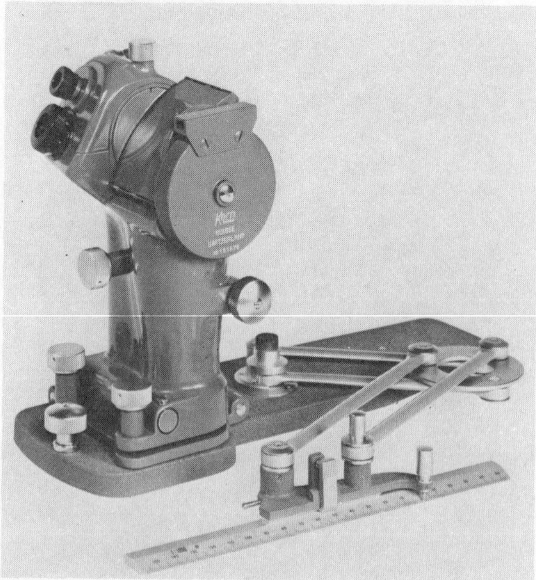

**Fig. 11-14 Self-reducing alidade**

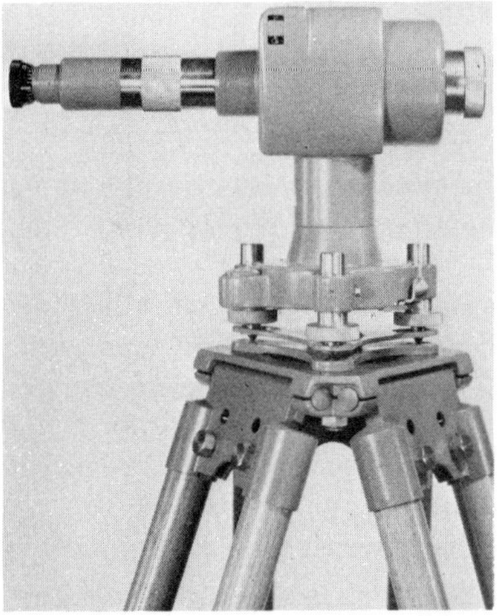

**Fig. 11-15 Zenith and nadir plummet**

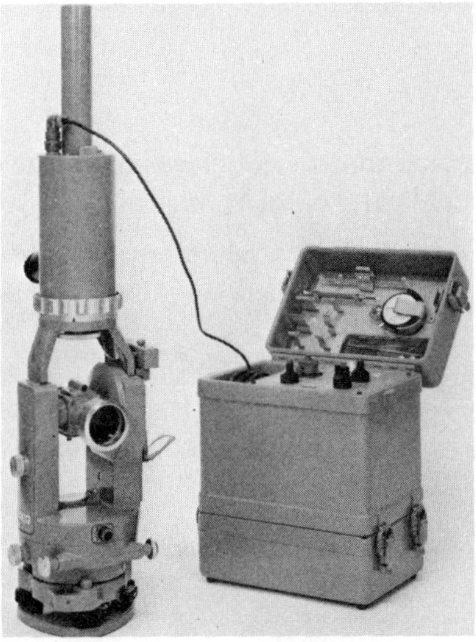

**Fig. 11-16 Gyro attachment**

## SUMMARY

State plane coordinate systems permit calculating the relationship of widely separated points without resorting to geodetic surveying methods. When state plane coordinates for a point are determined, permanent evidence of its location is established.

Zones for state plane coordinate systems are unlimited in length, but are restricted to widths of 158 miles. Where length is in an east-west direction, zones are on lambert conformal conic projections. Where length is in a north-south direction, zones are on transverse Mercator projections.

State plane coordinate systems in the continental United States are based on 67 lambert and 44 Mercator zones.

Standard parallels of latitude in the lambert projection, and standard meridian lines in the Mercator projection, are 1/6 of a zone's width from its edge.

A convenient point south and west of a zone is chosen for the origin of coordinates. That location keeps all coordinate values positive.

Grid lines are parallel to each other. Grid north coincides with true north at the central meridian of a zone. It diverges clockwise from meridians lying east of the central meridian and counterclockwise for meridians lying west. The angle of rotation is designated theta ($\theta$) for the lambert projection and alpha ($\alpha$) for the Mercator projection.

Surveyed distances are converted to grid distances by a sea level factor and a scale factor. The sea level factor reduces surveyed distances to sea level distances. The scale factor shrinks surveyed distances between standard parallels or standard meridian lines. It expands surveyed distances between the standard meridian lines and the sides of the projection.

## ACTIVITIES

1. Tie-in a tract of land to the state plane coordinate system, and compute the state plane coordinates of each corner.
2. Assume state plane coordinates values for one corner of a tract you have surveyed. Calculate the state plane coordinates for all other corners.

## REVIEW QUESTIONS

### A. Multiple Choice

1. Zone widths are restricted to
   a. 30 miles.
   b. 58 miles.
   c. 100 miles.
   d. 158 miles.
2. X-coordinate values of 0 are assigned to all points on the central meridian of a state plane coordinate system based on
   a. the lambert conformal conic projection.
   b. the transverse Mercator projection.
   c. both projections.
   d. neither projection.
3. Grid north diverges clockwise from meridians lying in what direction from the central meridian?
   a. North
   b. South
   c. East
   d. West

4. Grid north coincides with true north
   a. at standard meridian lines.
   b. at the standard parallels of latitude.
   c. at the central meridians.
   d. only at the equator.

5. The lambert conformal conic projection has standard parallels of latitude how far in from the edge?
   a. 1/12
   b. 1/6
   c. 1/3
   d. 1/2

6. For a transverse Mercator projection, the x-coordinate of the central meridian is usually assigned a value of
   a. 0
   b. 250,000
   c. 500,000
   d. 1,000,000

## B. Short Answer

7. The state plane coordinate system achieves projections with what amount of accuracy?

8. What two factors are considered for converting surveyed distances to grid distances?

9. What two Greek letters identify angles of rotation between true north and grid north?

10. Where are the grid and surveyed distances equal in length?

## C. Problems

11. Convert the following surveyed course to grid bearing and distance:
    - Length 865.02
    - True Bearing N43°32′ E
    - Elevation 3500 ft.
    - Earth's radius 20,902,600 ft.
    - $\theta$ angle is 1°27′ clockwise
    - Scale factor is 0.999908.

# CHAPTER 12

# Horizontal and Vertical Curves

## OBJECTIVES

After studying this chapter, the student will be able to:

- calculate curve data necessary to measure and lay out horizontal curves.
- lay out horizontal curves with a transit and tape.
- calculate grade lines for vertical curves.

## TYPES OF CURVES

Highways, railroads, and other engineering works are constructed along curved as well as straight lines. Land surveyors encounter property lines abutting curved rights-of-way. They also lay out subdivisions involving curves. Topographic surveyors locate and map curving features. Construction surveyors lay out facilities requiring curved alignment.

Most horizontal curves are circular, although railroads and some main highways also have spiral curves. *Spiral curves* are transition curves with a constant change of radius. They are used between straight lines and circular curves, or between adjoining circular curves.

Vertical curves are parabolic. For certain applications such as landscaping, driveways, and curb returns at intersections, some horizontal curves are parabolic.

## CIRCULAR CURVES

Highway and railroad engineers refer to the sharpness of a curve more often by its degree than by its radius. In highway surveying, the *degree of curve* is the number of degrees subtended by a 100-foot arc. In railroad surveying, it is the number of degrees subtended by a 100-foot chord. For long radius curves, the difference is slight. In this chapter, all references are based on the highway definition.

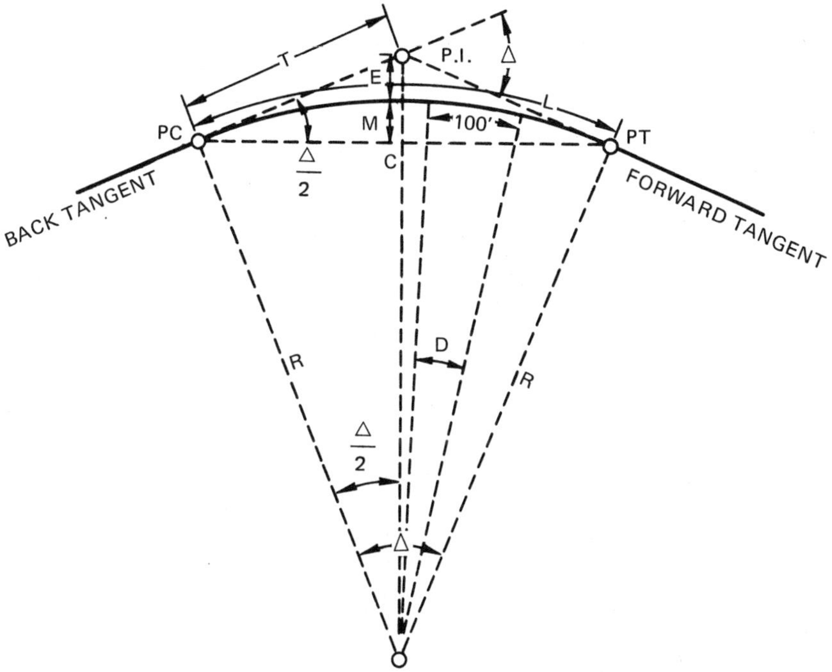


**Fig. 12-1 Elements of horizontal circular curve**

The names of curve elements are shown in figure 12-1. Straight lines connecting two curves, and tangent to both, are called *tangents.* Each curve has a *back tangent,* a line from which a curve is being approached. It also has a *forward tangent* which extends to the curve ahead. Extension of these two tangents brings them to a *point of intersection* (PI). The curve's *radius* is designated R. Radial lines at each end of a curve are at right angles to the tangents. The radial lines intersect at the curve's center to form a *central angle.* The central angle subtends the full length of curve (L), and equals the deflection angle at the PI. Both the central angle and the deflection angle are given the same designation, Δ. A curve begins at a *point of curvature* (PC) and ends at a *point of tangency* (PT). Distances from the PC to the PI, and from the PT to the PI, are the same length. They are called *tangent distances* (T).

A *long chord* (C) is a straight line between the PC and the PT. An angle at either the PC or the PT measured from the PI to the long chord has a value of $\Delta/2$. A distance from the PI to the midpoint of a curve is an *external distance* (E). A distance from the midpoint of a curve to the midpoint of a long chord is a *middle ordinate* (M).

The circumference of a circle is $2\pi r$. The circumference of a circle with a 100-foot radius is 628.32 feet, or 6.2832 arcs of 100 feet. Each 100-foot arc has a central angle of 57.2958°. A circle of 1000-foot radius has ten times that circumference and one-tenth of that central angle, figure 12-2. The relationship is that central angle D for a 100-foot arc is always:

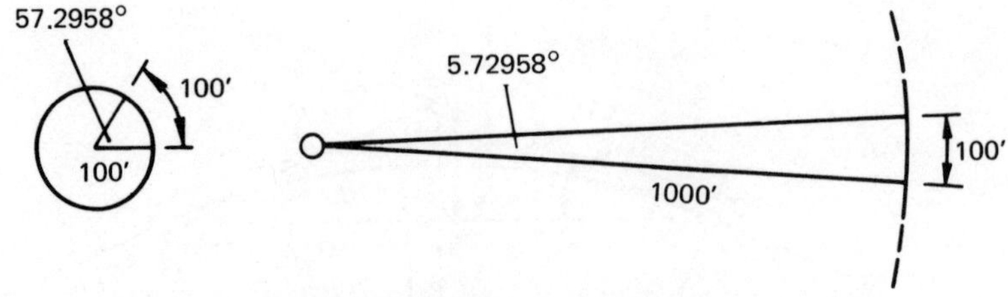


**Fig. 12-2 Central angle diminishes as radius increases**

$$D = \frac{360^\circ}{2\pi R} \times 100$$

$$= \frac{57.2958 \times 100}{R}$$

(A)

$$\text{or} \qquad D = \frac{5729.58}{R}$$

For most work, the value of 5729.58 is taken as 5730. Therefore:

(B)

$$D = \frac{5730}{R}$$

The central angle of any circular curve is proportional to the length of the curve it subtends. Central angle $\Delta$ equals the number of 100-foot arcs times the degree of curve. Expressing the number of 100-foot arcs as L/100,

(C)

$$\Delta = \frac{(L)}{(100)} D$$

Continuous stationing is carried out along tangents and around curves, figure 12-3. Station 14 + 50 at a PC indicates that it is 1450 feet from the beginning of a line. If the curve's length is 450 feet, the station of the PT of that curve is 19 + 00. The station of the PI is always obtained by adding the tangent distance to the station of the PC. It is never obtained by subtracting the tangent distance

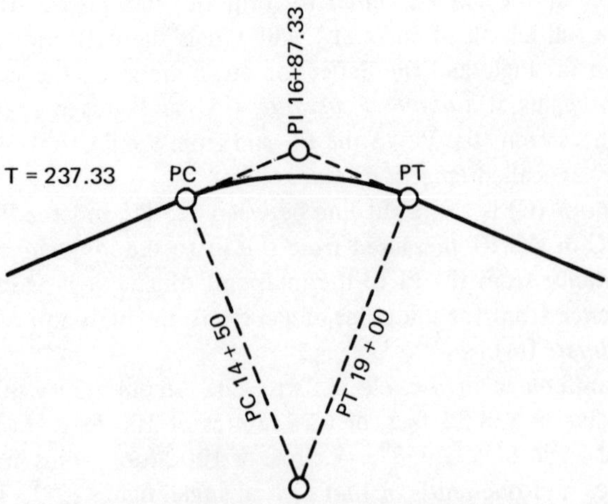


**Fig. 12-3 PI station reckoned from PC station**

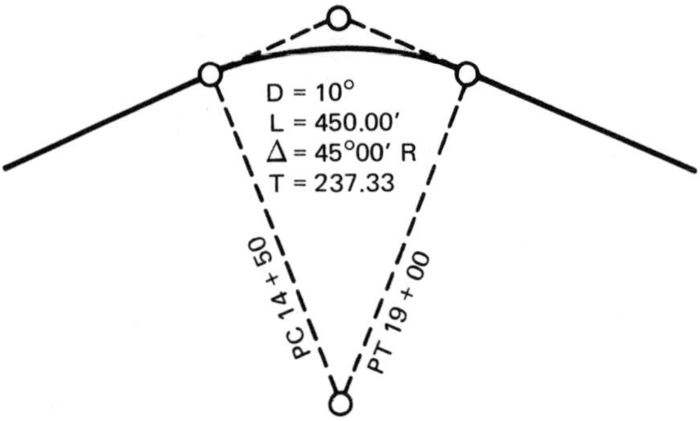


Fig. 12-4 Conventional curve data

from the station designating the PT. As shown in figure 12-3, PC station 14 + 50 plus 237.33 feet = PI station 16 + 87.33.

An intermediate transit station on a curve is called a *point on curve* (POC). An intermediate transit station on a tangent is called a *point on tangent* (POT).

Curve information entered on plans and in field notebooks normally include only values for D, L, Δ, and T, figure 12-4. The center is not shown except for very short radius curves.

## EQUATIONS FOR CIRCULAR CURVES

In addition to Equations (A) and (C), four other equations are needed to calculate the values of elements of curves.

(D) $T = R \tan \frac{\Delta}{2}$ Figure 12-5

(E) $C = 2R \sin \frac{\Delta}{2}$ Figure 12-6

(F) $M = R - R \cos \frac{\Delta}{2}$ Figure 12-7

(G) $E = \frac{R}{\cos \frac{\Delta}{2}} - R$ Figure 12-8

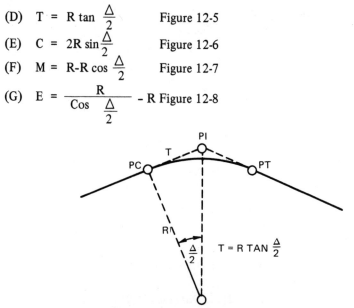

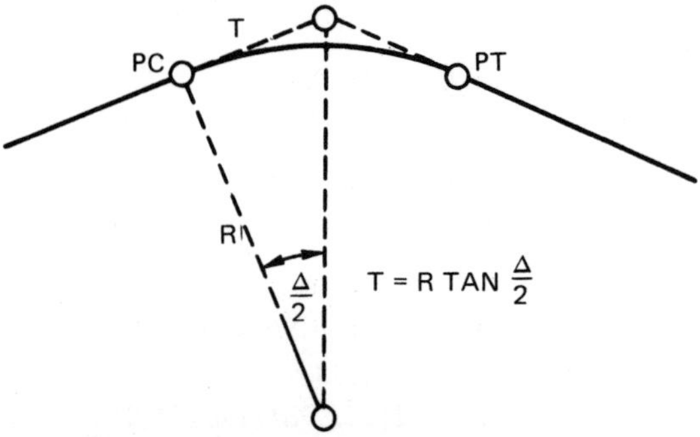


Fig. 12-5 Derivation of tangent distance

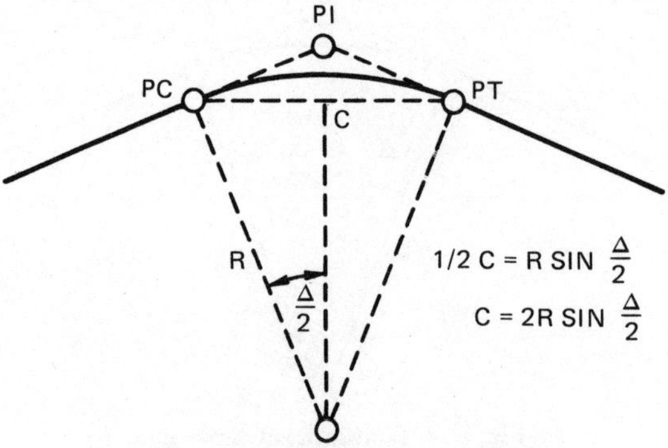


Fig. 12-6 Derivation of long chord

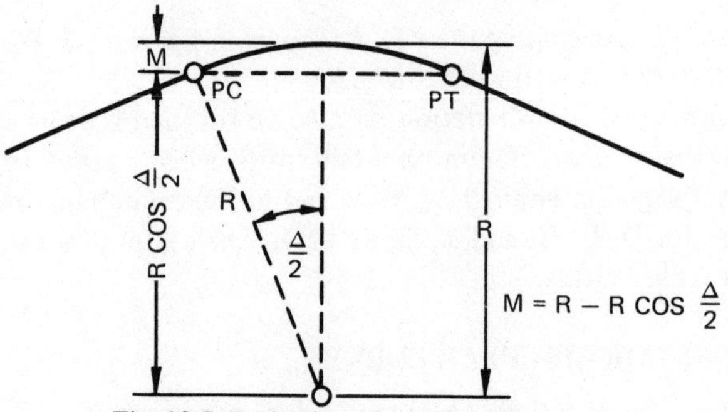


Fig. 12-7 Derivation of middle ordinate

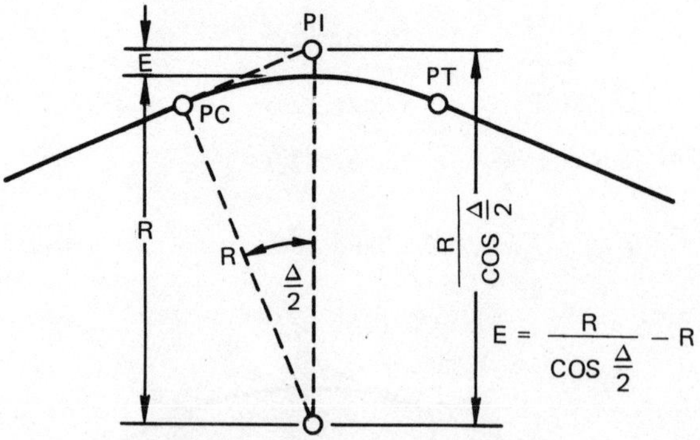


Fig. 12-8 Derivation of external

From these formulas and the curve data from figure 12-4, the following examples are given:

$$T = (572.96)(\tan 22^\circ 30') = 237.33 \text{ feet}$$

$$C = 2(572.96)(\sin 22^\circ 30') = 438.52 \text{ feet}$$

$$M = 572.96 - (572.96)(\cos 22°30')$$
$$= 43.61 \text{ feet}$$
$$E = \frac{572.96}{\cos 22°30'} - 572.96$$
$$= 47.21 \text{ feet}$$

## LAYOUT OF A CIRCULAR CURVE

### Alignment

Layout of a highway curve begins by setting up a transit at the PC. Backsight along the back tangent. For each stake set along the curve, the proper deflection angle is calculated and turned. Stakes are set at regular intervals, usually no greater than 50 feet apart.

A deflection angle between any two points on a curve is easily calculated. Its value is one-half the central angle subtending an arc between them. It is equal to the degree of curve multiplied by the arc length divided by 200. For a point on a 10-degree curve at 250 feet along the curve, it is 12°30′. At 350 feet along the curve it is 17°30′. Figure 12-9 shows several deflection angles with the transit at the PC. Figure 12-10 shows two deflection angles with the transit at a POC.

Good control of alignment cannot be maintained if deflection angles exceed 25 degrees. Before deflection angles reach that value, it is good practice to move

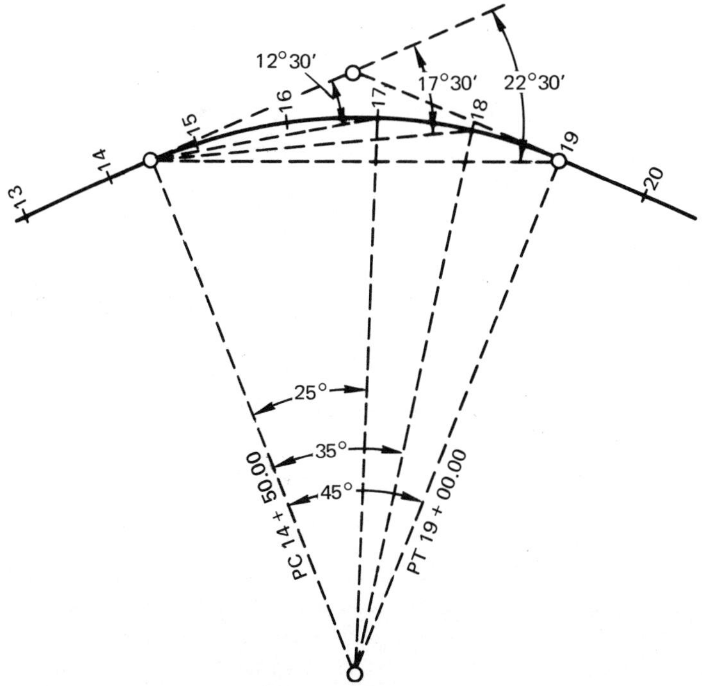


**Fig. 12-9 Deflection angles at PC**

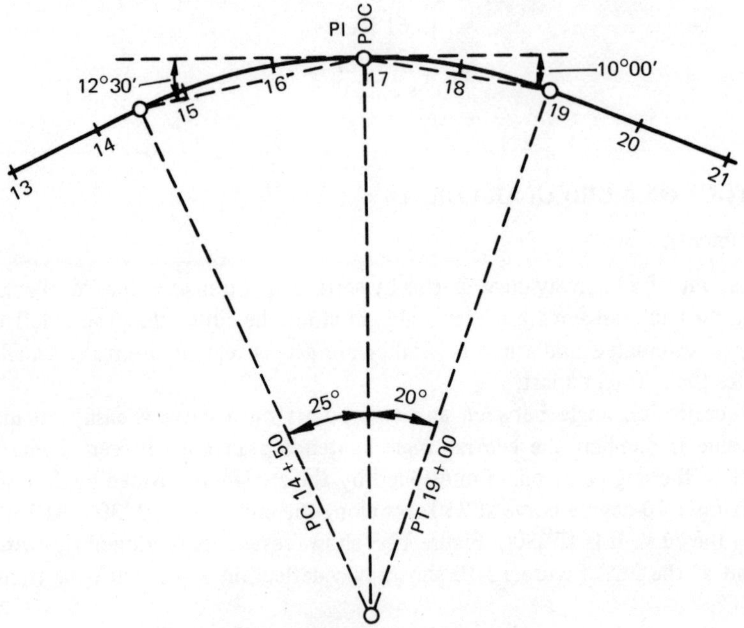


Fig. 12-10 Deflection angles at POC

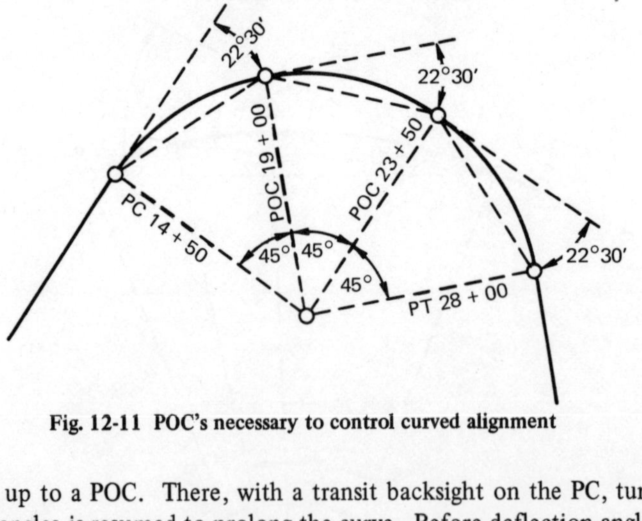


Fig. 12-11 POC's necessary to control curved alignment

the transit up to a POC. There, with a transit backsight on the PC, turning of deflection angles is resumed to prolong the curve. Before deflection angles again reach the 25 degree limit, another POC is set, and a backsight is taken on the previous POC, figure 12-11. This procedure is repeated as many times as necessary. For example, curves having a Δ angle of 90 degrees are laid out in two segments. Those with Δ 's of 180 degrees have at least four segments.

POC's are needed not only to avoid loss of control of alignment. They are also needed at hill crests and wherever topography or other existing features block lines of sight. In very restricted places, such as in dense woods, reasonable effort is made to clear lines. Intervals between POC's should exceed 100 feet wherever possible to avoid loss of control resulting from short backsights.

Deflection angles are doubled when setting all transit points. Hubs or other serviceable and stable devices are used to identify them.

There are two ways of setting the transit's plates for backsights and for deflection angles to align stakes on a curve. Values of deflection angles under Procedure A are as calculated by reference to the curve's back tangent. Under Procedure B they are calculated by reference to a tangent to the point the transit occupies. Backsights are taken as follows:

**Procedure A.** At the first POC, the upper plate is set at 0 degrees. Backsight on the PC with the telescope inverted. At the remaining POC's, the upper plate is set at the deflection angle used to align the previous POC. Backsight on the previous POC with the telescope inverted.

**Procedure B.** At any POC, the upper plate is set at a value equal but algebraically opposite to a deflection angle to the point backsighted; clockwise for counterclockwise, and vice versa. Backsight with the telescope inverted. The instrument is then always oriented to make plates read 0 degrees when aimed tangent to the curve. Calculated bearings of tangents are easily compared with observed compass bearings.

For the 1350-foot curve illustrated in figure 12-12, upper plate settings for backsights are:

| Transit At Station | Procedure A | Procedure B |
|---|---|---|
| 14 + 50 PC | 0° | 0° |
| 19 + 00 POC | 0° | -22°30′ |
| 23 + 50 POC | 22°30′ | -22°30′ |
| 28 + 00 PT | 45°00′ | -22°30′ |

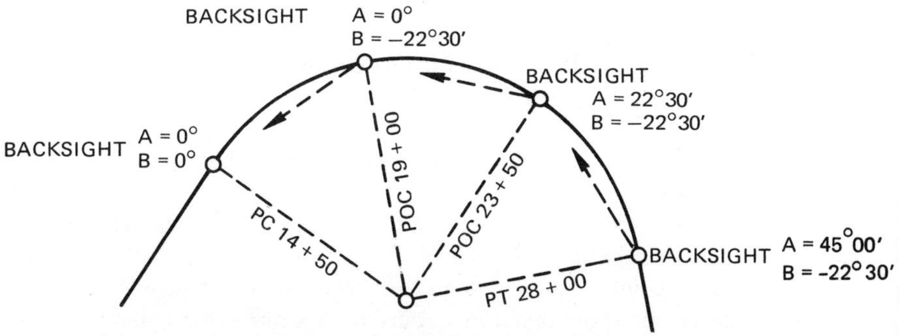


**Fig. 12-12 Plate settings for backsights at POC's**

Deflection angles for forward points are listed in figure 12-13.

| Transit At Station | Forward Point | Deflection Angles Procedure A | Procedure B |
|---|---|---|---|
| 14 + 50 | 15 + 00 | 2°30′ | 2°30′ |
| | 15 + 50 | 5°00′ | 5°00′ |
| | 16 + 00 | 7°30′ | 7°30′ |
| | 16 + 50 | 10°00′ | 10°00′ |
| | 17 + 00 | 12°30′ | 12°30′ |
| | 17 + 50 | 15°00′ | 15°00′ |
| | 18 + 00 | 17°30′ | 17°30′ |
| | 18 + 50 | 20°00′ | 20°00′ |
| | 19 + 00 | 22°30′ | 22°30′ |
| 19 + 00 | 19 + 50 | 25°00′ | 2°30′ |
| | 20 + 00 | 27°30′ | 5°00′ |
| | 20 + 50 | 30°00′ | 7°30′ |
| | 21 + 00 | 32°30′ | 10°00′ |
| | 21 + 50 | 35°00′ | 12°30′ |
| | 22 + 00 | 37°30′ | 15°00′ |
| | 22 + 50 | 40°00′ | 17°30′ |
| | 23 + 00 | 42°00′ | 20°00′ |
| | 23 + 50 | 45°00′ | 22°30′ |
| 23 + 50 | 24 + 00 | 47°30′ | 2°30′ |
| | 24 + 50 | 50°00′ | 5°00′ |
| | 25 + 00 | 52°30′ | 7°30′ |
| | 25 + 50 | 55°00′ | 10°00′ |
| | 26 + 00 | 57°30′ | 12°30′ |
| | 26 + 50 | 60°00′ | 15°00′ |
| | 27 + 00 | 62°30′ | 17°30′ |
| | 27 + 50 | 65°00′ | 20°00′ |
| | 28 + 00 | 67°30′ | 22°30′ |

**Fig. 12-13**

The advantage of Procedure A is that deflection angles are computed sequentially for all points on the curve. The advantage of Procedure B is its flexibility where:

- backsights are needed on almost any point on the curve, either forward or backward.
- transit points occur at a point of change from a circular curve of one radius to another.
- transit points occur at a point of change from one kind of curve to another.

Experienced surveyors prefer Procedure B. Beginners often prefer Procedure A, especially where no more than one POC is needed.

**Distances**

Distances from transits to stakes on curves are measured only to stakes adjacent to the transit stations. Beyond adjacent stakes, all distances are measured from stake to stake around the curve. Both head and rear ends of the tape are moved forward for each measurement. The person holding the rear of the tape always stays on the outside of curve to avoid obstructing the line of sight.

By definition, distances between points on curves are arc lengths. In practice it is impractical to bend a tape along the arc's path. Measurement is along the paths of chords. For long radius curves, differences between 50-foot arcs and chords are so slight that they are ignored. For shorter radius curves, stakes are set at 25-foot intervals. For still shorter curves where significant differences occur even for 25-foot lengths, measured distances are corrected to chord lengths, a procedure called *short chaining.*

For 50-foot intervals between stakes on curves, differences in chord and arc lengths are:

| Degree of Curve | Difference |
|---|---|
| $8^\circ$ | 0.01 ft. |
| $11^\circ$ | 0.02 ft. |
| $14^\circ$ | 0.03 ft. |

For 25-foot intervals between stakes on curves, differences in chord and arc lengths are:

| Degree of Curve | Difference |
|---|---|
| $15^\circ$ | 0.005 ft. |
| $22^\circ$ | 0.01 ft. |
| $32^\circ$ | 0.02 ft. |
| $57^\circ 18'$ (100 ft. radius) | 0.065 ft. |

An error of 1/5000 is introduced when a difference of 0.01 feet is tolerated for 50-foot arcs. Errors greater than 1/5000 occur where differences are ignored for 50-foot arcs on curves exceeding 8 degrees. For 25-foot arcs errors greater than 1/5000 occur where differences are ignored for curves exceeding 15 degrees.

All distances between stakes on curves are measured horizontally. On sloping ground, intermediate stakes are set wherever necessary to keep within the capability of measuring horizontally. Where slopes change abruptly between regularly spaced stakes, supplemental stakes are often needed to meet job requirements.

**Circumventing Obstacles**

When obstacles are encountered in running a curve, one solution is to run segments from each end. Both tangents are prolonged to the PI. The tangent distance, T, is measured off to the PC and the PT. The curve is run forward from the PC to the obstacle, and backward from the PT to the obstacle.

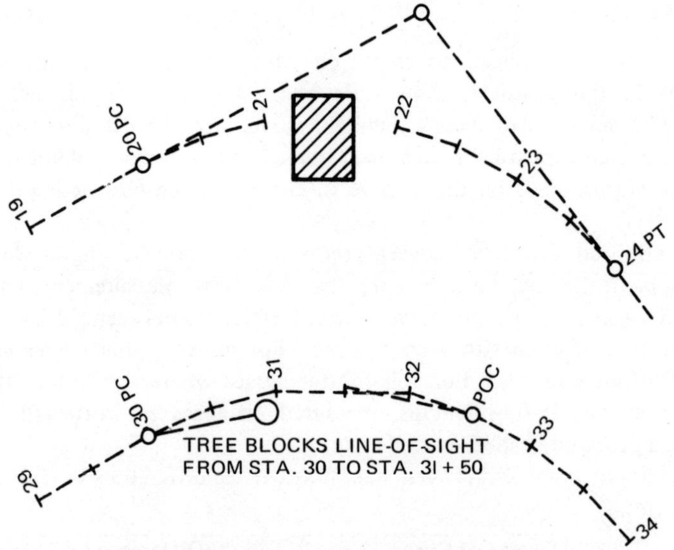


**Fig. 12-14 Bypassing stakes because of obstacles**

Another method is to simply bypass one or two of the regularly spaced stakes. Distances to stakes beyond omitted stakes are calculated as chords either from the transit or from any previous stake set. Bypassed stakes are often set easily from the next POC, figure 12-14.

One method that always works is to calculate rectangular coordinates for all key points. This is done by considering all straight lines between PI's as lines of an open traverse. Compute the latitudes, departures, and coordinates of the PI's as for any traverse. Compute the latitudes, departures, and coordinates of the PC's, POC's, and PT's. Compute coordinates of points obstructed by obstacles. Locate obstructed points by methods employed wherever coordinates are known.

### Notekeeping

Field notes for a horizontal curve are kept in the form shown in figure 12-15. It is customary to enter the lowest station number at the bottom of the page and work up the page. This arrangement keeps all points in the same order as the notekeeper sees them while looking forward.

## COMPOUND AND REVERSE CURVES

Two or more curves sometimes join (the PT of one curve being the next curve's PC). They are components of a *compound curve* when both are in the same direction. They are components of a *reverse curve* if they are in opposite directions. Reverse curves are no longer constructed on main highways. Compound curves are avoided whenever possible. Both are found in subdivisions and low speed, lightly traveled park and farm roads. A point common to two circular curves is called a *point of compound curve* (PCC) or a *point of reverse curve* (PRC), figure 12-16.

Another class of curve seldom found acceptable is a *broken-back curve.* It consists of two curves in the same direction connected by a short tangent, figure 12-16. Modern practice calls for the substitution of a single curve, such as that shown by the dashed line.

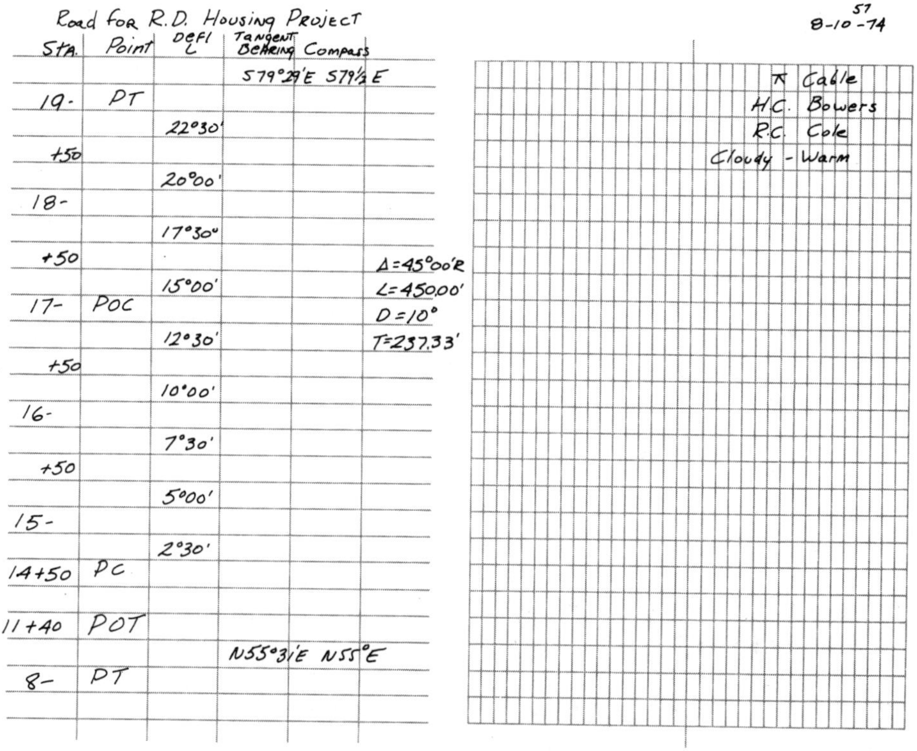
Road for R.D. Housing Project

57
8-10-74

| Sta. | Point | Defl ∠ | Tangent Bearing | Compass | |
|---|---|---|---|---|---|
| | | | S79°29'E | S79½E | |
| 19- | PT | | | | |
| | | 22°30' | | | |
| +50 | | | | | |
| | | 20°00' | | | |
| 18- | | | | | |
| | | 17°30' | | | |
| +50 | | | | | Δ=45°00'R |
| | | 15°00' | | | L=450.00' |
| 17- | POC | | | | D=10° |
| | | 12°30' | | | T=237.33' |
| +50 | | | | | |
| | | 10°00' | | | |
| 16- | | | | | |
| | | 7°30' | | | |
| +50 | | | | | |
| | | 5°00' | | | |
| 15- | | | | | |
| | | 2°30' | | | |
| 14+50 | PC | | | | |
| 11+40 | POT | | | | |
| | | | N55°31'E | N55°E | |
| 8- | PT | | | | |

Cable
H.C. Bowers
R.C. Cole
Cloudy - Warm

**Fig. 12-15 Field notes for horizontal curves**

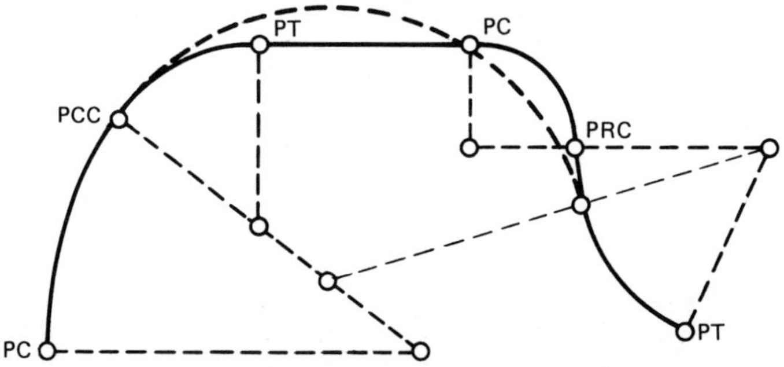


**Fig. 12-16 Compound, reverse, and broken-back curves**

## SPIRAL CURVES

An abrupt change from a tangent to a circular curve is undesirable for railroads and high-speed highways. Horizontal alignment for such facilities is eased by a *spiral curve.* It matches paths of vehicles during the interval their steering controls are operating to accomplish gradual changes in turning radius. Its departure from a tangent begins with an infinitely long radius curve. It gradually sharpens until the desired radius is reached. When spirals are used, centers of circular curves are forced inward. Inward displacement of a curve is shown in figure 12-17. The amount of *offset* from tangents is identified by letter "o". The point of departure from the back tangent is a *point of spiral* (PS). The point where the spiral meets the circular curve is a PSC. The point where the circular curve meets the forward spiral is a PCS. The point where the spiral meets the forward tangent is the ST.

The length of a spiral is twice the displaced circular arc's length. The simplest way of locating spirals is to run the entire length of the circular curve from the displaced PC to the displaced PT. Establish midpoints of spirals at the middle of offsets "o" opposite both the displaced PC, and PT. Establish other points along the spirals as needed by offsets from the tangent or the displaced circular arc.

Between the PS and the midpoint of the spiral, perpendicular offsets from the tangent to the spiral are proportional to the cubes of their distances from the PS. Between the PSC and the midpoint of the spiral, radial offsets from the displaced circular arc to the spiral are proportional to the cubes of their distances from the PSC. Offsets from either the tangent or the displaced arc reach their maximum values of o/2 at the spiral's midpoint. The same method applies to the forward spiral.

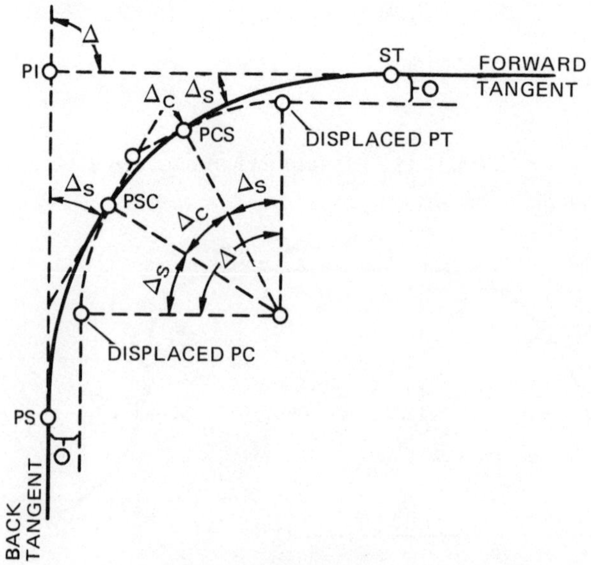


**Fig. 12-17 Spiral curve**

An experienced surveyor sets points along spirals by deflection angles, based on the following characteristics of spiral curves:

- The spiral angle, $\Delta$ s, equals the central angle of the displaced circular arc. The remaining portion of the circular arc, from the PSC to the PCS, has a central angle $\Delta$ c. It equals $\Delta$ for the entire curve, less twice $\Delta$ s.
- The deflection angle at the PS for the entire spiral curve is $\Delta$ s/3. The deflection angle at the PSC for the entire spiral curve is 2 $\Delta$ s/3.
- The deflection angles at the PS for all points on the spiral curve are proportional to the squares of distances from the PS.
- The instantaneous degree of curve at any point on a spiral is proportional to the point's distance from the PS.

Transits are set with the upper plates reading 0 degrees when the instrument is aimed tangent to a spiral curve.

Distances from the PI to points on back and forward tangents opposite the displaced PC and PT equal (R + o) tan $\Delta$/2. From these points, distances to the PS or ST are only slightly less than half the spiral length. These distances are given exactly in published spiral tables and on construction plans. When laying out spirals by the offset method, half spiral lengths are used without significant error.

## VERTICAL CURVES

Ground elevations along the line of proposed highways or railroads are plotted in profile form. Survey stations are plotted horizontally, and elevations are plotted vertically on grid paper. Vertical dimensions are usually exaggerated by a factor of 10:1. Popular scales are one inch equals 100 feet horizontally, and one inch equals 10 feet vertically. Tangent sections of grade lines are shown on the profile as a succession of straight lines. These lines represent elevations to which the facility is designed. Differences in elevation per hundred feet horizontally identify the *percent of grade.* Grades change with major changes in ground profile. Where grade changes occur, vertical parabolic curves are used. The term *grade line* applies to both tangent and curved sections of a grade used for construction.

A vertical *point of intersection* of adjacent tangent sections is called the VPI. Points of departure of a curve from vertical tangents are the vertical point of curvature (VPC) and the vertical point of tangency (VPT), figure 12-18. The *length* of a vertical curve, L, is the horizontal distance from the VPC to the VPT. A straight line between the VPC and the VPT is a *chord.* The vertical curve's center lies midway between the chord's center and the VPI. The vertical distance between the curve's center and the vertical point of intersection is the *external distance,* E.

Positions of points along a vertical curve are defined by survey station designation and elevation. Elevations of the VPC, VPT, and VPI are fixed by tangents. The average elevation of the VPC and the VPT is the elevation of the chord's center. The average elevation of the VPI and the chord's center is the elevation of the curve's center.

Vertical offsets between tangents and curves are proportional to the squares of the horizontal distances from the VPC (or the VPT), to the VPI. Offsets reach a maximum at the VPI where the offset value is E. Figure 12-19 shows examples of two vertical curves.

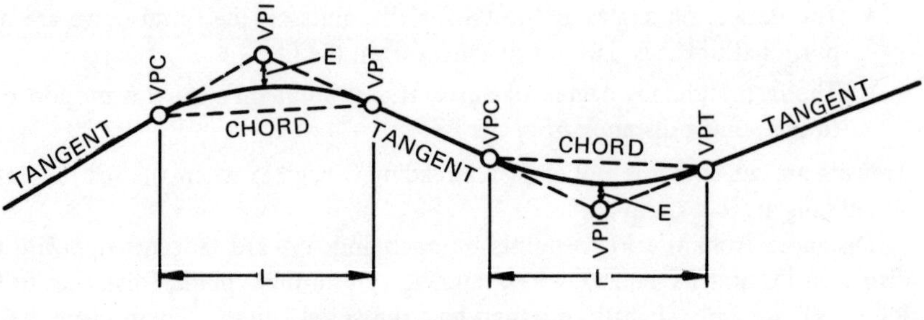


**Fig. 12-18 Vertical curves**

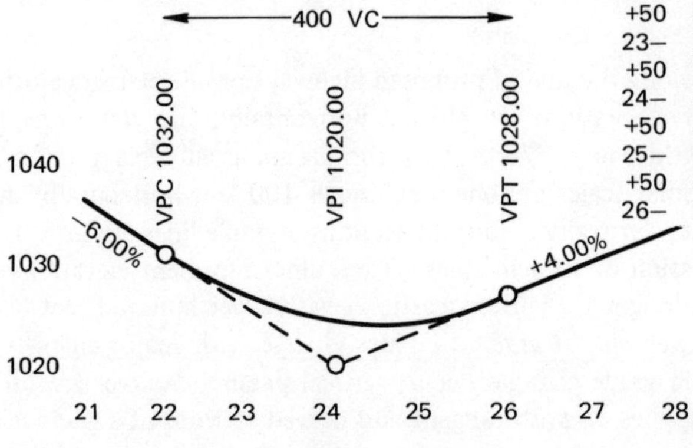


| STA. | TANG. GRADE | OFF-SET | CURVE GRADE |
|---|---|---|---|
| 22— | 32.00 | 0.00 | 32.00 |
| +50 | 29.00 | 0.31 | 29.31 |
| 23— | 26.00 | 1.25 | 27.25 |
| +50 | 23.00 | 2.81 | 25.81 |
| 24— | 20.00 | 5.00 | 25.00 |
| +50 | 22.00 | 2.81 | 24.81 |
| 25— | 24.00 | 1.25 | 25.25 |
| +50 | 26.00 | 0.31 | 26.31 |
| 26— | 28.00 | 0.00 | 28.00 |

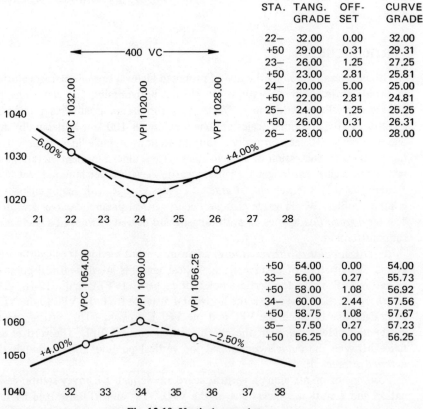

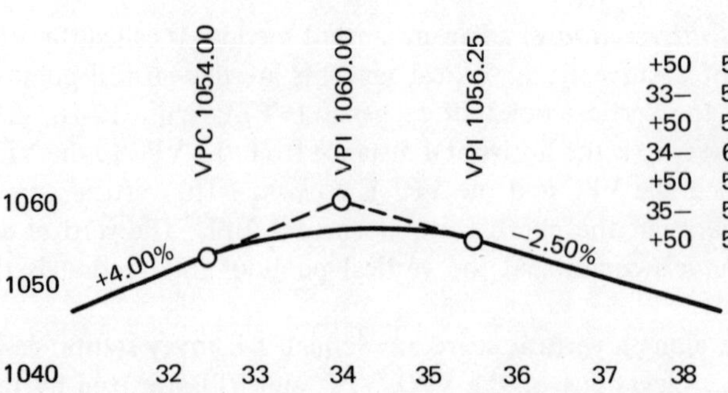


| | | | |
|---|---|---|---|
| +50 | 54.00 | 0.00 | 54.00 |
| 33— | 56.00 | 0.27 | 55.73 |
| +50 | 58.00 | 1.08 | 56.92 |
| 34— | 60.00 | 2.44 | 57.56 |
| +50 | 58.75 | 1.08 | 57.67 |
| 35— | 57.50 | 0.27 | 57.23 |
| +50 | 56.25 | 0.00 | 56.25 |

**Fig. 12-19 Vertical curve data**

## VALUES OF ELEMENTS OF CURVES

The art of selecting degrees of curvature, fitting highway locations to terrain, or choosing the lengths of spirals or vertical curves, are rarely left to the surveyor to decide. Surveyors are governed by design standards adopted by agencies responsible for the construction of facilities requiring curves. Such standards are developed following years of study, training, and experience by specialists in those fields.

## SUMMARY

The layout and measurement of curves is an essential part of surveying. Most horizontal curves are circular. Vertical curves are parabolic. Railroads and highways also have spiral curves, usually in combination with circular curves.

The sharpness of horizontal circular curves is generally defined by degree rather than by radius. In highway work, the degree of a circular curve is the number of degrees subtended by a 100-foot arc. The degree multiplied by the curve's radius in feet always equals a constant, 5729.58.

Common abbreviations of important elements of horizontal curves are PC, PI, PT, POC, D, L, T, Δ, C, E, and M. Surveyors should know what each represents.

A deflection angle from one POC to another is one-half the central angle of an arc between them. POC's are set where necessary to prevent deflection angles accumulating to more than 25 degrees.

Orientation of a transit at a POC gives maximum flexibility if the upper plate reads 0 degree with the telescope aimed tangent to the curve.

Stakes are set at regular intervals around a curve, usually not more than 50 feet apart. Distances are measured from stake to stake.

Where differences between chord and arc lengths cannot be made insignificant by shortening the intervals between the stakes, short chaining is used.

In rough topography, supplemental stakes are often set to meet horizontal measurement or other job requirements.

When obstacles are encountered, segments of a curve are run from each end. Another method is to bypass the obstructed points, and set stakes at such points from the next POC. If these methods cannot be used, the obstructed points are set from calculated coordinates.

In notekeeping for curves, stations are listed with the lowest numbered station at the bottom of the page.

Adjoining curves in the same direction are compound curves. The degree of one curve should be at least twice the degree of the other.

Adjoining curves in opposite directions are reverse curves.

Spiral curves are used between tangents and circular curves on railroads and high-speed highways. They are laid out by offsets without much difficulty.

Vertical parabolic curves are used where grade changes occur on profiles of highways and other facilities. Common abbreviations of important elements are VPC, VPT, VPI, L, and E. Their lengths are in terms of horizontal distance.

The difference in elevation between the curve's center and the VPI is the external distance, E. Vertical offsets between tangents and curve are proportional to the squares of the horizontal distances from the VPC or VPT to the VPI.

## ACTIVITIES

1. A long tangent section of a new highway location intersects a section of an existing highway, figure 12-20. The PI is at station 83 + 56.29 on the new location. Deflection angles are 80°00′ L and 100°00′ R. A y-intersection is needed, beginning curves to left and right at the same PC station. The curve to left is a 20-degree curve. Establish lines in a field representing the intersecting tangents. Calculate curve data for both curves and lay out centerlines for the y-intersection, using 25-foot chords. Locate the PTs of each curve by running in the curves. Check their locations by measuring tangent distances out from the PI.

2. Lay out two tangents intersecting with deflection angle of about 40°00′. Designate the PI as station 15 + 00. Stake a curve connecting the two tangents with an external of 20.00 feet.

3. Fit a 3-degree curve 500.00 feet long to two intersecting tangents which are established with the proper $\Delta$.

4. Fit a 150-foot radius curve 250.00 feet long to two intersecting tangents which are established with the proper $\Delta$.

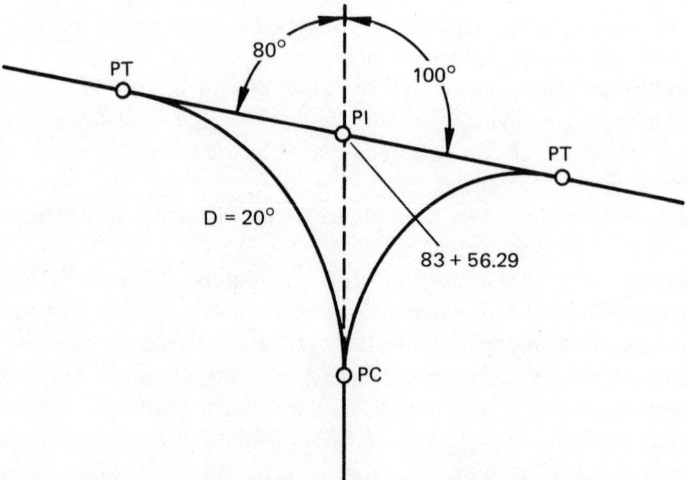


**Fig. 12-20 Curve layout for activity 1**

## REVIEW QUESTIONS

### A. Short Answer

1. How does the field work involved in circumventing obstacles on curves compare to the technique for straight lines?
2. What does the practice of measuring by short chaining refer to?
3. What does the product of a circular curve's degree multiplied by its radius equal?
4. What does the product of a circular curve's degree multiplied by its number of 100-foot station lengths equal?
5. What do each of the following abbreviations represent?

| | | | |
|---|---|---|---|
| PC | POC | T | E |
| PI | D | Δ | M |
| PT | L | C | |

### B. Problems

6. Convert the degree of each of the following curves to its radius, using formula (B).

| Degree |
|---|
| 5° |
| 10° |
| 15° |
| 20° |

7. Convert the radius of each of the following curves to its degree using formula (B).

| Radius (ft.) |
|---|
| 100 |
| 200 |
| 500 |
| 1000 |

8. What are the lengths of the following curves?

| Degree | Δ |
|---|---|
| 5° | 20° |
| 10° | 30° |
| 15° | 60° |
| 20° | 60° |

9. Station 8 + 25 is the PC of a 10-degree curve having a Δ of 40 degrees. What is the station of the PT? Of the PI?
10. Station 11 + 00 is the PC of a 9-degree curve. Station 26 + 00 is its PT. List the deflection angles for stakes to be set at 50-foot intervals as far as the first POC at station 16 + 00.

| Station |
|---|
| 16 + 00 |
| 15 + 50 |
| 15 + 00 |
| 14 + 50 |
| 14 + 00 |
| 13 + 50 |
| 13 + 00 |
| 12 + 50 |
| 12 + 00 |
| 11 + 50 |

11. For curves cited in review questions 9 and 10, calculate the radii, tangent distances, long chords, middle ordinates and externals.

12. A number of curves are to be staked for a subdivision. Errors no greater than 1/2500 may be introduced because of differences in chord and arc lengths. Short chaining is to be avoided. What is the maximum degree of curve for which stakes may be set at 50-foot intervals? What is maximum degree for 25-foot intervals?

13. Stakes are being set for a 10-degree curve beginning with the transit at station 10 + 50. Line of sight to station 12 is obstructed by a tree 2-feet in diameter. Lines of sight are clear to station 11 + 50 and 12 + 50. Station 12 + 50 is to be set by the normal deflection angle, and a measured chord length from station 11 + 50. What is that chord length? Another way is to measure a chord length from the transit to station 12 + 50. What would that chord length be?

14. Continuing review question 13, the transit is moved forward to a POC at station 12 + 50. Line of sight from 12 + 50 back to station 12 is clear. After taking a backsight on PC, the telescope is left inverted for setting the stake at station 12. Using procedure A, what upper plate reading is needed to set the proper deflection angle? Using procedure B, what upper plate reading is needed?

15. A 16-degree curve with PC at station 7 + 50 is being staked at 25-foot intervals in rough topography. Intermediate stakes are found desirable at stations 7 + 85, 8 + 12, 8 + 40, 8 + 64, and 8 + 88. With the transit at PC, what deflection angles are required?

| Station |
|---|
| 9 + 00 |
| 8 + 88 |
| 8 + 75 |
| 8 + 64 |
| 8 + 50 |
| 8 + 40 |
| 8 + 25 |

8 + 12
8 + 00
7 + 85
7 + 75

16. The bearing of a forward tangent to a 5-degree curve to the right is N 60° 00′ E. What is the bearing of a tangent to the curve at a POC 200 feet behind the PT?

17. Compute vertical offsets including external distance, E, and vertical curve grade elevations for the following vertical curve:

| | Station | Tangent Grade Elevation | Vertical Offset | Vertical Curve Grade Elevation |
|---|---|---|---|---|
| VPC | 12 + 00 | 120.00 | 0.00 | 120.00 |
| | 12 + 50 | | | |
| | 13 + 00 | | | |
| | 13 + 50 | | | |
| VPI | 14 + 00 | 112.00 | | |
| | 14 + 50 | | | |
| | 15 + 00 | | | |
| | 15 + 50 | | | |
| VPT | 16 + 00 | 116.00 | 0.00 | 116.00 |

18. The PI of two tangents is at station 10 + 50. The $\Delta$ is 39°54′. A curve is to be run between the tangents with an external of 18.00 feet. What is the radius of the curve? What are the stations of the PC and PT?

# CHAPTER 13

# Stadia Measurements

## OBJECTIVES

After studying this chapter, the student will be able to:

- make stadia measurements with a transit, or with an alidade and plane table.
- reduce stadia measurements to horizontal distances and differences in elevation.
- establish adequately controlled stations for stadia measurements.

## USES OF STADIA MEASUREMENTS

Stadia methods produce quick measurements of horizontal distances and differences in elevation where high precision is unnecessary. Stadia measurements make it easy to locate topographic details and measure across deep ravines and bodies of water. They are also used for rough traverses and checking for major mistakes in taped distances.

Stadia measurements are made either by transit and stadia rod, or by alidade and plane table.

## TRANSIT-STADIA

Transits equipped for stadia measurements have telescopes with three horizontal cross hairs. The upper and lower ones are *stadia cross hairs.* They are equally spaced above and below the central horizontal cross hair, figure 13-1.

With the transit sighted on a vertical graduated rod, the length of rod appearing between stadia cross hairs is measured. That length is called a *stadia interval, s.* It is proportional to the distance between the transit's focal point and the rod, figure 13-2.

Manufacturers usually space the cross hairs to make distances 100 times the stadia intervals when the telescopes are level. The ratio of the distance compared to the stadia interval is called a *K factor.* Its value is given in instrument manufacturers' specifications. For external focusing transits, the focal

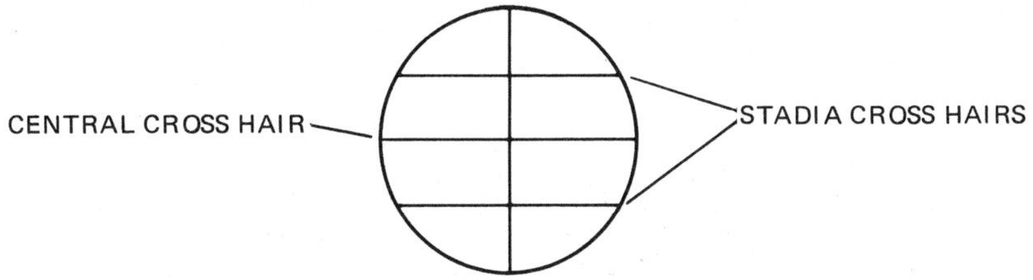


**Fig. 13-1 Stadia cross hairs as seen through telescope**

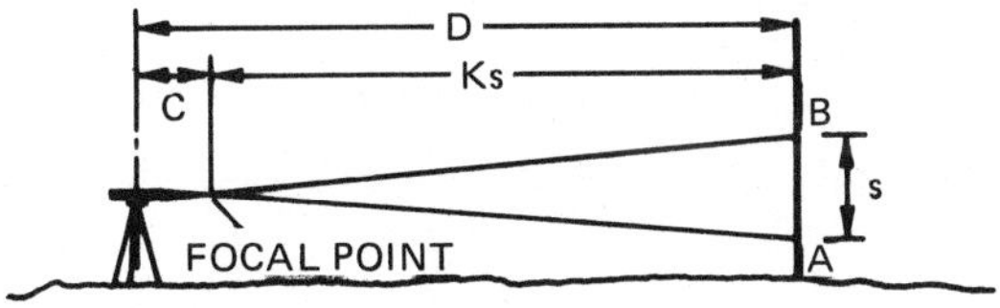


**Fig. 13-2 Stadia interval, s**

points are about a foot ahead of the centers of the instruments. A *focal correction,* C, also furnished by manufacturers, is necessary for all observed distances, D. Therefore:

(A) $$D = Ks + C$$

For internal focusing instruments, C = O. When manufacturer's values of K and C are unavailable, they are determined by observing stadia intervals at known distances.

Any self-reading leveling rod can serve as a stadia rod for short distances. Special stadia rods with graduations that are more easily read at longer distances are manufactured.

## Inclined Lines of Sight

Usually there are differences in elevation between transit stations and points being located. Wherever inclined lines of sight occur, the calculations of horizontal distance and difference in elevation involves performing two steps. Determine the slope distance from the transit station to the point. Then convert the slope distance to a horizontal distance and a difference in elevation.

It is customary to hold the rod vertically rather than perpendicular to the inclined lines of sight, figure 13-3. The observed stadia interval AB is reduced mathematically to right angle interval A'B'. Calling the right angle interval s', the slope distance is Ks' + C. The angle between AB and A'B' is the same as the line of sight's inclination. This angle, vertical angle alpha ($\alpha$), is measured at the instrument. To calculate s' and slope distance D':

$$s' = s \cos \alpha$$

(B) $$D' = Ks \cos \alpha + C$$

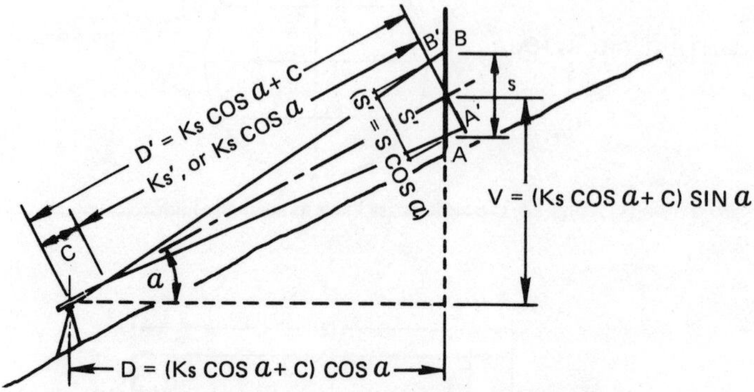


**Fig. 13-3 Basic relationships for an inclined line-of-sight**

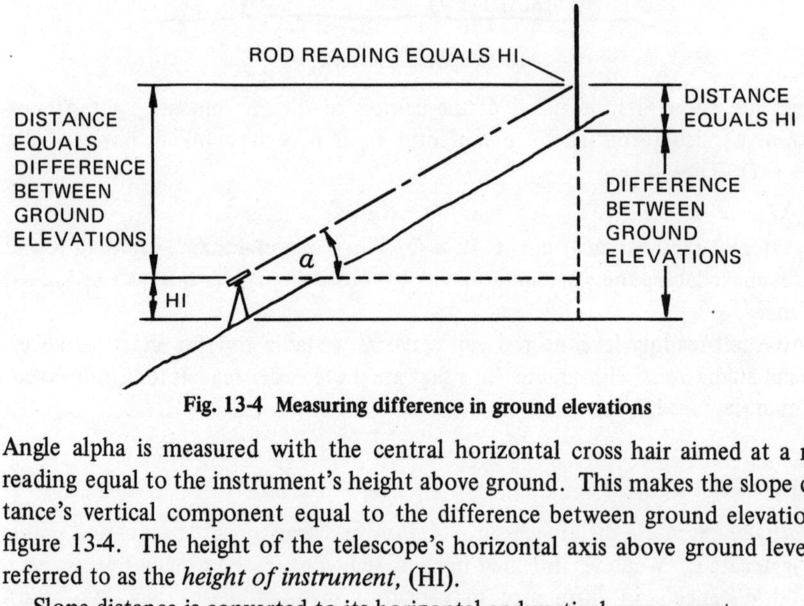


**Fig. 13-4 Measuring difference in ground elevations**

Angle alpha is measured with the central horizontal cross hair aimed at a rod reading equal to the instrument's height above ground. This makes the slope distance's vertical component equal to the difference between ground elevations, figure 13-4. The height of the telescope's horizontal axis above ground level is referred to as the *height of instrument,* (HI).

Slope distance is converted to its horizontal and vertical components.

(C) Horizontal distance

$$D = (Ks \cos a + C) \cos a$$

(D) Difference in elevation

$$V = (Ks \cos a + C) \sin a$$

Assume, using figure 13-3, that C = 1.0, K = 100, A = 3.12 feet, B = 5.66 feet, and alpha = $10^\circ 30'$. The slope distance D' is equal to:

D' = $Ks \cos a + C$

D' = (100) (2.54) (.98325) + 1

D' = 250 + 1

D' = 251 feet

The difference in elevation equals:

$V = (Ks \cos a + 1) \sin a$

$V = (100 \times 2.54 \times .98325 + 1)(.18224)$

$V = 250.7 \times .18224$

$V = 45.7$ feet

The horizontal distance equals:

$D = (Ks \cos a + C) \cos a$

$D = (100 \times 2.54 \times .98325 + 1)(.98325)$

$D = 251 \times .98325$

$D = 247$ feet

Dealing with the value C is inconvenient. However, its omission leaves horizontal distances and vertical components significantly short. It is usually included as part of an adjusted value of s. When K = 100 and C = 1 foot, the value of s is then increased by 0.01 foot. This treatment causes only insignificant shortening of the horizontal distances and differences in elevation. Entering the adjusted values of s or Ks in the field notebook at the time of observation is recommended.

## Stadia Reduction

The reduction of stadia observations to horizontal distances and differences in elevation are seldom made by formulas. Tables, diagrams, stadia slide rules, or stadia arcs on instruments provide easier solutions. Appendix E gives reduced values in tabular form.

A transit equipped with a *stadia arc* eliminates the need for reading vertical angles. Stadia intervals are reduced by multiplying them by horizontal and vertical scale readings on the arc. Scales are made in several designs and their use is explained in instructions furnished by equipment manufacturers. One type of stadia arc has scale readings that are applied as percentage factors to stadia intervals.

Ordinarily horizontal distances obtained by stadia measurements are expressed to the nearest foot. Differences in elevation are reduced to the nearest 0.1 foot.

## Field Work

Surveys are usually controlled by local networks of transit stations and levels. For accurate work, transit stations are triangulation points or transit-tape traverse points. Stadia measurements are used only for locating details. Elevations at transit stations or nearby temporary bench marks are established by regular differential leveling methods. Occasionally, low orders of precision for local horizontal and vertical control networks are acceptable. Locations and elevations of supplementary transit stations, and even entire local control networks are established by stadia measurement alone.

From transit stations, the locations of other points are determined by direction, distance, and difference in elevation. The turning of horizontal angles to determine a direction is performed as for radiation. The stadia method is used to determine the distance and difference in elevation.

When distance alone is determined by stadia, vertical angles are not read as close as usual. The number of minutes for vertical angles up to 20 degrees is estimated to the nearest 15 minutes. Angles of less than 3 degrees are ignored entirely. When differences in elevation are also being determined, vertical angles are usually read to the nearest minute.

Level rod targets are clamped at a reading equal to the height of the instrument. If a target is unavailable, a rubber band placed around the level rod is sufficient. These devices make it easier to aim the central cross hair at the proper level for vertical angle readings.

The following sequence of stadia observations is recommended:

1. Aim and focus the vertical cross hair on the rod.
2. Bring the central horizontal cross hair to the rod reading approximating the height of the instrument. Then bring the lower cross hair to the nearest even foot mark.
3. Read the upper cross hair and subtract the lower cross hair reading to obtain the stadia interval.
4. Bring the central cross hair to the rod reading equaling the height of the instrument.
5. Read the horizontal angle.
6. Read the bearing if necessary.
7. Read the vertical angle, or stadia arc.

Instrument orientation is confirmed by rechecking the backsight on an adjoining transit station following each dozen consecutive observations. Also, upon completion of all observations at a station.

Field notes for transit-stadia surveys are illustrated in figure 13-5. Horizontal angles are recorded as azimuths to simplify plotting. When C = 1.0 foot, each entry in the stadia distance column is K times the stadia interval plus 1. Horizontal distances and elevations are entered in the field notebook whether reduced in the field or later in the office. A sketch and numbering system for points is used for topographic details.

## ALIDADE AND PLANE TABLE

The location of details by stadia measurement is also accomplished with plane-table equipment. This method permits map plotting in the field. It allows a more perfect representation of ground topography with fewer observations and a minimum of notekeeping. Map making at the site makes the omission of data from fieldwork unlikely.

Field measurement methods are similar to the transit-stadia method. The difference is that an alidade and plane table are substituted for a transit.

A *plane table* is a drawing board, covered with drawing paper, mounted on a tripod. The *alidade* is an instrument which works substantially the same way as

Topography Upper Turkey Creek (Transit-Stadia)

50
6-8-74

| Point | Az. | Stadia Dist. | Vert. ∠ | Hor. D | Diff. in Elev. | Elev. | | |
|---|---|---|---|---|---|---|---|---|
| | Instrument @ A | | H.I.=4.6' | 0°Az. is South | | 2025.1 | Ridge | ⊼ Somers. |
| B | 4°30' | 79 | -0°15' | 79 | -0.3 | 24.8 | Ridge | ɸ Thompson |
| 1 | 32°15' | 57 | -14°20' | 53 | -13.7 | 11.4 | Stream Jct. | Notes Hensley |
| 2 | 67°30' | 62 | +1°12' | 62 | +1.3 | 26.4 | Ridge | Clear-Cool |
| 3 | 109°15' | 142 | +9°42' | 138 | +23.6 | 48.7 | Ridge | |
| 4 | 136°00' | 86 | -2°35' | 86 | -3.9 | 21.2 | Stream | |
| 5 | 204°45' | 87 | +8°12' | 85 | +12.3 | 37.4 | Ridge | |
| 6 | 241°30' | 98 | -3°14' | 98 | -5.5 | 19.6 | Stream | |
| 7 | 305°15' | 148 | +6°17' | 146 | +16.1 | 41.2 | Ridge | |
| | | | | | | | | |
| | Instrument @ B | | H.I.=4.7' | 0°Az. is South | | 2024.8 | Ridge | |
| 8 | 294°45' | 153 | -0°53' | 153 | -2.4 | 22.4 | Stream | |
| 9 | 318°15' | 72 | -6°55' | 71 | -8.6 | 16.2 | Stream | |
| 10 | 73°15' | 118 | -10°26' | 114 | -21.0 | 03.8 | Stream Jct. | |
| 11 | 101°45' | 136 | -2°37' | 136 | -6.2 | 18.6 | Side Slope | |

Fig. 13-5 Transit-stadia survey notes

a transit in a vertical plane. It has a telescope equipped with a striding level, a vertical arc, a vertical clamp and tangent screw, horizontal and vertical crosshairs, and stadia crosshairs, figure 13-6. It often has an auxiliary vernier control level and a stadia arc. The telescope is supported by a post mounted on a horizontal rectangular base plate called a *blade*. The blade has beveled edges parellel to the telescope's line of sight. On the blade there is a circular level and a trough compass.

The drawing board, usually 18" x 24" or 24" x 31", screws onto the tripod head. Under the tripod head is a ball-and-socket joint with two independent clamps controlled by wing nuts. The upper wing nut controls the leveling clamp and the lower wing nut controls the orientation clamp.

The best practice is to set up plane tables only at stations already located by transit and tape traverses. The courses of the complete traverse are plotted on the plane table drawing paper before fieldwork begins.

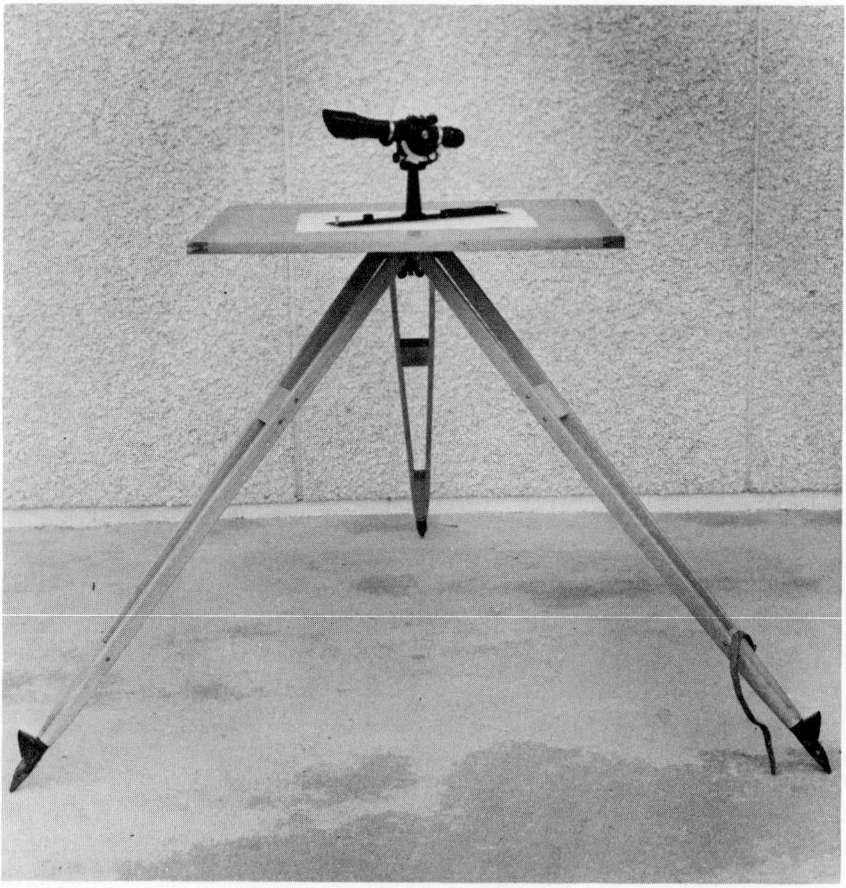

**Fig. 13-6 Plane table and alidade**

A plane table is set up at a height slightly below elbow level. The tripod is first placed in the approximate position at the station. Lateral shifting brings the point on the drawing paper representing the station over the station. Alignment is by eye from two directions at right angles to each other. Plumb bobs are unnecessary because positioning within a few tenths of a foot does not effect the accuracy of this method. Tripod legs are pressed firmly into ground.

Leveling begins after placing the alidade on the plane table. The upper wing nut is loosened. The table's position is adjusted until the alidade level bubble shows it to be level. The upper wing nut is then tightened. While the wing nut is loose, the table tilts easily and can spill the alidade to the ground. Keep a firm grip on the alidade post during the leveling operation.

The map scale and plotted location of the initial plane table station depend upon the size and shape of the survey area. The scale is chosen so that the survey area stays within the bounds of the paper. A needle is stuck vertically into the board at the initial point. The needle is moved forward to each station as it is occupied. One of the blade's beveled edges remains in contact with the needle for all backsights and foresights. Map scales are large enough that the width of the blade introduces no significant error.

**Initial Station Orientation**

Orientation of the plane table at the initial station begins by loosening the lower wing nut. The board is rotated to a position that confines the area being plotted within the paper's limits. Tightening the lower wing nut locks the table to prevent its rotation.

The alidade is placed near one edge of the paper and rotated to a north-south direction. The compass mounted on the blade indicates when the alidade is pointed correctly. A long north-south reference line is drawn along one of the blade's beveled edges.

The final step is to stick the needle vertically into the point representing the station.

**Subsequent Station Orientation**

Orientation of the plane table at other stations is accomplished by backsighting. The initial step is to stick the needle into the point representing the occupied station.

The blade's beveled edge is fitted to a line between the needle and the point being backsighted. The lower wing nut is loosened. The board is rotated to bring the vertical cross hair on the backsight. The lower wing nut is then clamped.

When traverse points have been previously plotted, the backsight is taken on an adjacent traverse station, figure 13-7. Orientation is then confirmed by also backsighting the other adjacent traverse station. When traverse points have not been previously plotted and only one station is available for backsighting, orientation is confirmed by the compass. It points north when the blade's beveled

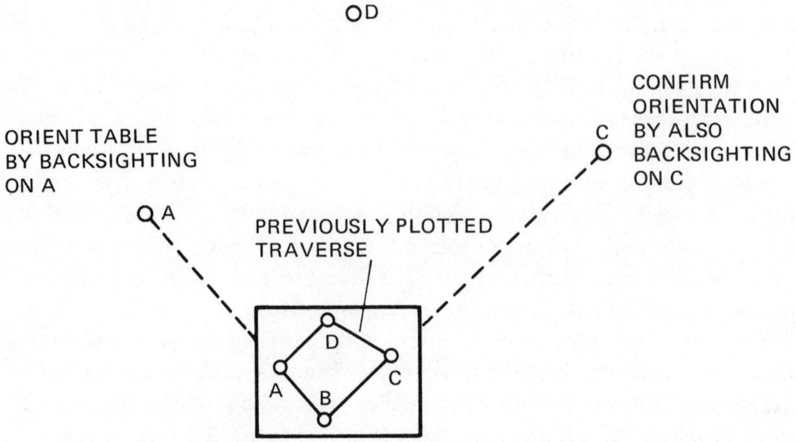


**Fig. 13-7 Backsighting adjacent stations**

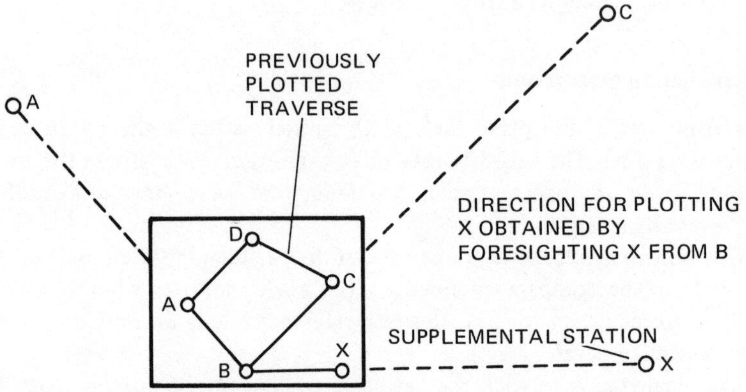


**Fig. 13-8 Locating a supplemental station**

edge is aligned along the north-south reference line. Where a prominent landmark is visible from several stations, orientation is confirmed by sighting on it from each station.

A plane table can be set up and oriented at a location omitted from the basic control network. A point for a supplemental station can be set from any regular plane table station. Distance to it is measured by stadia and plotted as is any other point, figure 13-8. Actually, it can be set up and oriented at any point without first knowing the map location. Distances and differences in elevation are obtained by stadia to two stations already plotted. Arcs with radii equal to the scaled distances are drawn from each of those stations. The point of intersection of the arcs on the map is the location occupied, figure 13-9, step 1. A needle is stuck in vertically at the occupied point. The drawing board is finally oriented by backsighting on one or both of the known stations, figure 13-9, step 2. This method of establishing the location of an accupied station is called *resection.*

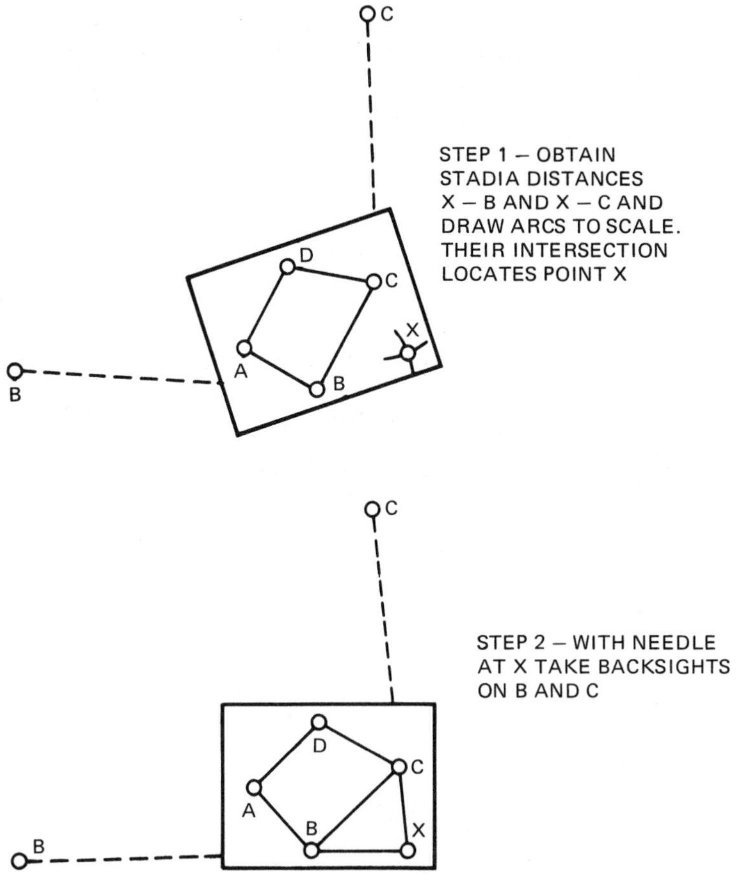
C
STEP 1 – OBTAIN
STADIA DISTANCES
X – B AND X – C AND
DRAW ARCS TO SCALE.
THEIR INTERSECTION
LOCATES POINT X
D
C
X
A
B
B
C
STEP 2 – WITH NEEDLE
AT X TAKE BACKSIGHTS
ON B AND C
D
C
A
B
X
B

**Fig. 13-9 Finding a station location by resection**

## Plane Table Plotting

Details visible from a station are usually located by *radiation.* With the blade's beveled edge in contact with the needle, the alidade is foresighted on a point being located, figure 13-10. A *ray,* a short line, is drawn along the blade's beveled edge to define the direction. Distance is determined by stadia, and scaled along the ray. Differences in elevation are normally determined by stadia. It is also possible to use the alidade for differential leveling the same as a transit or level.

A remote point difficult to reach is located by intersection. Foresights from each of two plane table stations are taken and rays from each are extended to an intersection, figure 13-11.

Lines are drawn with hard pencils, 6H to 9H. Rays begin a distance from the needle. They are drawn only long enough to bracket the estimated distance to the point. Where elevations are entered, the plotted point is also made to serve as the decimal point. Elevations are entered with two digits preceding and one digit following the decimal point.

Fig. 13-10 Drawing a ray

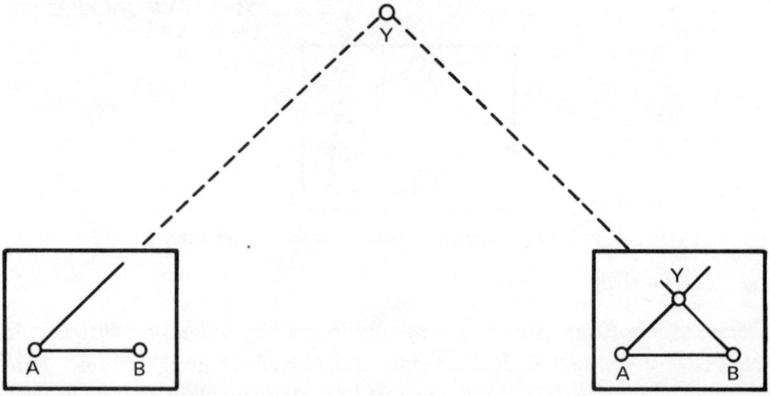


Fig. 13-11 Locating a point by intersection

## Field Operation

The recommended sequence of operations is the same as when making stadia observations using a transit and stadia rod. Since recording is done directly on the paper, drawing a ray takes the place of reading a horizontal angle. If the alidade is equipped with stadia arcs reading in percentage, the horizontal and vertical percentage factors are read instead of vertical angles. It is important that the auxiliary vernier control level bubble be centered when these readings are made.

An umbrella or awning supported by poles is set up over a plane-table station to shade the drawing board from excessive glare. Other forms of relief are sunglasses or the use of green paper. To avoid loss, detachable striding levels are always removed from alidades when moving the equipment between stations.

## SUMMARY

Stadia measurements of horizontal distances and differences in elevation are quick but lack the precision of tape measurements and leveling. They are effective for locating topographic features and working across inaccessible areas.

Stadia cross hairs are in the telescope. They are equally spaced above and below the central horizontal cross hair.

A stadia interval is the difference between the graduations on a vertical rod observed between the stadia cross hairs.

Stadia distance from a transit focal point to a vertical rod is proportional to the stadia interval. The ratio is called the K factor.

Stadia distances for inclined lines of sight are resolved into horizontal distances and differences in elevation. Horizontal distances are expressed to the nearest foot, and differences in elevation to the nearest 0.1 foot.

Vertical angles for inclined lines of sight are measured with the central cross hair aimed at a rod reading equal to the height of the instrument.

Stadia measurements are used for locating details. Stations from which they are made are located by methods more precise than stadia.

When large numbers of stadia measurements are made from a single station, the instrument orientation needs periodic confirmation.

Stadia measurements are made by either transit-stadia, or by alidade and plane table. The techniques are essentially the same for either system.

Horizontal direction by plane table is recorded by rays on a map drawn in the field. Horizontal direction when using transit-stadia is recorded by azimuth.

Orientation of a plane table at each station is accomplished by backsighting. It is confirmed by backsighting an additional station, by compass observation, or by sighting the same prominent landmark from several stations. For best control of a survey, the locations of primary stations are predetermined and plotted before other field measurements are made. Supplemental stations can be located by stadia methods.

The location of topographic details is ordinarily accomplished by radiation. For remote points, intersection is occasionally employed.

To avoid cluttering the map, rays are drawn a mininum length to bracket a point. The plotted point is made to serve as the decimal point for the elevation.

## ACTIVITIES

1. Set hubs at points A and B at the terminals of a base line 500.00 feet long. Between 400 and 500 feet off to one side of the base line, set a meandering line of five hubs. Assume a deep gorge exists between the base line and the meandering line. Locate the meandering hubs by intersection. Calculate the lengths of lines between A and each hub, and between B and each

hub. Make stadia observation on each hub from both points, A and B. Compare the horizontal distances derived from stadia observations with the calculated distances.

2. Run a circuit of differential levels to include points A and B and the five meandering hubs set in activity 1. Compare the differences in elevation derived from stadia observations in activity 1 with the differences in elevation obtained from differential leveling.

## REVIEW QUESTIONS

### A. Multiple Choice

1. A stadia rod is held
   a. horizontally.
   b. vertically.
   c. parallel to the line of sight
   d. Perpendicular to the line of sight.
2. Stadia cross hairs are
   a. vertical.
   b. horizontal.
   c. parallel to the line of sight.
   d. both located above the normal cross hairs in the transit.
3. Stadia measurements, when reduced to horizontal distances, are normally expressed to the nearest
   a. foot.
   b. inch.
   c. 0.1 foot.
   d. 0.01 foot.

### B. Short Answer

4. How high above ground is a plane table usually set up?
5. For level lines of sight, what is the formula for finding horizontal distance?
6. Under what condition does the slope distances vertical component equal the difference between its terminal ground elevations?
7. Name three methods by which stadia observations are reduced to horizontal distances and differences in elevation.
8. Manufacturers usually space the cross hairs to make the K factor equal to what number?
9. List the following operations in the order they are to be performed when using a transit and stadia rod.
   a. Read horizontal angle
   b. Read vertical angle
   c. Aim vertical cross hair on stadia rod
   d. Bring central cross hair to rod reading equaling the height of the instrument
   e. Determine stadia interval

10. When C=1.0, what would the entry for stadia distance be in a field notebook?
11. What is the main difference between a transit and an alidade used with a plane table?
12. Which wing nut under a plane table controls the leveling clamp? Which controls the orientation clamp?
13. What limits both map scale and plotted location of initial plane table station on map?
14. Why can either side of a blade be used for drawing rays?
15. The most common method of locating details by a plane table is radiation. What is another method?
16. What should the hardness be for pencils used in plane-table mapping?
17. What are three ways of avoiding excessive glare on a plane table map on sunny days?
18. How are locations of points identified on plane-table maps where elevations are entered?
19. A plane table is set up at a point not previously located on the map. Orientation is accomplished by observations on two stations already plotted. What is the name of the method employed?
20. A plane table is set up and oriented by backsighting on two adjacent preplotted stations. How can its orientation be further confirmed?

## C. Problem

21. Using Appendix E, find the horizontal distances and differences in elevation for the flollowing stadia interval readings. C = 1.0 and K = 100.

| s | Vertical angle | Horizontal distance | Difference in elevation |
|---|---|---|---|
| 1.11 | 0°00′ | | |
| 0.87 | - 3°15′ | | |
| 2.06 | + 6°12′ | | |
| 1.53 | - 8°37′ | | |
| 0.92 | + 12°20′ | | |

# CHAPTER 14

# Topographic Surveying

## OBJECTIVES

After studying this chapter, the student will be able to:

- trace contours on the ground.
- obtain elevations by practical means at points on the ground where significant changes in slope occur.
- map contours by interpolation between points of known elevation.
- recognize locations of terrain features by contour characteristics.
- make a topographic map.

## TYPES OF MAPS

Maps made to show the positions of objects in a horizontal plane are *planimetric maps.* Maps which add vertical dimensions to represent the shape of the earth's surface are *topographic maps.* City street maps, park trail maps, and highway route maps are planimetric maps. United States Geological Survey quadrangles showing the position of natural features, man-made objects, and contour lines are topographic maps. Topographic maps are an essential tool in the design of major construction projects. They are needed for planning land developments and subdivisions.

*Topographic surveying* is the occupation of obtaining the necessary field measurements for preparing topographic maps.

The difficult and time consuming work once performed in making ground measurements for extensive topographical mapping programs is no longer necessary. Photogrammetry today provides a superior means of defining location of objects and contours for most topographic mapping. There still remains, however, many situations needing ground measurements by the methods explained in this chapter.

*Photogrammetry* is stereoscopic interpretation of aerial photographs. It is a specialized field closely allied to surveying. Many surveyors make it a subject of advanced study.

## CONTOURS

*Contours* are lines connecting points of equal elevation, figure 14-1. Contours are at multiples of whole numbers of feet above a datum. This datum is usually sea level. A multiple of 5 means a differential of 5 feet in elevation between two adjacent contour lines. A multiple chosen for a particular map is called a *contour interval.* It can be 1, 2, 5, 10, 25, 50, or even 100 feet or more. The interval depends upon the purpose and scale of the map. The smaller the contour interval, the more detailed and precise are the requirements for topographic surveys.

Study of figure 14-1 reveals the following characteristics of contours:

- Steepness of slopes are indicated by the spacing of contours. The steeper the slope, the closer is the spacing.
- Contours are perpendicular to ridge lines, drainage lines, and lines of steepest slope.
- V-shaped contours indicate stream crossings, with the point of the V upstream; and ridge lines, with the point of the V downhill.

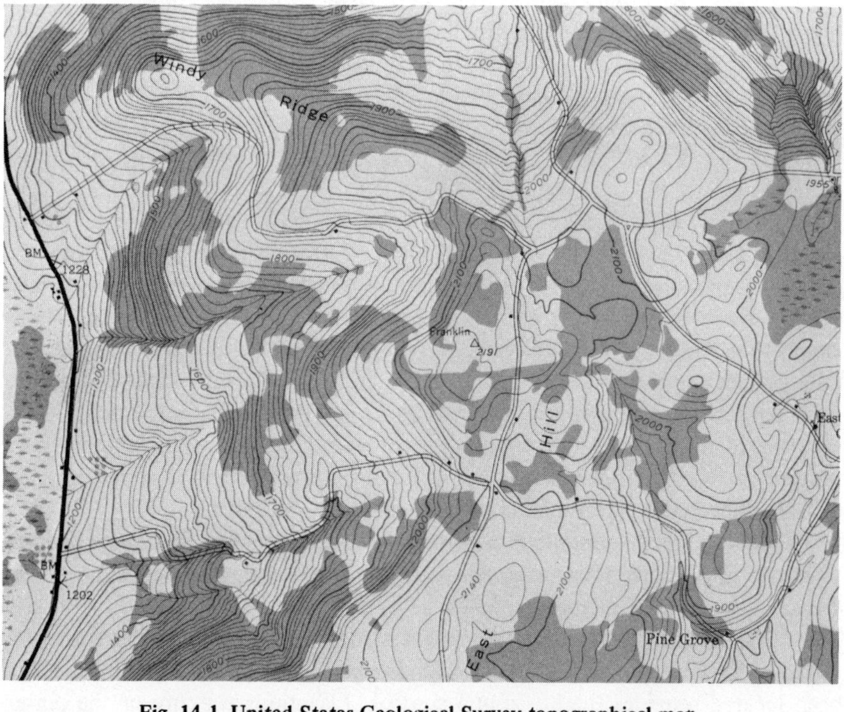


**Fig. 14-1 United States Geological Survey topographical map**

- Contours encircle summits. All contours close upon themselves either within map limits or beyond. Contours encircling dished depressions are called *depression contours.*
- Contours do not merge or cross each other except where vertical or overhanging cliffs exist.

It is possible to *trace a contour.* By this method, successive points are selected where identical rod readings are read. As such points are found, they are located by radiation. If the height of instrument is 1618.30 feet above sea level, all points with rod readings of 3.3 are then on contour 1615. This method is usually used for locating contours at close intervals in relatively flat terrain. Elsewhere, more rapid and economical methods are employed.

For surveys of large areas, particularly in rough terrain, the *controlling-point method* is usually used. This method requires that only the elevations and locations of prominent points be plotted. Contours are then drawn through points located at interpolated distances between the plotted points. Success with this method depends upon the surveyor's skill in choosing the right number and location of controlling points. These points are selected wherever it is necessary to define the land forms. Such points are at summits, along ridge lines, along drainage lines, and at significant points in between. The plane table method requires fewer controlling points than transit-stadia. By use of the plane table, it is easier to give the proper form to irregularities and slopes of various shapes. The surveyor benefits by observing the terrain peculiarities while sketching its contours. Figure 14-2 shows a map in the preliminary stage of development. Courses of drainage lines, elevations of plotted controlling points, and points at interpolated contour intervals are identified. Figure 14-3 shows the same map after the contours have been drawn in.

Other topographic survey methods include the grid method and the cross-section method.

In the *grid method,* stakes are driven at the corners of squares or rectangles of uniform size. Ground elevations are obtained at each stake, and between the stakes where major changes of slope occur. A logical notation system for identifying stakes is adopted. Letters in alphabetical order are assigned to parallel lines connecting stakes in one direction. Figures in numerical order are assigned to lines connecting the stakes at right angles. Each stake is now identified by the letter and figure line intersecting at its location. Intermediate points, if necessary, are identified by a reference to nearby stakes. Elevations of all points are plotted to a suitable scale on grid paper. Contours are drawn by connecting the points established at interpolated contour intervals, figure 14-4.

In the *cross-section method,* cross-section lines are established perpendicular to a base line. Elevations are obtained where changes of slope occur along each cross-section line. This method is ideal where topography is needed for a strip 50 to 200 feet wide. Typical strips occur along a tentative center line for a proposed highway, railroad, airport runway, or canal. A center line that has already been located provides the horizontal control. Elevations at points on the center

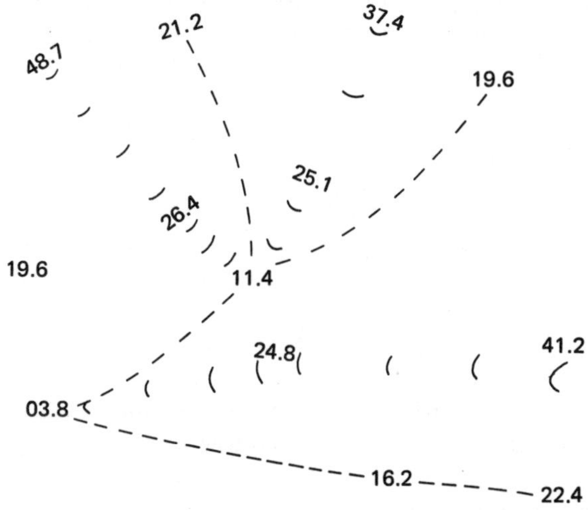


**Fig. 14-2 Contours by controlling point method, preliminary stage**

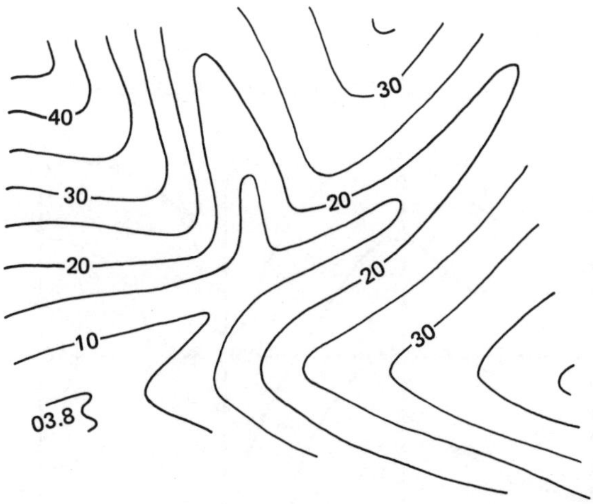


**Fig. 14-3 Contour by controlling point method, completed**

line at each cross section are determined by differential leveling. Perpendicular direction for cross-section lines is estimated by eye. Distances between successive cross sections are as much as 100 feet for preliminary work. They are ordinarily closer for earthwork calculations. Contours are drawn by connecting points at interpolated contour intervals. Interpolations are made along dotted lines as shown in figure 14-5, as well as along the center line and cross-section lines.

Regardless of the method, local horizontal and vertical control networks are always established. Their precision depends upon the specified scale of the map and the contour interval. Local control stations are normally established by triangulation, traversing, and leveling. They are tied into national or state control networks whenever possible.

## FIELD METHODS FOR OBTAINING ELEVATIONS

Stadia measurement in conjunction with differential leveling is effective for tracing contours. Differential leveling is an effective means of obtaining elevations by the grid method. Conventional differential leveling is practical for the controlling point method and the cross-section method only in flat terrain. Stadia measurement is ideal for the controlling point method.

For the cross-section method, forms of differential leveling and slope measurement not previously described are suitable. This method deals with distances which are relatively short from point to point where elevations are needed. Vertical measurements are made with sufficient precision using a hand level or an abney level and level rod. Horizontal measurements are made with a woven tape or horizontal rod. Slope measurements are also practical utilizing an abney level.

A hand level or abney level is placed at an even foot mark on a vertical range pole. The range pole is held at either end of a horizontal taped distance. A level rod is read at another point along the tape to get the difference in elevation to that point. When the limit of the level rod is reached, the hand level is moved forward. Backsight and foresight level readings are taken as in regular differential

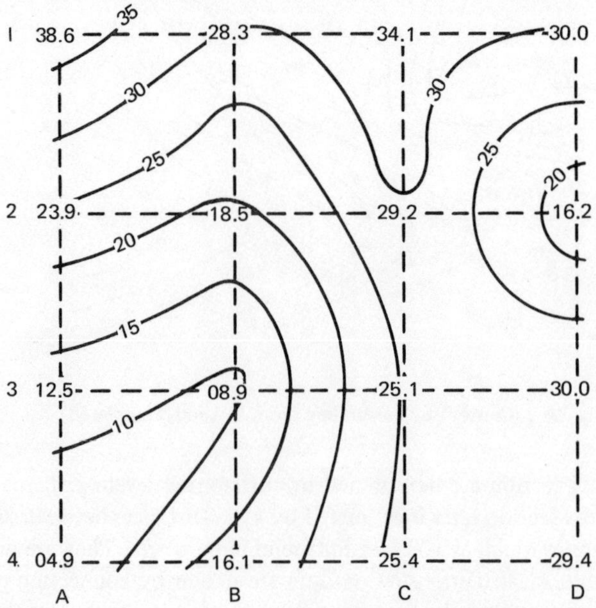


Fig. 14-4 Contours by grid method

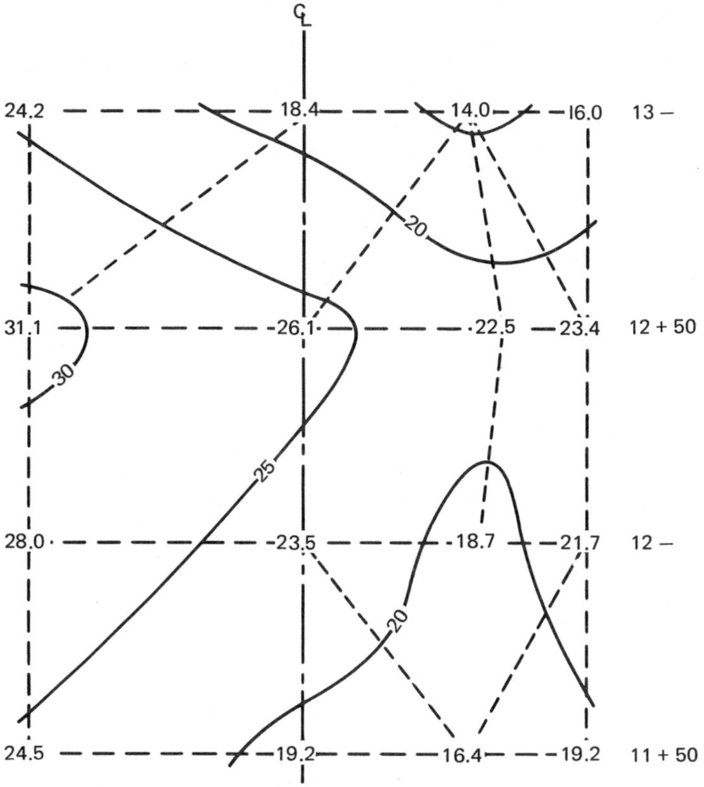


**Fig. 14-5 Contours by cross sections**

leveling. Measurement of differences in elevation continue out to the edge of the strip being plotted.

Where cross-section slopes are steep, a 14-foot one-piece rod offers advantages for horizontal measurements. It is especially useful where the rise or fall of the ground exceeds five feet in a horizontal rod length. To measure uphill, the toe of the horizontal rod is placed at the uphill point, with the rod graduations to one side. The rod is leveled by placing a hand level on its top surface. A conventional rod is held vertically on the lower point and against the horizontal rod, figure 14-6, page 210. The vertical distance is read up to the horizontal rod's bottom surface. Supplemental rod readings are recorded for intermediate changes in slope. The horizontal rod is kept in place until the vertical rod is moved forward. This operation is repeated for each horizontal rod length.

Time is often saved by using an abney level to read the percent of slope. It is especially helpful for the last measurement to the edge of a strip. It is also effective for full widths of strips where the slopes are relatively uniform. To measure slope, the abney level is sighted on a rod at a height equal to the abney's height of instrument. The distance is often paced or estimated.

## INTERPOLATION FOR CONTOUR INTERVALS

The location of contours between two points of known elevation is estimated by experienced topographers. Until this skill is acquired, mathematical or graphical interpolation is necessary. *Interpolation* is simply a more accurate way of estimating an unknown value when it is between two known values.

Mathematical interpolation divides the map distance between two points of known elevation in proportion to the differences in their elevation. Assume point A has an elevation of 64.2 and is 3.79 scale inches from point B, elevation 97.3, figure 14-7. The 3.79 inches divided by 33.1 (the difference in elevation) is

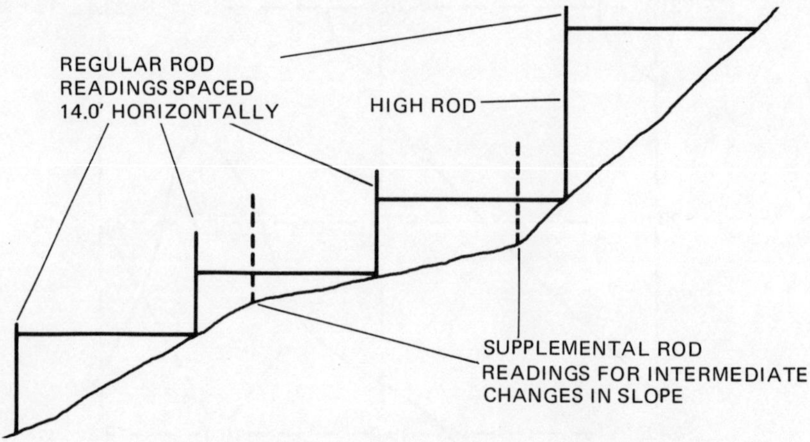


**Fig. 14-6 Ground elevations by horizontal and vertical rods**

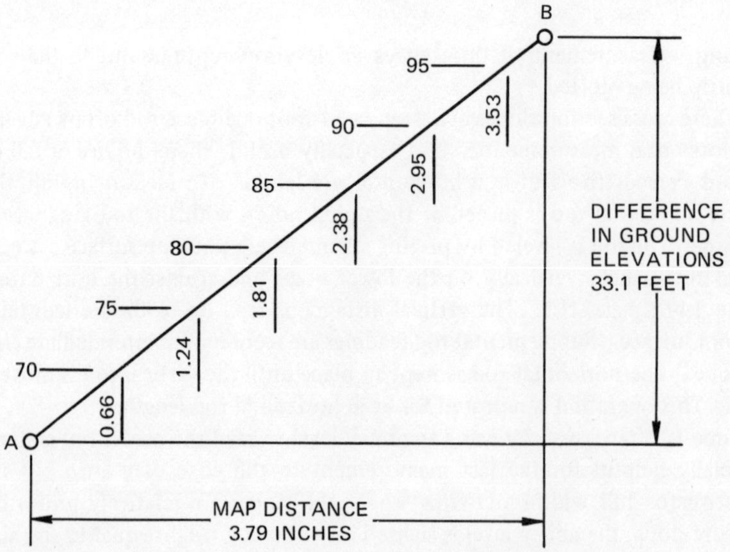


**Fig. 14-7 Example for mathematical interpolation**

0.115, the number of horizontal scale inches per foot of rise. This factor times the height above point A to any contour gives the contour's distance from point A. Calculated values are:

| Point | Height Above Point A | Map Distance in inches from Point A |
|---|---|---|
| Point A. Elevation 64.2 | 0.0 | 0.00 |
| Contour 65 | 0.8 | 0.09 |
| Contour 70 | 5.8 | 0.66 |
| Contour 75 | 10.8 | 1.24 |
| Contour 80 | 15.8 | 1.81 |
| Contour 85 | 20.8 | 2.38 |
| Contour 90 | 25.8 | 2.95 |
| Contour 95 | 30.8 | 3.53 |
| Point B Elevation 97.3 | 33.1 | 3.79 |

Graphical interpolation is accomplished with a triangle and an engineer's triangular scale. Of the six scales on an engineer's triangular scale, one is chosen. The length of the scaled elevation differential chosen is slightly less than the map distance between the terminal points. A 10 scale has a length of 3.31 inches covering a 33.1 foot elevation differential. That length is slightly less than the 3.79 inches of map distance between the terminal points.

Each inch on the scale is made to represent a 10 foot difference in elevation. The useful span of scale readings range between values one-tenth of the terminal elevations 64.2 and 97.3. The 6.42 mark on the scale is pivoted on point A, figure 14-8, page 212. The 90-degree corner of the triangle is placed opposite the 9.73 mark. The triangle's bottom edge is kept in contact with the scale. Both are rotated until the triangle's perpendicular edge passes through point B. The scale is then held firmly in place. The triangle is shifted along the scale until its corner is opposite the 9.5 mark. A mark is made where the triangle's perpendicular edge intersects line BA. Marks are likewise made on BA with the triangle's corner opposite scale marks, 9.0, 8.5, 8.0, 7.5, 7.0, and 6.5. Each mark on BA is a position of one of the desired interpolated contours at a 5-foot interval.

## POSITION OF OBJECTS

Most of this chapter concerns locating and plotting contours. They are features that distinguish topographic maps from planimetric maps.

Complete topographic maps also show the positions of significant objects along with the contour of the earth's surface. Topographic surveying includes locating and plotting natural and man-made objects. The variety and number of objects depends on the maps purpose. Figure 14-9 shows the topographical map symbols used on United States Geological Survey maps.

Where stadia Measurements are used to obtain contours, they are usually employed for other details also.

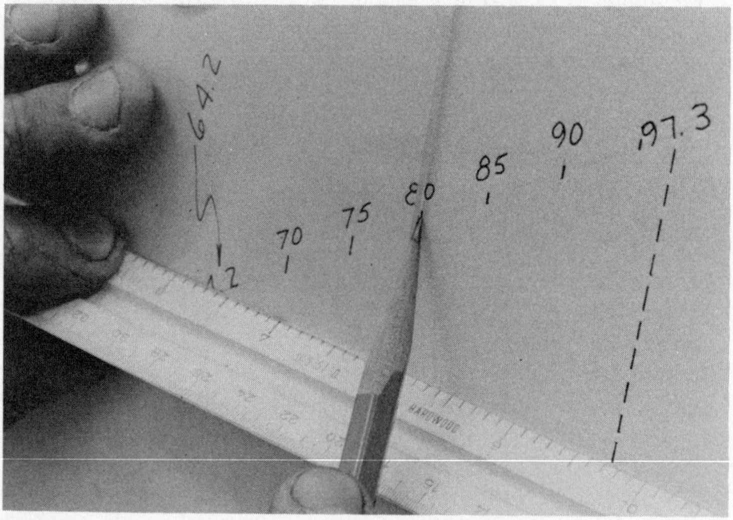


**Fig. 14-8 Graphical Interpolation**

## SUMMARY

Topographic maps show the shape of the earth's surface and the positions of natural and man-made objects. Photogrammetric and topographic surveying methods produce data for preparing topographic maps.

Contours are lines connecting points of equal elevation. The process of locating a contour on the ground is called tracing a contour.

Contours are drawn on maps at intervals of whole numbers of feet. Points through which they are drawn are generally identified by interpolating between points whose elevations have been measured.

It is important that elevations be obtained at all points where significant changes in ground slope occur. Most important are points at summits, along ridge lines, and along drainage lines.

Common methods for obtaining the elevations of whole groups of points are called grid, cross section, and controlling point. All of the methods require adequate local horizontal and vertical control networks.

Stadia measurement is most suitable for topographic mapping by the controlling-point method. Differential leveling generally serves best for the grid method. Hand or abney levels and woven tapes are common for the cross-section method. A combination of horizontal and vertical rods is also suggested for cross sections in rough terrain.

Interpolation for contour locations is performed mathematically, graphically, or by estimation.

Contours have characteristics which reveal the direction and steepness of slopes. They also indicate the location of ridges, streams (including the direction of flow), summits, cliffs, and depressions.

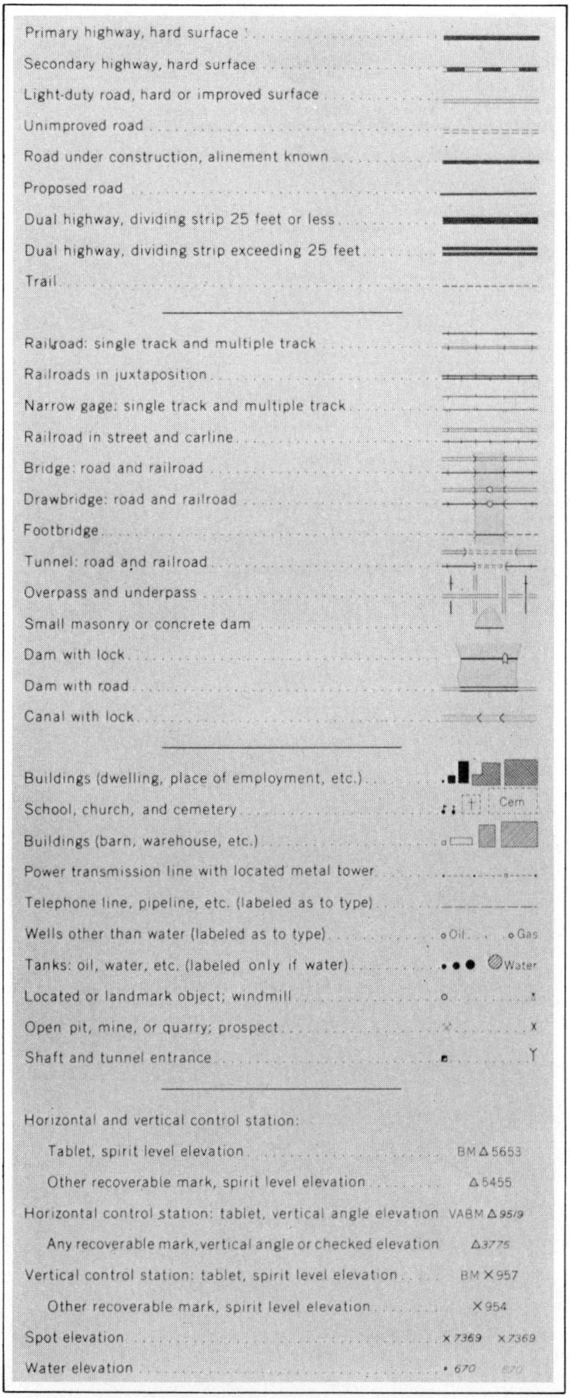
Primary highway, hard surface
Secondary highway, hard surface
Light-duty road, hard or improved surface
Unimproved road
Road under construction, alinement known
Proposed road
Dual highway, dividing strip 25 feet or less
Dual highway, dividing strip exceeding 25 feet
Trail
Railroad: single track and multiple track
Railroads in juxtaposition
Narrow gage: single track and multiple track
Railroad in street and carline
Bridge: road and railroad
Drawbridge: road and railroad
Footbridge
Tunnel: road and railroad
Overpass and underpass
Small masonry or concrete dam
Dam with lock
Dam with road
Canal with lock
Buildings (dwelling, place of employment, etc.)
School, church, and cemetery
Cem
Buildings (barn, warehouse, etc.)
Power transmission line with located metal tower
Telephone line, pipeline, etc. (labeled as to type)
Wells other than water (labeled as to type)
Oil
Gas
Tanks: oil, water, etc. (labeled only if water)
Water
Located or landmark object; windmill
Open pit, mine, or quarry; prospect
Shaft and tunnel entrance
Horizontal and vertical control station:
Tablet, spirit level elevation
BM Δ 5653
Other recoverable mark, spirit level elevation
Δ 5455
Horizontal control station: tablet, vertical angle elevation
VABM Δ 9519
Any recoverable mark, vertical angle or checked elevation
Δ 3775
Vertical control station: tablet, spirit level elevation
BM × 957
Other recoverable mark, spirit level elevation
× 954
Spot elevation
× 7369
× 7369
Water elevation
670
670

**Fig. 14-9 U.S.G.S. topographical map symbols (continued)**

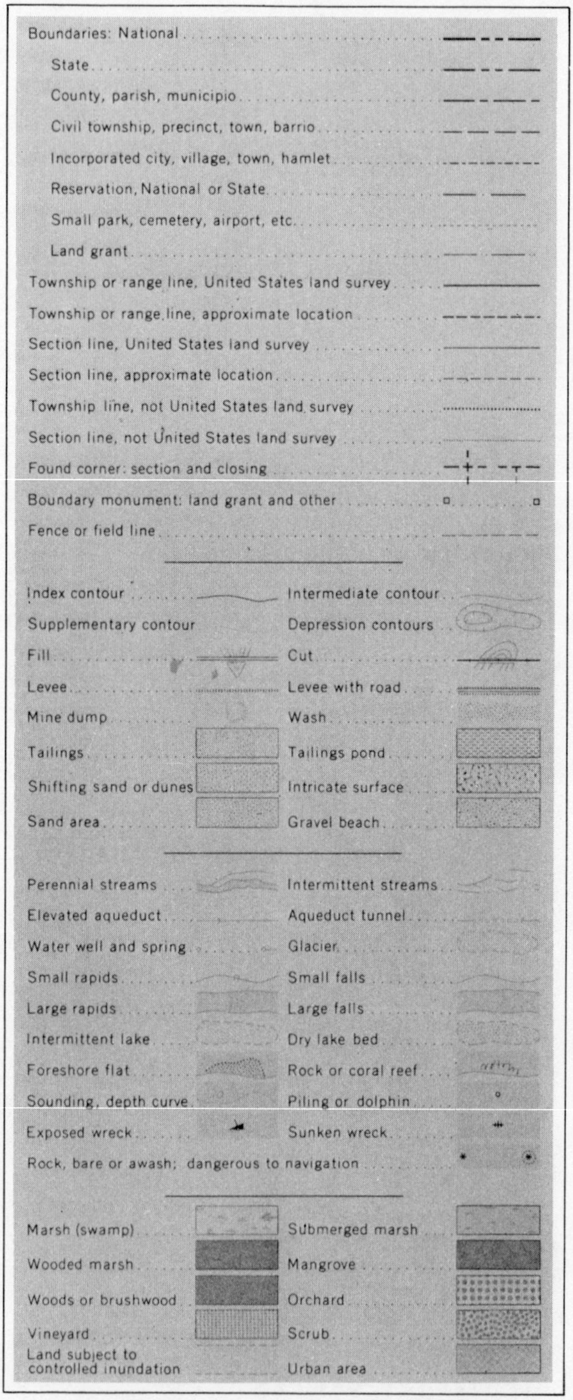


**Fig. 14-9 U.S.G.S. topographical map symbols**

## ACTIVITIES

1. Locate ground features and obtain sufficient elevations using the transit-stadia method to prepare a topographic map. Using the controlling-point method, make a topographic map with a 2-foot contour interval on tracing paper. Occupy the same stations with a plane table and make a topographic map of the same area also with 2-foot contour intervals. Again use the controlling-point method. Compare the results by placing the first map over the plane-table map.
2. Lay out grid lines 20 feet apart within a 100-foot square. Set stakes at the intersections of all grid lines. Obtain the ground elevations at each stake by differential leveling. Plot the results and draw in contours at 2-foot intervals.
3. Run cross-section lines at 50-foot intervals extending 50 feet each way from a base line 250 feet long. Obtain ground elevations at the base line, edges, and intermediate points where the slope changes significantly. Plot the results and draw in contours at 2-foot intervals.

## REVIEW QUESTIONS

### A. Multiple Choice

1. Which of the following is not classified as a field method for obtaining ground elevations?
   a. Controlling point
   b. grid
   c. Interpolation
   d. Cross-section

### B. Short Answer

2. How much work is involved in obtaining ground elevations for contour intervals of 10 feet as compared to 2 feet?
3. Stream crossings are denoted by V-shaped contours with the point of the V pointing in which direction?
4. Steep slopes are denoted by what type of spaced contours?
5. A contour interval of 10 feet means that the contours are 10 feet apart in what direction?
6. What important information appears on topographic maps but not on planimetric maps?
7. What is the most suitable kind of measurement for obtaining the elevations of points by the controlling-point method?
8. List three terrain features where the most significant changes in slope occur.
9. What do adjacent concentric circular contours indicate?

10. Name three ways by which interpolation for contour locations isperformed.

11. Name two principal uses of topographic maps.

**C. Problems**

12. Point A is at elevation 56.7 Point B is at elevation 21.2. The map distance from A to B is 2.84 inches. What are map distances from B to the following contours?

    Contours
    25
    30
    35
    40
    45
    50
    55

13. A contour map shows four contour lines merging at the location of a cliff. The contour interval is 40 feet. How high is the cliff?

14. The height of instrument above sea level is 164.70. What rod readings would be used for tracing 5 feet contours?

# CHAPTER 15

# Land Surveying

## OBJECTIVES

After studying this chapter, the student will be able to:

- recognize inconsistencies and deficiencies in property descriptions.
- adapt field methods to meet the immediate needs of the land being surveyed.
- make complex land surveys including those abutting curving highway rights-of-way.
- calculate the positions of corners for partitioned tracts of specified size and shape.
- assist adjoining property owners to settle boundary disputes.

## THE SURVEYOR'S RESPONSIBILITIES

During colonial days, some individuals became owners of vast tracts of land. Others acquired town lots. Some land was held in public ownership. This process extended westward as the population grew and spread. Down through the years, as the population continued to grow, smaller estates were carved from larger ones. Boundaries of farm and town properties were altered. Towns and cities grew outward creating many new small parcels of land with individual ownership.

Occasionally, new large estates were formed by the consolidation of a number of small parcels.

Whenever land boundaries are created or altered, land surveys are necessary. Unaltered parcels of land are often resurveyed when ownership changes. Land values increase with time or by change of use. The greater the land value, the greater the precision of the survey of land boundaries must be. Ever-growing numbers of land parcels, coupled with the need for higher precision, intensifies the demand for land surveys.

Persons engaging in land surveying today are registered surveyors or work under the direction of a registered surveyor. A land surveyor is responsible for finding boundary locations as near as possible to their original locations. The first step is to obtain all available evidence of the original boundary locations. This includes:

- studying deeds describing not only the client's property but also deeds for adjoining tracts.
- examining all natural and man-made monuments marking important boundaries.
- obtaining and considering the statements of knowledgeable persons as necessary to reinforce visual evidence.

Surveyors employed by public agencies usually have legal rights to enter private property to make a survey. Private surveyors do not. They should not enter private property without permission. Both public and private surveyors or their employers are liable for damage done to trees, crops, etc.

A land surveyor does not fix disputed boundaries. Settlement is made by an agreement of adjoining owners or by court decision. Nevertheless, a surveyor often assists disputing owners to reach agreement out of court by contributing information about local practices of defining boundaries. Surveyors also assist them by explaining the relationship between the locations of points on the ground and points as described in deeds.

Customs and practices affecting land titles vary widely. A land surveyor is obliged to learn what influences in the community affect a number of matters, including:

- history of land ownership
- principles of law and boundary control
- basis of land titles
- peculiarities of deed descriptions and terminology
- precedence of conflicting elements of measurement
- regulations governing subdivisions, lot sizes, road and street widths, and monuments
- highway and utility right-of-way procedures

By study and experience, a professional land surveyor becomes skilled in these matters. Many changes in custom and practices make the reading of up-to-date publications essential. The many legal aspects of land surveying and boundary control are of particular importance.

## PROPERTY DESCRIPTIONS

By 1784, land titling practices in the United States were recognized as inadequate. Boundary lines were vaguely defined and poorly described. Locations of many corners could only be established by a compromise of claims. For opening new lands to development, a grid system of land subdivision was adopted for *public land states.* Now included are all the states except the original thirteen colonies, Kentucky, Tennessee, Texas, West Virginia, Maine, Vermont, and Hawaii. Since 1784, the United States General Land Office, now the Bureau of Land Management, has been setting monuments marking grid corners. The system is based on townships as close to six miles square as converging meridian lines permit. Each township is divided into 36 *sections,* each being approximately one mile square. Each section is identified by a system of numbers and letters. It is now possible to describe land by reference to township, section, and smaller subdivisions. The system is fully described in the Bureau of Land Management's *Manual of Instruction for the Survey of the Public Lands of the United States.*

There remain twenty states not surveyed under the public land system. There, rural land is generally described by metes and bounds, and urban land by lot and block. Descriptions by state plane coordinates are often being made now in both rural and urban areas. This system is required in some jurisdictions under certain circumstances.

In the term metes and bounds, the word *metes* relates to description by measurement. It involves a recitation of measured distances and bearings with reference to a selected point of beginning. The word *bounds* relates to owners of adjoining tracts and the physical features which bound a tract, such as a stream, ridge, or road.

Maps showing subdivision lots are frequently recorded in a county registry of deeds. They permit descriptions to be made simply by lot and block number. County regulations establish standards for subdivision maps.

In preparing deeds, lawyers may choose to write property descriptions themselves or request surveyors to furnish descriptions. If they do it themselves, they rely upon drawings or other material furnished by a surveyor. Descriptions can be in terms of any combination of several methods already described.

Good descriptions define corners and courses in a manner which enables competent surveyors to retrace the boundaries. Directions given by magnetic or grid north are accompanied by an explanation of the amount of rotation from true north.

Copies of deeds transferring land are recorded and filed in the county registry of deeds or the city or town clerk's office. Subdivision plats and property maps are often recorded at the same places. Such documents are open to public view and copies can be made. File indexes are usually by years, and within years by listings of names of *grantors* (sellers) and *grantees* (buyers). From the recorded deeds, a surveyor obtains the information for the tract being surveyed, and the adjoining tracts. The wording of descriptions and terms of reference reflect local influences. It is dependent upon history, law, traditions, customs, regulations, and even the usage of the units of measurement. A surveyor adapts to these local influences for better understanding and to be understood.

## INTERPRETATION OF PROPERTY DESCRIPTIONS

Property descriptions in many old deeds suffer deficiencies. They are vague and may contain gross mistakes. Inconsistencies exist in descriptions of adjoining properties. There are other problems encountered in the field. Many monuments marking corners are obliterated. Compass declinations change. A surveyor must remain keenly aware that the original boundary locations are what is being sought. All available evidence is evaluated and interpreted accordingly.

Undisturbed monuments are the best evidence of the positions of boundary lines. *Natural monuments,* such as streams, ridges, and distinctive rocks are generally the best evidence. They are superior to *artificial monuments,* such as fences, iron pipes, or walls. Where physical evidence of boundaries cannot be found, the testimony of reliable residents is helpful. They can often point out the locations of monuments marking corners that are uncovered by digging. They may know which stump is the remains of a tree marking a corner.

Where boundaries are vague and adjoining deeds inconsistent, it is good practice to survey and map the prominent features in the vicinity of likely locations of corners. Make overlays on tracing paper showing the boundaries as called for in deed descriptions for the property being surveyed and the adjoining properties. Demonstrate to the property owners alternate ways of orienting the overlays on the map. Once they understand the true relationship between property descriptions and familiar physical features on the ground, solutions are often found.

Land bounded by a road or a street is presumed to extend to its center line unless otherwise indicated. Where deed discrepancies occur between acreage and calls for direction and distance, the acreage is usually disregarded. Other rules for the interpretation of descriptions exist but they are not consistent through the United States. They are learned by practice in a local area.

## FIELD METHODS

In rural areas, open land is often fenced along boundary lines. It is sometimes marked by lines of trees or brush. Corners may be marked by trees. Along such lines, and in some wooded areas, it is impractical to run transit lines along the boundaries. In open country, some traverse lines are offset and run parallel to a fence. It is usually more practical both in open and wooded country to run a traverse from one point of convenience to another. Found corners that are not easily occupied are then located by sideshots. This procedure saves much brush cutting and minimizes tree damage. It takes less time in the field. When coordinates are used for calculating and plotting, office time often works out about the same.

Where the exact location of corners cannot be identified, traverse points are established in their approximate area. Such corners themselves are bypassed. The traverse is then balanced. Coordinates are computed for the missing corners using called directions and distances for lines between preceding and succeeding found corners. Missing corners that cannot then be found are re-set from nearby traverse points. The direction and distance to such a corner is calculated from the differences in coordinates.

Magnetic bearings in old property descriptions require adjustments to offset changes in local declination since the date of survey. Some old surveys were performed with insufficient precision to warrant such refinement.

In urban areas, where points on property lines cannot be occupied, parallel offset lines often offer the best solution. Precision is dependent upon property values, high precision being essential for expensive land. Cities have their own networks of control monuments and bench marks to which city surveys are tied. When control monuments are within a reasonable distance, it is a good practice to tie-in to more than one. Cities maintain maps of developed areas, showing the dimensions and other features of lots. This information makes a surveyor's research in connection with re-surveys easier.

The act of cutting out rectangular lots from a large tract is not an adequate layout of a subdivision. Skillful layout of subdivisions produces attractive, well-drained lots along well-graded streets carefully fitted to the terrain. Their patterns provide greater measures of convenience and safety. It is false economy to perpetuate monotonous grid patterns of streets and rectangular lots. The cost of skillful planning, engineering, and surveying is repaid many times over by enhanced land values. Books are available on city and suburban planning and landscape architecture for surveyors interested in this field.

Surveyors place objects in the ground to mark corners they set. Temporary markers are nails or hubs. They are later replaced by permanent monuments such as iron pipes. In some cases, they are concrete cylinders with cross marks or metal plugs marking exact points. While setting permanent monuments, four stakes are driven in a square pattern around the point. Strings are stretched between the two diagonally opposite stakes to intersect exactly above the point. They hold the location while the permanent monument is being substituted for the temporary one, figure 15-1.

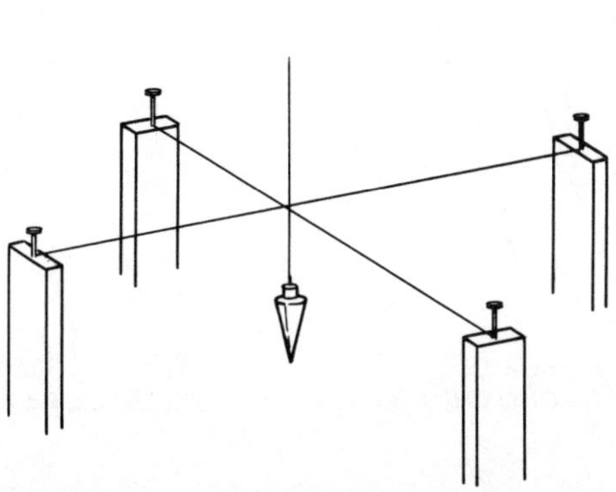

**Fig. 15-1 Arrangement of stakes and strings for holding the location of a point**

Occasionally, corners fall at locations which cannot be marked. They may be in a stream, or over the edge of a cliff. Monuments are then placed at *witness corners* along one or both boundaries at recorded distances from the corners.

Corner monuments are often referenced to make it easier to find or replace them. Bearings and distances are measured to references, such as points on trees, buildings, and rocks.

Sufficient information is obtained in the field to make good descriptions and plats of surveyed properties. The location of improvements are included when desired by the owner. Where the land abuts a curved highway right-of-way, the straight-line perimeter survey includes either a long chord between the ends of the curving right-of-way, or tangents along the right-of-way lines extended to their point of intersection, figure 15-2. If curve data are unavailable in deeds, or at

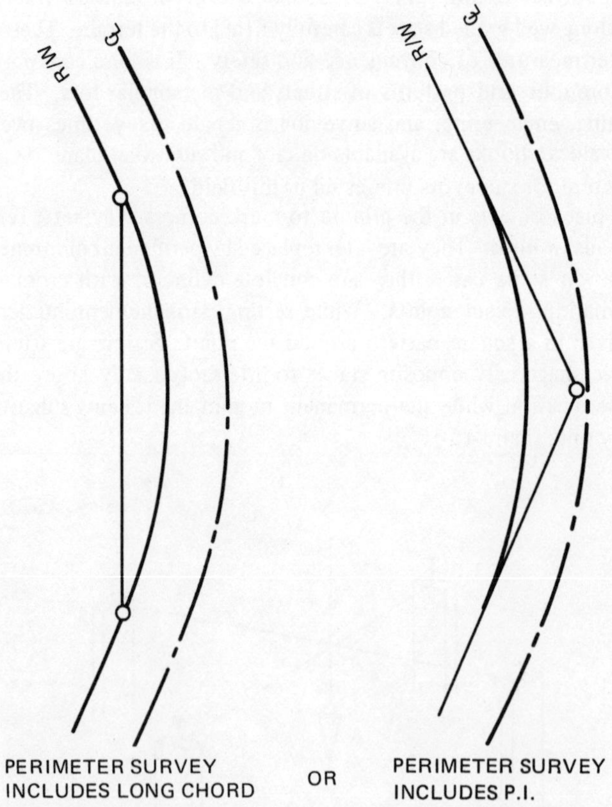


**Fig. 15-2 Perimeter points at boundary abutting curving highway**

the highway department, the radius is determined by plotting two or more chords. Perpendiculars are erected at their midpoints and extended to an intersection at the curve's center. The radius is then scaled, figure 15-3. All other curve data is then derived by scaling or calculation. Calculation is accomplished by determining the rectangular coordinates of all points including the terminal and midpoints of plotted chords.

## CALCULATION OF LAND AREAS BOUNDED BY CURVED HIGHWAYS

Areas immediately adjacent to curved boundaries are in one of two geometric shapes. One is a circular segment between the curve and a chord. The other is an external area lying between the curve and two tangents. In figure 15-4, page 224, a highway with a 66-foot right-of-way separates three properties; A, B, and C. The full length of a 20-degree curve with $\Delta = 50$ degrees borders A. Portions of the curve border B and C. The centerline radius of a 20-degree curve is $\frac{5730}{20}$, or 286.5 feet. The boundary radius on A is 253.5 feet and boundary radius on B and C is 319.5 feet.

The area of property A is equal to polygon 1, 2, 3, 4, X, Y, plus the cross-hatched segment XY in figure 15-5, page 225. The area of polygon 1, 2, 3, 4, X, Y is surveyed and calculated as any figure with straight-line boundaries. The area of segment XY equals the area of sector XY minus triangle OXY. The area of sector XY is $\left(\frac{50^\circ}{360^\circ}\right) \pi R^2$. A sector's area is in the same ratio to a circle's area as its central angle is to 360 degrees. In other words, a piece of pie cut with a central angle of 60 degrees is $\left(\frac{60^\circ}{360^\circ}\right)$ or 1/6th of the pie. A piece cut with a central angle of 50 degrees is $\left(\frac{50^\circ}{360^\circ}\right)$ or 5/36ths of the pie. Since R = 253.5 feet:

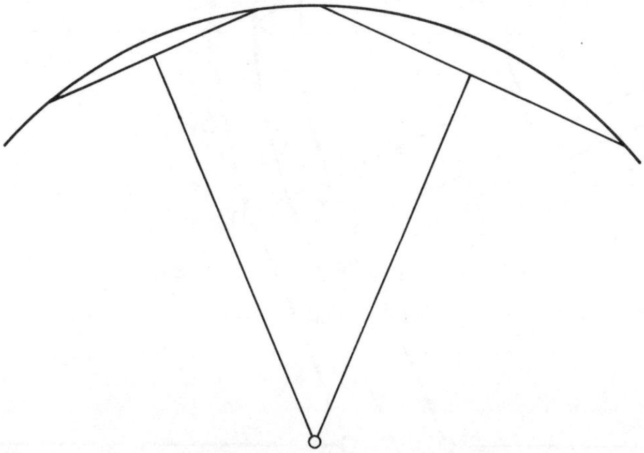

**Fig. 15-3 Determination of a curve's center and radius**

$$\text{Area of sector XY} = \left(\frac{50^\circ}{360^\circ}\right) \pi (253.5)^2$$
$$= 28040 \text{ square feet}$$

In triangle OXY

$$\text{Base XY} = 2(R \sin 1/2\ \Delta)$$
$$= 2(253.5)(\sin 25^\circ)$$
$$= 214.27 \text{ feet}$$
$$\text{Altitude OM} = R \cos 1/2\ \Delta$$
$$= (253.5)(\cos 25^\circ)$$
$$= 229.75 \text{ feet}$$
$$\text{Area triangle OXY} = 1/2\ (XY)(OM)$$
$$= 1/2\ (214.27)(229.75)$$
$$= 24614 \text{ square feet}$$
$$\text{Area segment XY} = \text{area sector XY} - \text{area triangle OXY}$$
$$= 28040 - 24614$$
$$= 3426 \text{ square feet}$$

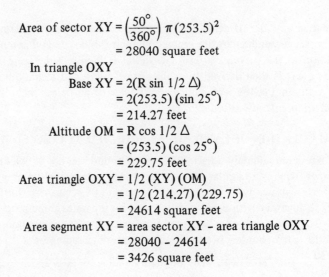

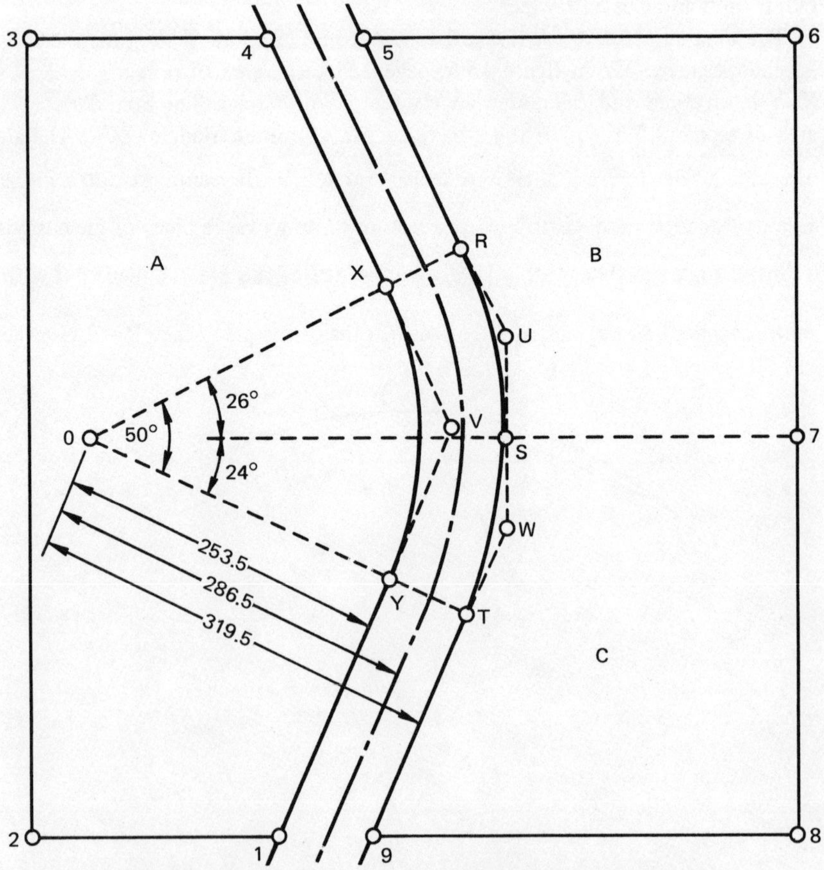


**Fig. 15-4 Three pieces of property bordering a curved highway**

Area property A = area 1,2,3,4,X,Y + area segment XY
= area 1,2,3,4,X,Y + 3426 square feet

The area of property A is also equal to polygon 1,2,3,4,V minus cross-hatched external area in figure 15-6. The area of polygon 1, 2, 3, 4, V is surveyed and calculated as any figure with straight-line boundaries. External area XY equals

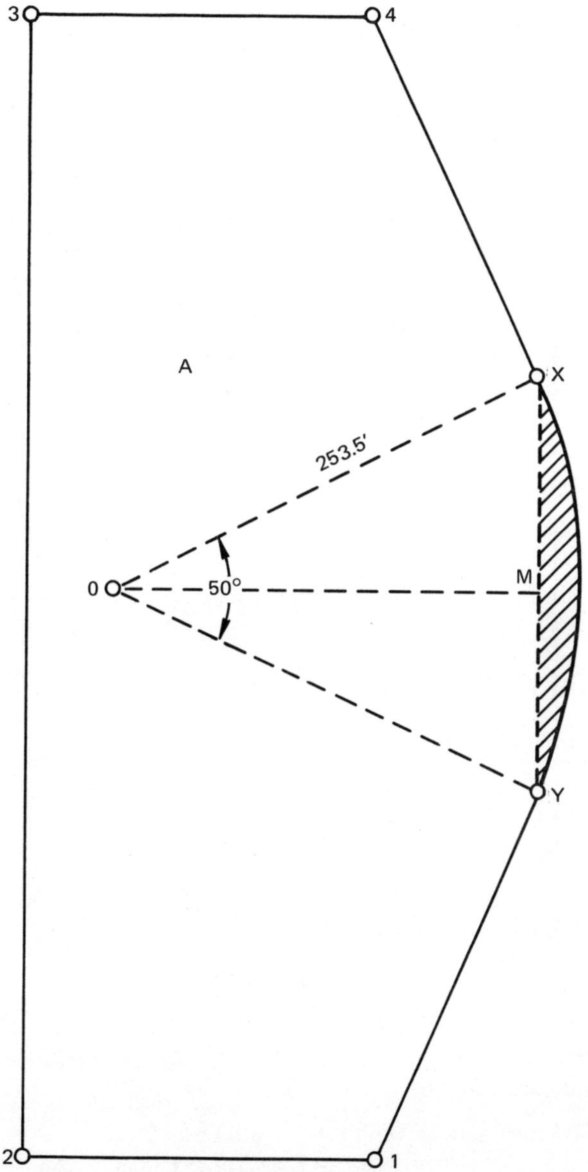


**Fig. 15-5 Property A segment**

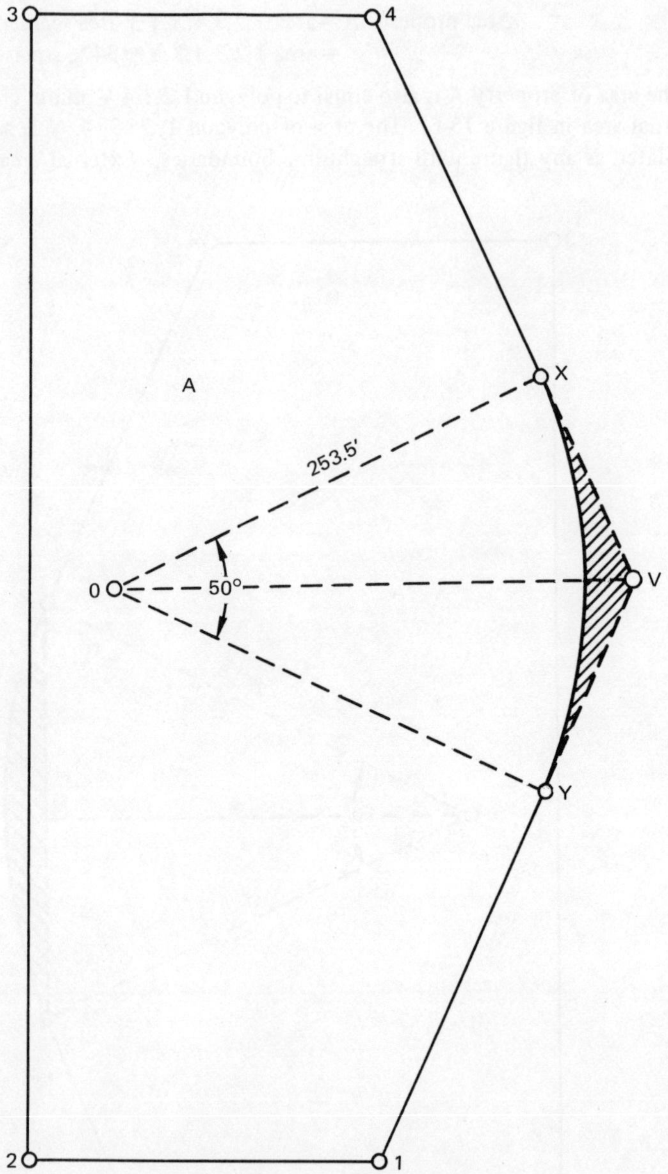


**Fig. 15-6 Property A external area**

the area of quadrilateral OXVY, minus sector XY. Quadrilateral OXVY consists of two right triangles, OXV and OYV. Each triangle has a base equal to radius R and an altitude equal to tangent distance T.

$$\text{Area quadrilateral OXVY} = 2(1/2\ RT)$$
$$= (253.5)\,(253.5 \tan 25^\circ)$$
$$= 29966 \text{ square feet}$$

$$\text{Area external XY} = \text{area quadrilateral OXVY - area sector XY}$$
$$= 29966 - 28040$$
$$= 1926 \text{ square feet}$$
$$\text{Area property A} = \text{area 1,2,3,4,V - external area XY}$$
$$= \text{area 1,2,3,4,V} - 1926 \text{ square feet}$$

In areas B and C, the same method is applied. However, subtangent distances are used instead of the full tangent distance T. Also, algebraic signs of segments and externals are reversed. R = 319.5 feet, $\Delta$ for property B is 26 degrees and $\Delta$ for property C is 24 degrees. Calculations for property B, figures 15-7 and 15-8 are:

$$\text{Area sector RS} = \left(\frac{26^\circ}{360^\circ}\right) \pi\ (319.5)^2$$
$$= 23161 \text{ square feet}$$
$$\text{Area triangle ORS} = 2\,\frac{[(319.5)(\sin 13^\circ)]\,[(319.5)(\cos 13^\circ)]}{2}$$
$$= 22375 \text{ square feet}$$
$$\text{Area segment RS} = 786 \text{ square feet}$$
$$\text{Area property B} = \text{Area 5,6,7,S,R} - 786 \text{ square feet}$$

or alternatively

$$\text{Area quadrilateral ORUS} = (319.5)\,(319.5 \tan 13^\circ)$$
$$= 23567 \text{ square feet}$$
$$\text{Area external RS} = 406 \text{ square feet}$$
$$\text{Area property B} = \text{Area 5,6,7,S,U,R} + 406$$

Calculations for property C, figures 15-9 and 15-10:

$$\text{Area sector ST} = \left(\frac{24^\circ}{360^\circ}\right) \pi\ (319.5)^2$$
$$= 21380 \text{ square feet}$$
$$\text{Area triangle OST} = 2\,\frac{[(319.5)(\sin 12^\circ)]\,[(319.5)(\cos 12^\circ)]}{2}$$
$$= 20760 \text{ square feet}$$
$$\text{Area segment ST} = 620 \text{ square feet}$$
$$\text{Area property C} = \text{Area 7, 8, 9, T, S} - 620 \text{ square feet}$$

or alternatively

$$\text{Area quadrilateral OSWT} = (319.5)\,(319.5 \tan 12^\circ)$$
$$= 21698 \text{ square feet}$$
$$\text{Area external ST} = 318 \text{ square feet}$$
$$\text{Area property C} = \text{Area 7,8,9,T,W,S} + 318 \text{ square feet}$$

## LAND PARTITIONING

*Land partitioning* is dividing a large tract into two or more parts. A common requirement is that one or more part is to contain a specified area. Often a point is designated on a boundary line through which the dividing line must pass. Sometimes the direction of the dividing line is specified. Land partitioning occurs

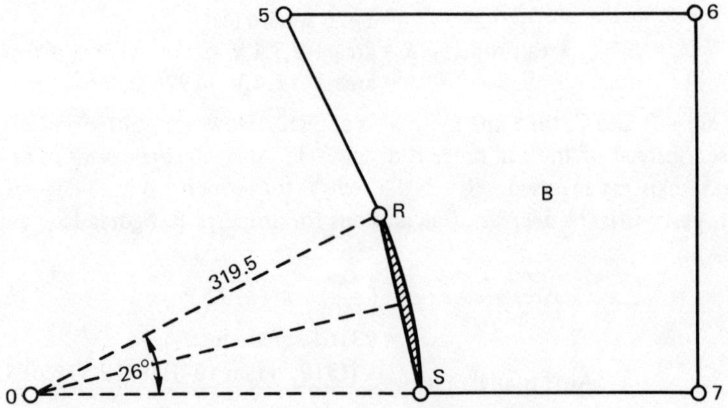


Fig. 15-7 Property B segment

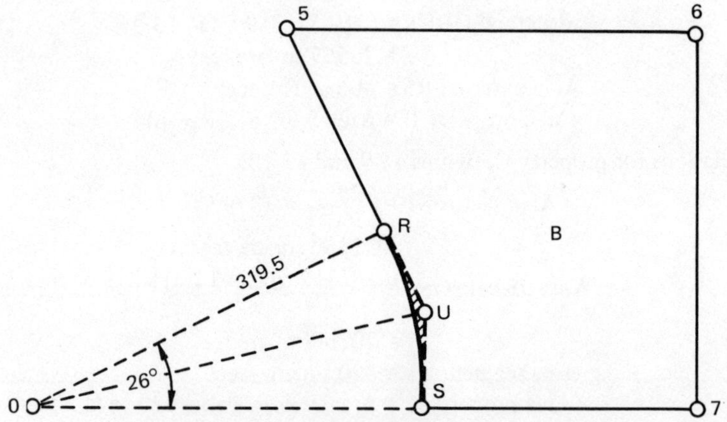


Fig. 15-8 Property B external area

for inheritance purposes, or to bring more favorable prices in disposing of property. It differs from laying out subdivisions. There may be no interest in producing building lots oriented for access, views, utilities, or drainage. Generally, partitioned parts of large tracts remain larger than lot size. The process need not involve road layouts. Different considerations influence the size and orientation than those which shape building lots.

**Dividing Line Through a Specified Point**

It is desired to cut off a specified area from tract ABCDE by a line through M, figure 15-11, page 230. It appears that the line strikes side CD at a point nearer C than D. A reference line is therefore drawn from M to C rather than to D.

Coordinates of points A, B, C, and M are listed in sequence. Area ABCM is computed by the coordinate method. It is found to be smaller than the specified area desired. An additional amount of land in the shape of a triangle MCN is needed. Point N is located so that the triangle's area adds just enough. The area of triangle NCM is one-half of base CM times its altitude h. The length and bearing of CM is computed from the coordinates of C and M. Angle NCM is the difference between the bearings of CM and CD. Altitude h equals CN times the sin NCM.

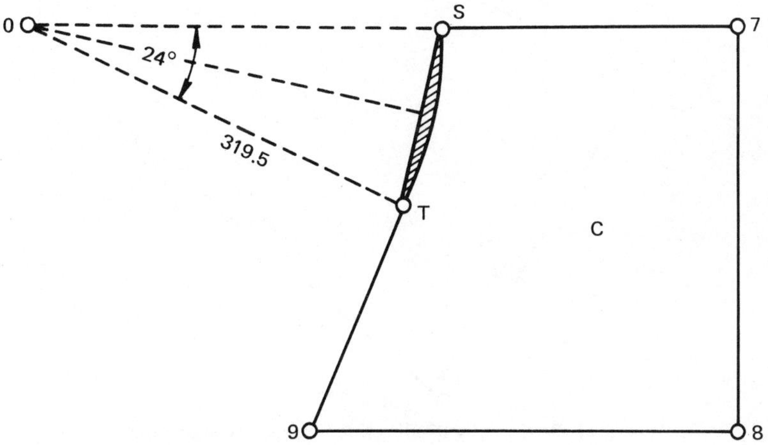


**Fig. 15-9 Property C segment**

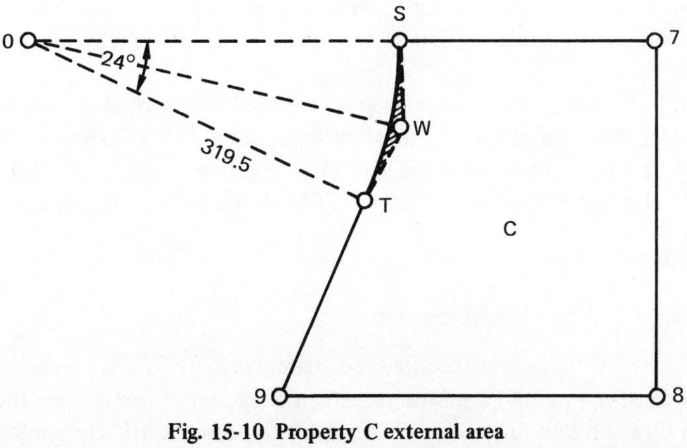


**Fig. 15-10 Property C external area**

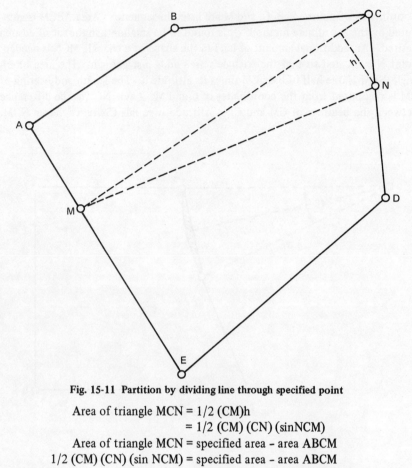

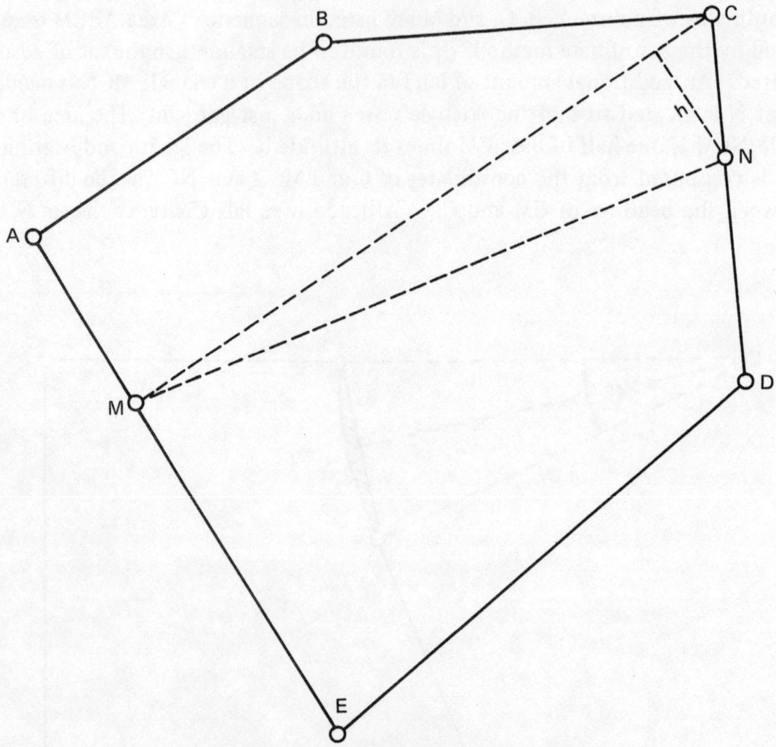


**Fig. 15-11 Partition by dividing line through specified point**

$$\text{Area of triangle MCN} = 1/2\,(\text{CM})h$$
$$= 1/2\,(\text{CM})\,(\text{CN})\,(\sin\text{NCM})$$
$$\text{Area of triangle MCN} = \text{specified area} - \text{area ABCM}$$
$$1/2\,(\text{CM})\,(\text{CN})\,(\sin \text{NCM}) = \text{specified area} - \text{area ABCM}$$
$$\text{(A)} \qquad \text{CN} = \frac{2\,(\text{specified area} - \text{area ABCM})}{\text{CM} \sin \text{NCM}}$$

All factors on the right side of equation (A) are known. Line CN and the coordinates of N are calculated. Points M and N are set in the field by any convenient method.

It should be noted that the reference line could have been drawn from M to D. The area ABCDM is then larger than the specified area desired. A triangular area MDN is then subtracted from ABCDM. Elements of triangle MDN are calculated in the same manner as triangle MCN. DN is derived in the same manner as CN.

### Dividing Line in a Specified Direction

It is desired to cut off a specified area from tract ABCDE by a line in a specified direction, figure 15-12. A line through C appears to be nearer the desired line than one through any other corner. A reference line CP′ is therefore drawn in the specified direction from C. It intersects boundary AE at P′. In triangle CAP′, the length and direction of CA are computed from the known coordinates of C and A. Angle ACP′ is the difference in bearings of CA and CP′. Angle CAP′ is the difference in bearings of CA and AP′. By the law of sines,

$$CP' = \frac{(\sin CAP')}{(\sin AP'C)} AC$$

$$AP' = \frac{(\sin ACP')}{(\sin AP'C)} AC$$

Bearings and lengths of CP′, AP′, AB, and BC are listed on a traverse computation sheet in sequence. Coordinates are calculated for P′, and the area is calculated for ABCP′.

Line PQ is established parallel to CP′ at a perpendicular distance x. Area CP′PQ must equal the difference between the specified area and area ABCP′. Area CP′PQ is equal to the area of rectangle CP′RS minus triangle PP′R minus triangle QCS. In triangle PP′R, angle theta ($\theta$) is the difference in bearings of AE and a perpendicular to the specified bearing of the dividing line. Side P′R is x, and side RP is x tan $\theta$. In triangle QCS, angle phi ($\phi$) is the difference in bearings of CD and perpendicular to the specified bearing of the dividing line. Side SC is x, side SQ is x tan $\phi$.

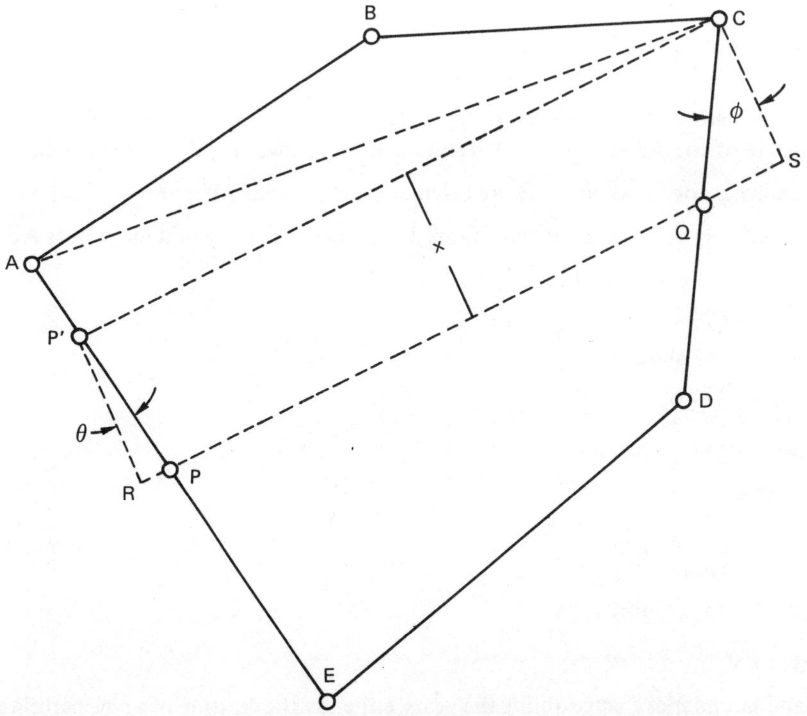


**Fig. 15-12 Partition by dividing line in specified direction**

$$\text{Area of rectangle } CP'RS = (CP')\,x$$

$$\text{Area of triangle } PP'R = \frac{(x)\,(x \tan \theta)}{2}$$

$$\text{Area of triangle } QCS = \frac{(x)\,(x \tan \phi)}{2}$$

$$\text{Area of } CP'PQ = (CP')x - \frac{(x)\,(x \tan \theta)}{2} - \frac{(x)\,(x \tan \phi)}{2}$$

(B) $$= (CP')x - \frac{x^2}{2}\,(\tan \theta + \tan \phi)$$

When arranged as a quadratic in the form $ax^2 + bx + c = 0$ where $x = \dfrac{-b \pm \sqrt{b^2 - 4ac},}{2a}$ equation (B) becomes:

$$\left(\frac{\tan \theta + \tan \phi}{2}\right) x^2 + (-CP')\,x + (\text{area } CP'PQ) = 0$$

$$x = \frac{CP' - \sqrt{(CP')^2 - 4\left(\dfrac{\tan \theta + \tan \phi}{2}\right)\quad (\text{Area } CP'PQ)}}{2\left(\dfrac{\tan \theta + \tan \phi}{2}\right)}$$

If $CP'PQ = 200{,}000$ sq ft, $CP' = 1000.00$ ft, $\theta = 10°00'$ and $\phi = 30°00'$

$$x = \frac{1000.00 \sqrt{1{,}000{,}000 - 2(.17633 + .57735)\,(200{,}000)}}{.17633 + .57735}$$

$$= \frac{1000.00 - \sqrt{698{,}529}}{.75368}$$

$$= 217.89 \text{ ft}$$

Algebraic signs of $\tan \theta$ and $\tan \phi$ depend upon whether angles $\theta$ or $\phi$ are inside or outside of the polygon. Correct algebraic signs are always evident from the plat.

After solving for x, AP and CQ are calculated. AP equals AP′ plus $\dfrac{x}{\cos \theta}$. Line CQ equals $\dfrac{x}{\cos \phi}$. Lines AP and CQ are then measured along boundary lines AE and CD to set points P and Q.

$$PQ = CP' - x(\tan \theta + \tan \phi)$$

Concluding the above example:

$$P'P = \frac{217.89}{.98481}$$

$$= 221.25 \text{ feet}$$

$$CQ = \frac{217.89}{.86603}$$

$$= 251.60 \text{ feet}$$

$$PQ = 1000 - 164.22$$

$$= 835.78 \text{ feet}$$

## SUMMARY

Land boundaries change during the years, either by the creation of a new parcels or the alteration of old ones. Land surveys are necessary when boundaries change.

Persons engaging in land surveying are registered surveyors or work under a registered surveyor's direction. Their responsibility on resurveys include finding boundary locations conforming to the original locations.

Unlike surveyors employed by many public agencies, private surveyors lack the authority to enter private property without permission.

Surveyors do not decide the locations of boundaries in dispute. They often assist disputing land owners to reach agreement.

There are thirty public land states where a grid system of land subdivision exists. In the other twenty states, land is generally described by metes and bounds or lot and block. In recent years, descriptions by state plane coordinates have been introduced.

Surveyors often prepare descriptions of land boundaries. Sometimes lawyers prepare descriptions using information supplied by surveyors.

Descriptions of land contained in deeds become public information by recording the deeds in local public places. From recorded deeds, surveyors obtain information for tracts to be resurveyed and for adjoining tracts.

Where boundaries are vague, and adjoining deeds inconsistent, surveyors prepare maps showing important physical features. These maps help adjoining property owners reach agreement on boundaries.

Where it is impractical to run transit lines along property boundaries, traverses are run nearby. Found corners which are not occupied as transit stations, are located by sideshots. Coordinates of all corners are calculated. Corners not found are set by measuring to their calculated positions.

Traverses upon which land surveys are based are tied-in to control monuments whenever posssible.

Witness corners are set where it is impractical to set permanent markers at corner locations.

Corner monuments are often referenced to nearby points to make it easier to find them.

In addition to boundary locations, land surveys include the location of improvements and curve data for abutting highways.

Calculations of land areas abutting curved highways requires calculation of either curcular segments or external areas.

For land partitioning, a common requirement is that one part or more is to contain a specified area. A point on a boundary line through which the dividing line must pass may be given. In other instances, the direction of the dividing line is specified. In either case, the approach is to lay off on a plat a trial dividing line. The area partitioned by the trial line is calculated. Necessary adjustment of the line is made to bring the partitioned area to the specified amount. Positions of intersections of the desired line with existing boundaries are calculated. Points are set in the field at those new corners.

## ACTIVITIES

1. Make a survey of a tract of land bordering a curved highway. Make a plot acceptable under local requirements, compute its area, and prepare the boundary description.
2. Using one of the land surveys previously made, partition off a portion of about 1.00 or 2.00 acres, selecting a point on one of the boundaries through which a dividing line must pass. Partition off another portion, selecting a direction for the dividing line.
3. Determine the radius of a curved highway by locating points at regular intervals around the curve. Following the office procedure, determine the radius by the graphical method described. From the rectanglar coordinates of terminals and midpoints of selected chords, calculate the radius.

## REVIEW QUESTIONS

### A. Multiple Choice

1. Which of the following statements is true?
   a. Private surveyors usually have legal rights to enter private property adjoining a tract they have contracted to survey.
   b. Surveyors employed by public agencies are not liable for any damage they do to trees, crops, etc. while surveying.
   c. Neither public nor private surveyors should enter property without permission.
   d. Surveyors employed by public agencies usually have legal rights to enter private property to make a survey.
2. How many public land states are there?
   a. 9
   b. 18
   c. 30
   d. 37

### B. Short Answers

3. Would a surveyor undertake a survey for a client who requested setting corners strictly as called for in a deed? Why?
4. List five matters on which land surveyors are obliged to learn the effect of influences in their local community.
5. List three ways of obtaining evidence of land boundaries.
6. In public land states, land is described by reference to townships, section, and smaller subdivision. How many square miles are contained in a township? How many square miles are in a section?
7. In states not surveyed under the public land system, how are rural lands generally described? How are urban lands generally described?

8. What type of property description consists of measured distances and bearings with reference to a selected point of beginning?
9. What type of property description lists owners of tracts adjoining a property?
10. What is the most important part of a good property description?
11. What is a grantor and grantee?
12. What are natural monuments?
13. What are artificial monuments?
14. What two survey methods are used where traversing boundaries is impractical?
15. What field survey procedure is followed when no physical evidence of a corner's location is found?
16. When it is impractical to set a permanent marker at a corner, what kind of monument is used?

## C. Problems

17. In figure 15-4, assume the highway right-of-way width to be 100 feet instead of 66 feet. Calculate the areas of the segments and the externals for Properties A, B, and C.

**Area Designation**

XY

RS

ST

18. In figure 15-11, assume the following values:

Specified area = 56,180 square feet

Area ABCM = 39,120 square feet

CM = 432 feet

Angle NCM = $62°30'$

What is the value of CN?

# CHAPTER 16

# Construction Surveying

## OBJECTIVES

After studying this chapter, the student will be able to:

- identify and apply surveying techniques required to meet construction needs.
- establish and maintain local survey controls and reference points for construction projects.
- stake alignment, grades, and slopes for the layout of construction projects.
- make field measurements to calculate the quantities of cut or fill.

## CONSTRUCTION DEMANDS

Years before construction begins, major construction programs and projects create needs for new surveys and maps. Construction planning, location, and design is a lengthy, complex process. It embraces legal, economic, and engineering considerations. In the engineering field, many matters are dealt with by contributing professionals. They determine the effects of geology and soils on the stability and design of foundations, slopes, and retaining walls. Steepness of slopes, types of soils, and vegetation are studied to see how they affect the run-off of rain, and designs of canals; dams; tunnels; culverts; or bridges. Vehicular speeds, traffic needs, and safety factors are considered in determining the design capacity, standards for curvature, and grades of transportation facilities.

Decisions on locations are preceded by studies of land boundaries, geology, and the shape of the earth's surface. The land owned by a project sponsor is evaluated. Requirements for the acquisition of new lands are determined. Existing land and topographic surveys and maps are supplemented with more specific and detailed surveys. Large projects requiring the acquisition of new lands often generate extensive land surveying. Boundaries of thousands of parcels may have to be defined for a single construction program.

Where national and state systems of monuments and bench marks are inadequate, control must be extended. Additional control traverses, base lines, photogrammetric control points, and bench marks become necessary. As locations of specific construction projects are decided, surveyors tie them to control monuments and land boundaries.

The design process includes the selection of center lines of proposed sewers, highways, tunnels, and railroads. It includes decisions on location, dimensions, and lines of buildings, bridge foundations, and dams. Grades are established wherever earth is to be moved. Elevations are specified for all kinds of structures. Adequate clearances between highways, railroads, airports, and power lines are worked out. Relocation of some of these existing facilities becomes necessary. Channels of streams are altered. Existing structures and other improvements are removed from lands to be used for reservoirs. New and more detailed topographic surveys and maps become necessary for the investigation of such impacts.

In urban areas, the survey and mapping of utilities is a tremendous undertaking. The greatest impact is felt where underground mass transit and expressway projects develop.

Preliminary planning work ends with the preparation of project plans and specifications which show all work details. Before construction begins, earthwork, alignment, and elevations of structures are staked out as designed. Most measurements require plane surveying equipment and techniques. Some major tunnels, bridges and dams involve geodetic surveying principles and methods to achieve the necessary precision.

After construction is in process, additional stakes are set for the control of successive stages. When rough grading is completed, stakes are set for fine grading. Lines of structures are defined and redefined. As structures are built, additional points for line and grade are given. Accuracy of erection is checked. Periodic measurements of earth removed and materials put in place provide a basis for payments due.

Surveyors must learn to read construction plans and to understand specifications if they stake out the work. Engineers must learn surveying techniques if they stake out the work. Such work is actually performed by surveying firms, engineering firms, contractors, city and county engineers, surveying or public works departments, and state and federal highway, reclamation, and other public works agencies. All use plane surveying techniques.

Construction places many demands on plane surveying during planning, location, design, and land acquisition phases. Some demands are met by methods and techniques covered in previous chapters. Others, peculiar to construction, require methods and techniques explained in this chapter.

## SITE MAPS

Surveys needed for the preparation of site maps call for new applications and combinations of techniques already explained. Topography in more detail is needed, including the location and elevation of improvements within likely construction limits. For proposed buildings in urban areas, the location of utilities and the elevation and size of sewers is important. For bridges and dams, the locations of borings and soundings are spotted. The characteristics of stream flow are determined. For airports, the tops of surrounding hills, tall buildings, towers, and anything else creating potential hazards to aircraft are identified. For grade separation structures, sufficient details are obtained for the redesign of existing facilities that may require relocation.

## ROUTE SURVEYING

Route surveying for proposed highways and railroads requires the selection of control points between the route terminals. The alignment between the control points is fitted to the terrain. *Control points* are mountain passes, favorable stream crossing, points at or near towns, and similar strategic features. The ideal alignment depends upon the class of highway or railroad. There are main line railroads and spur lines. There are expressways, primary highways, secondary roads, and land-access roads. Adopted standards for each class govern the maximum grades and degree of curvature. Within tolerable working limits of the standards, there are alternate feasible alignments. Selection of the best alternate requires considerable experience, knowledge, and skill.

Control points are generally selected by map study. Alignment is defined by photogrammetric methods, or field survey. In preparation for field surveys, the best available topographic maps are studied to identify possible alignments. For hilly country, a pair of drafter's dividers are used to step-off uniform distances between contours, figure 16-1. They are set to a scaled length proportional to a

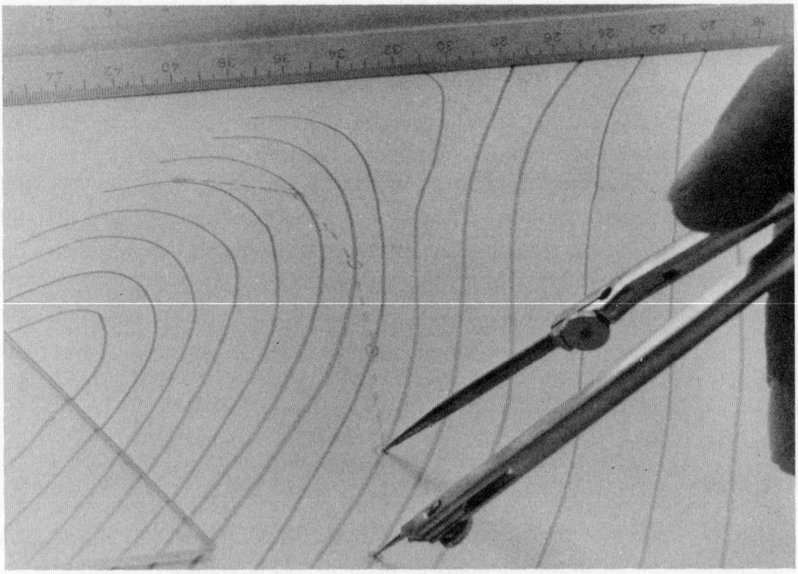

**Fig. 16-1 Map reconnaissance for feasible routes**

trial percent of grade for the proposed road. The distance set on the dividers is that required for the road to rise or fall an amount equaling the contour interval. Expressed mathematically, the length equals the contour interval divided by the trial percent of grade times 100 feet. Lines connecting points stepped off by dividers for each of several trial grades reveal possible alignments. Where available topographic maps are inadequate, field reconnaissance lines are run with a compass and abney hand level. The abney is set to a trial percent of grade, and points are staked at intervals of 100 feet or so. Field location surveys then develop some combination of tangent and circular curve alignment. That alignment follows approximately along the most favorable general alignment revealed by the topographic map study or from field reconnaissance.

After a satisfactory centerline (℄) is established, ground elevations are obtained by differential leveling. Readings are taken along the centerline at regular intervals of 50 or 100 feet and at significant breaks. From this information, a profile is plotted. Topography for design purposes is then obtained by the cross-section method.

When the design reaches an appropriate stage, the locations of property boundaries are determined. Right-of-way maps are prepared showing the proposed land acquisitions.

Route surveying is a subject deserving considerable study by those wishing to specialize in this field.

## EARTHWORK

Excavated earth is *cut*. Earth used to raise ground levels is *fill*. Excavated earth which cannot be used effectively on a project is *waste*. Earth which is imported from beyond project limits is *borrow*. Excavated earth is considered waste even though sold to nearby land owners needing borrow. Contractors engaging in excavation, movement, and placement of earth are said to be engaged in grading operations. The term *grade* also refers to the percent of grade as it applies to highways. It also refers to elevation to which land is brought by grading operations. Stakes indicating such elevation are called *grade stakes*. A graded surface is a level, curved, or sloping surface. The multiple meanings of the word grade are accepted by the construction industry with surprisingly little confusion.

In excavating for structures, pipelines and sewers, earthwork is often left standing vertically. Elsewhere it is seldom left standing vertically in cut, and cannot be in fill. Earthwork at the edges of graded surfaces slope upward in cut and downward in fill to natural undisturbed ground levels. Slopes are designed in accordance with rock and soil conditions, and with standards adopted by the project sponsors. The slope is expressed in terms of the ratio of the horizontal to vertical distance. A 1:1 cut slope, figure 16-2 means that for each foot beyond the graded area, the slope rises a foot. A 1½:1 fill slope, figure 16-3, means that for each 1½ foot beyond the graded area, the slope drops a foot. Slopes such as 4:1, 6:1, or 10:1 are justified where cuts and fills are small. Points where slopes

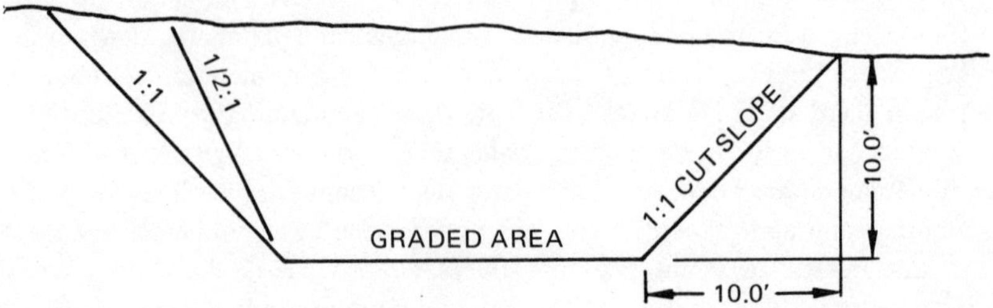


Fig. 16-2 Cut slopes

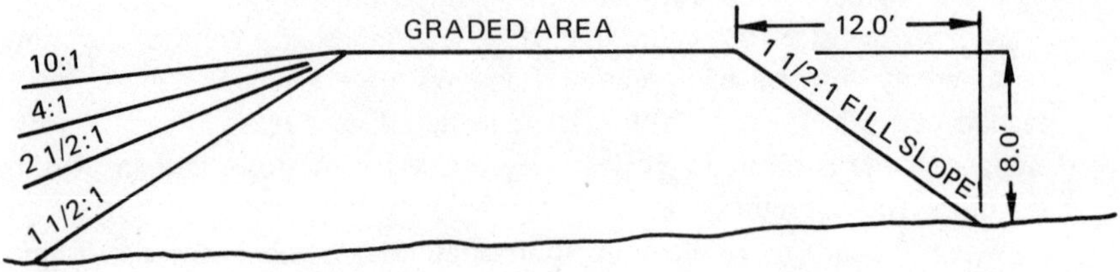


Fig. 16-3 Fill slopes

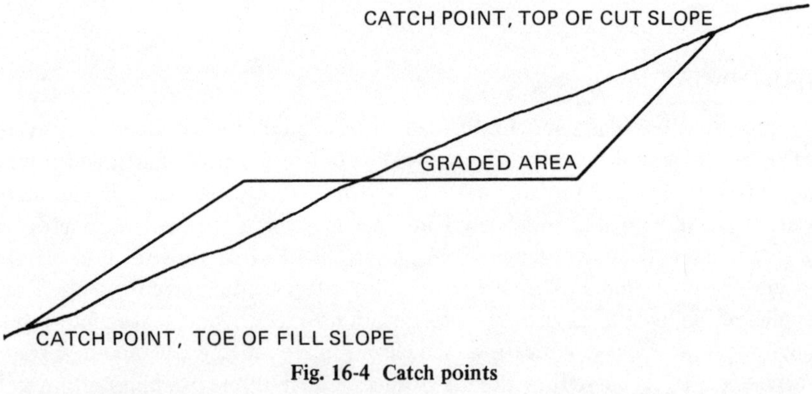


Fig. 16-4 Catch points

intersect natural undisturbed ground levels are sometimes referred to as *catch points.* They are usually described as the top of the cut slope and the toe of the fill slope, figure 16-4.

Landscape architects design finished grades to enhance the landscape. Highway designers try to set up grades so that cut and fill quantities balance to avoid waste and borrow. Occasionally, cut is hauled long distances for disposal in a needed fill. When the distance is too great, it is cheaper to waste or borrow than to balance cut and fill. On earthen dams, where soils are suitable, cut from reservoir areas is used as fill for dam structures.

One cubic yard of excavated rock makes more than one cubic yard of fill. One cubic yard of excavated soil makes less than one cubic yard of fill, figure 16-5.

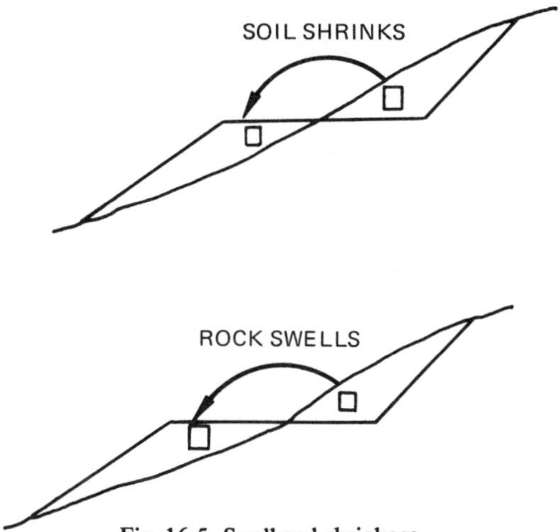


**Fig. 16-5 Swell and shrinkage**

The amount that rock swells and soils shrink varies. As construction proceeds, supervising engineers observe how close actual swell and shrinkage compare with assumed design values. They may authorize an adjustment in a line or grade to avoid unanticipated waste or borrow. Such changes require restaking of any section changed.

Earth from a borrow pit can be removed carelessly, leaving an unsightly hole. On the other hand, the pit can be designed, staked, and excavated to pleasing lines. Waste can be dumped in unsightly piles; or staked, placed, and finished to pleasing lines. It is in the public interest to stake all cuts, fills, waste, and borrow. Stakes are placed to show a contractor the alignment and grade called for in construction plans. They are in known relationship to local monuments, property boundaries and bench marks or some other datum. They are at points convenient for construction workers. They are placed around the perimeters of cut and fill areas, and at interior points as needed. Construction standards govern the intervals and precision at which grade stakes are set.

## Pipe Line or Sewers

Hubs, nails, or spikes, each protected by guard stakes, are set in line parallel to the centerline. They are offset an even number of feet from the centerline where they are out of the way of construction operations, figure 16-6. Their elevations are determined by differential leveling. The difference between their elevations and the grade elevations is the amount of cut. The station number is marked on the back of the guard stake, the side away from line. The cut is marked on its front side facing the line. The cut is expressed in feet, inches, and fractions. Below that, the offset distance is marked enclosed by a circle, figure 16-7. Sometimes the hubs, nails, or spikes are set at an even number of feet above grade elevations.

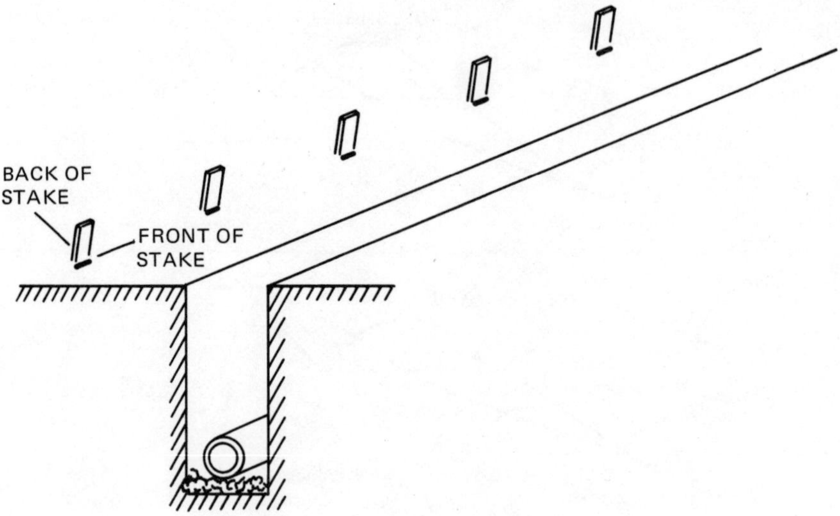


Fig. 16-6 Location of stakes for underground pipe

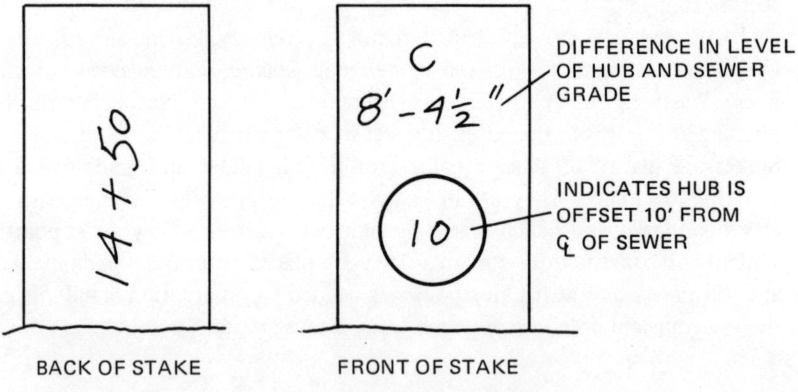


Fig. 16-7 Marking of stakes for underground pipe

Occasionally, the alignment for the construction of a sewer is maintained by means of a laser beam aimed inside the center of the pipe. A target centered at the opposite end of the pipe helps make the beam visible.

**Highway**

The centerline is staked at intervals shown on the plans, and at intermediate points if the topography requires. Ground elevations are obtained at each center-line stake by differential leveling. Differences between these elevations and the centerline grade elevations shown on the profile is the cut or fill, figure 16-8. Stationing is shown on the back face of the stake. Cut or fill expressed in feet and tenths is marked on its front face, figure 16-9. Hubs with tacks are set at all transit points. Each hub is referenced by methods already described. All reference points are placed outside of the construction limits. This permits the reestablishment of lines after hubs have been removed or covered by grading operations, figure 16-10. Slope stakes are set to the right and left of each center-line stake. These stakes mark the tops of cut slopes and the toes of fill slopes.

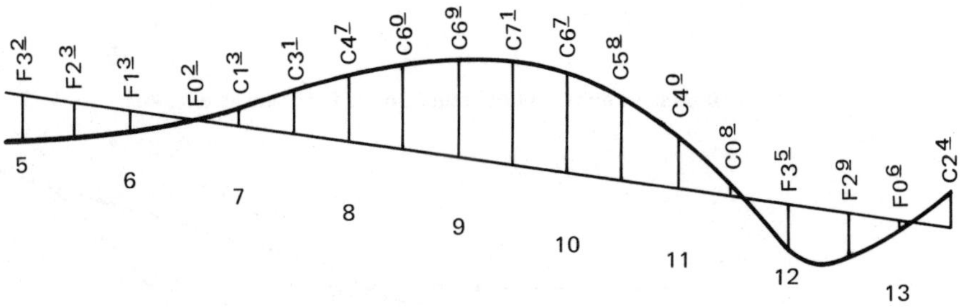


**Fig. 16-8 Highway profile**

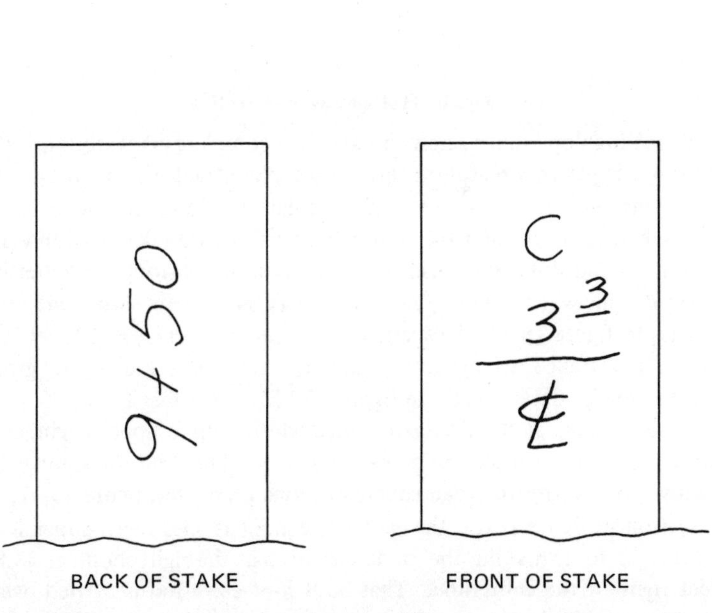


**Fig. 16-9 Marking of highway ℄ stake**

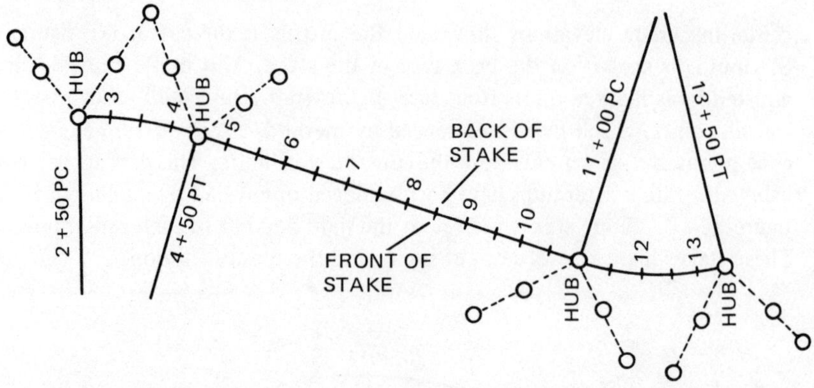


**Fig. 16-10 Stakes, hubs, and reference points for highway alignment**

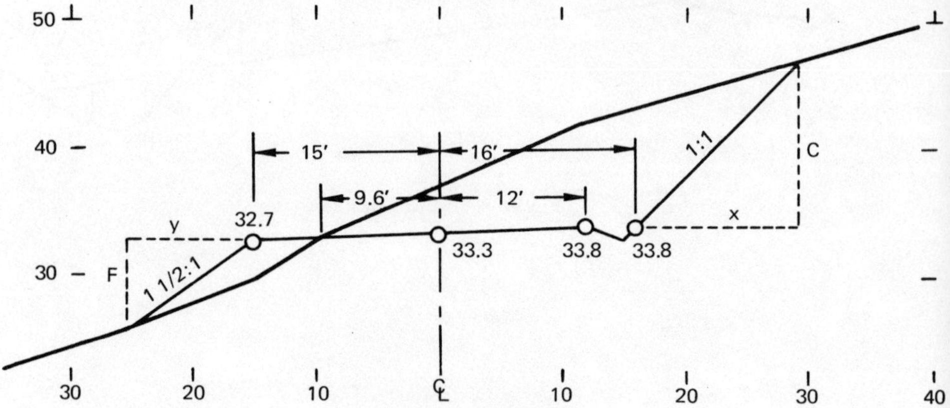


**Fig. 16-11 Highway cross-section**

While setting slope stakes, crews usually take complete cross sections. This time the purpose is for two reasons. One is for the contractor's guidance. The other is to provide data for a more precise calculation of the amount of earthwork, and for rebalancing cut and fill. Ground elevations are taken not only at breaks in natural ground slopes, but also at specified distances from the centerline. The specified distances are where grade lines break, and where slopes and grade lines intersect. In figure 16-11, the spcified distances are 15 feet L, 12 feet R, and 16 feet R. The distance to any point where the grade line and natural ground line intersect is always determined. In figure 16-12 it is 9.6 feet L.

Grade elevations at the limits of graded widths where slopes begin, are a little above or below the center-line grade elevations. They are different because of road crown on tangent or road superelevations on curves, figure 16-12. For the cross section in figure 16-12, the center-line grade is 33.3 feet. Superelevation is 0.04 foot per foot, making the grade elevation at the right shoulder 33.8 feet at 12 feet right of the centerline. That 33.8 foot elevation is carried over to the grade elevation ditch point at 16 feet right of the centerline. There a 1:1 cut slope begins to rise. The grade elevation at the left shoulder is 32.7 feet at 15 feet left of the centerline.

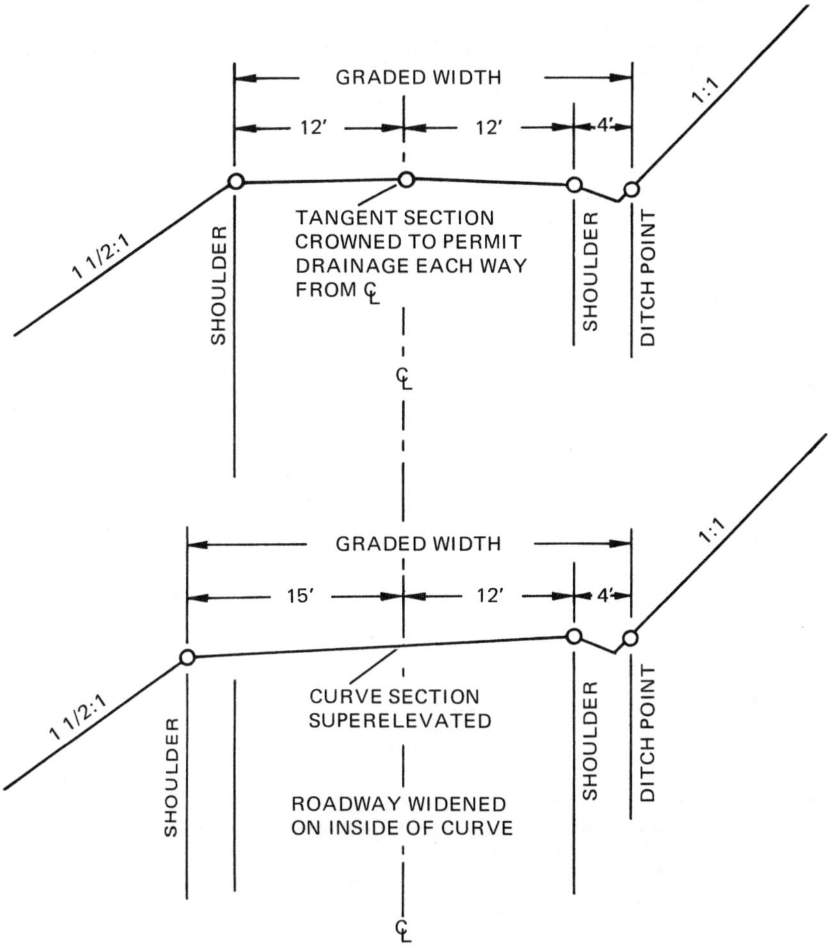


**Fig. 16-12 Highway crown and superelevation**

There a 1½:1 fill slope begins to drop. A point is found where the 1:1 cut slope intersects the natural ground. Its distance beyond 16 feet must equal the amount it rises. In figure 16-11, x must equal C. Another point is found where the 1½:1 fill slope intersects the natural ground. Its distance beyond 15 feet equals 1½ times the amount it drops. In figure 16-11, y must equal 1½F. These points are found by a cross-section crew using a trial and error method. Assume the crew's first trial discloses a rise C of 12.8 feet at horizontal distance x of 12.0 feet, figure 16-13. A second trial discloses a rise of 13.0 feet for horizontal distance of 12.7 feet. Finally, a third trial shows a 13.1 foot rise matching a 13.1 foot distance. Assume a first trial on the fill side gives a drop of 7.4 feet at 12.0 feet, figure 16-14. A second trial shows a drop of 6.9 feet at 10.1 feet. A third trial shows a 7.0-foot fill matching a 10.5-foot distance in the proportion desired. Experienced crews usually find catch points in two or three tries.

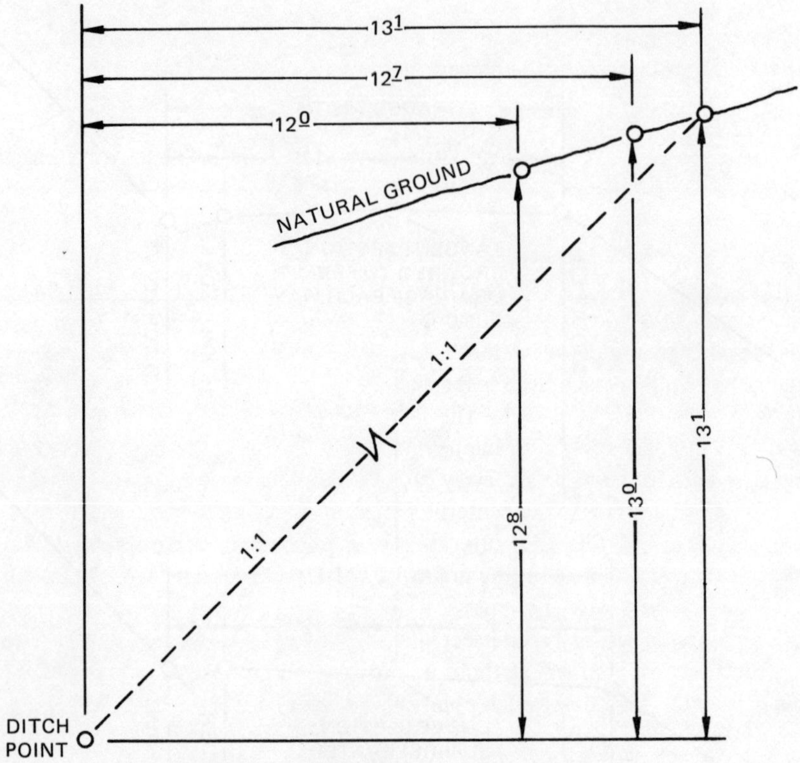


Fig. 16-13 Trial points for top of cut slope

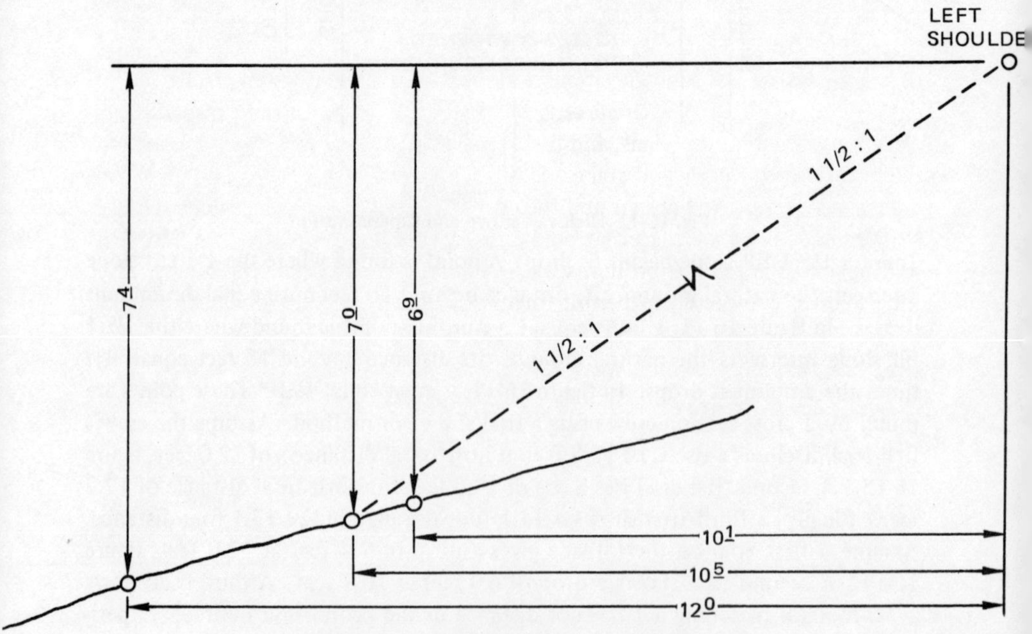


Fig. 16-14 Trial points for toe of fill slope

In notekeeping, the amount of cut or fill at each distance is noted, rather than the elevations. For the preceding example, notes appear as follows:

10–

9 + 50 $\frac{10^{\underline{1}}}{35^{\underline{5}}}$ $\frac{F7^{\underline{0}}}{25^{\underline{5}}}$ $\frac{F3^{\underline{1}}}{15^{\underline{0}}}$ $\frac{0^{\underline{0}}}{9^{\underline{6}}}$ $\frac{C\,3^{\underline{3}}}{℄}$ $\frac{C\,8^{\underline{2}}}{12^{\underline{0}}}$ $\frac{C\,9^{\underline{8}}}{16^{\underline{0}}}$ $\frac{C13^{\underline{1}}}{29^{\underline{1}}}$ $\frac{15^{\underline{6}}}{39^{\underline{1}}}$

9–

Six stakes are set for this cross section at distance 35.5 L, 25.5 L, 9.6 L, the centerline 29.1 R, and 39.1 R. Stakes at the extreme left and right are reference stakes 10 feet beyond the toe of the fill slope and the top of the cut slope. Total drop or rise from the shoulder elevation to the reference stakes is noted by the circles in the notes. In this example, a lost stake at the top of the cut slope is reestablished by measuring back 10.0 feet. The fact that it was 2.5 feet lower (15.6 - 13.1) is useful should a contractor happen to excavate beyond the staked top of slope. On the back of each stake, away from the centerline, the station is marked. Fronts of stakes facing the centerline are marked as follows, figure 16-15. Note the repetition of fills and cuts at the catch points on the reference stakes. Figures entered in circles in the notebook do not appear anywhere on the stakes.

Between stations where cross sections are taken, grade line of shoulders and ditch-points occasionally intersect natural ground. At those points, $0^{\underline{0}}$ stakes are set. Station numbers at those points are marked on the back of the stakes and entered in the notes. Such points are found by trial and error.

**Borrow Pits**

The grid method is effective for staking out borrow points. Stakes at each intersection of grid lines are marked with their letter and number designations as described earlier. Ground elevations at each stake are usually obtained by differential leveling. Cut at each stake is the difference between the designed grade elevation shown on the plans, and the actual ground elevation. Cut is marked on the front face of each grid stake. Slope stakes and reference stakes are set on extensions of the grid lines around the pit perimeter.

Reference hubs are placed at points that are on extensions of grid axial lines well beyond the excavation limits. Grid lines are then reestablished easily after excavation is completed to permit final measurement of the excavated material.

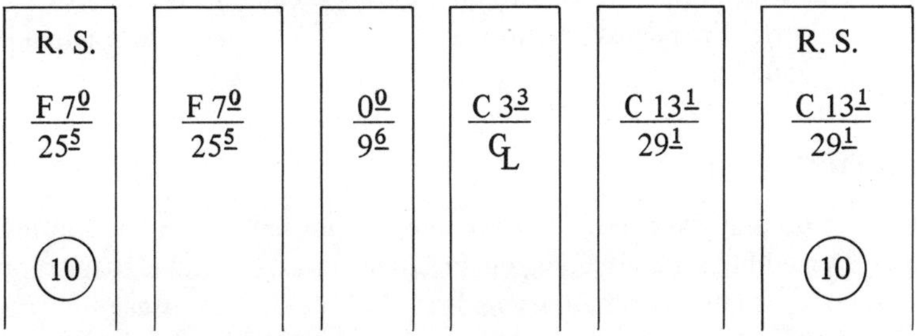


**Fig. 16-15**

## FINE GRADING

When rough grading is complete, new line and grade stakes are needed to govern the fine grading operations. On highway grades, hubs are reset at all transit points. The centerline is rerun. Pegs are driven to grade elevations, usually ± 0.03 feet, along the shoulder lines. Normal intervals are 50 feet on tangents and long radius curves, and 25 feet on the sharper curves. Since the tops of grade pegs are colored blue, they are referred to as *blue tops.* Guard stakes with station numbers marked on the back are set over them.

For work other than a highway; base lines, grid lines, or other control lines are similarly reestablished. Pegs are driven to grade level at necessary points for all kinds of grading operations.

## STRUCTURAL EXCAVATION

Excavated areas for structures extend far enough outside of the lines of the foundation to provide space for working. Hubs are placed either along parallel offset lines, or on extensions of the foundation lines. They are set at points safely beyond the excavation limits. Contractor's measurements are from string lines stretched between the hubs. Elevations of the tops of hubs are obtained by differential leveling. Amounts of cut down to grade are marked on the guard stake at each hub.

*Batter Boards,* figure 16-16, are horizontal planks nailed to upright 2 x 4's driven into the ground. Partially driven nails are set on line into the plank's top surfaces. Alignment is controlled by stretching a string between the nails from one batter board to the next. Workers plumb down from the string to the proper level. Planks are mounted horizontally, perferably at a fixed number of feet above the grade level to make the measuring easier.

For small structures, stakes are sometimes set at each corner of the excavation. The amount of cut is marked as measured from a line on the stake down to grade.

Stakes are set along road culvert centerline at each end, and at intermediate points as needed. The amount of cut is marked on each. After rough grading is completed, pegs are driven to grade at sufficient points to control the placement of the culvert.

After fine grading of large areas is completed, centerlines, base lines, and grid lines are established again. Deeper foundations if needed for portions of the area are also laid off. Frequently, additional structural excavation is required below the general grade line.

## STRUCTURES

After structural excavation is completed, lines and grades for footings are staked. Tacked hubs are set along, on an extension of, or parallel to the proposed foundation. In some cases, batter boards are not set until this stage.

After footings are in place, new line and grade points are again necessary to govern the erection of the structure itself. As large structures are erected, the

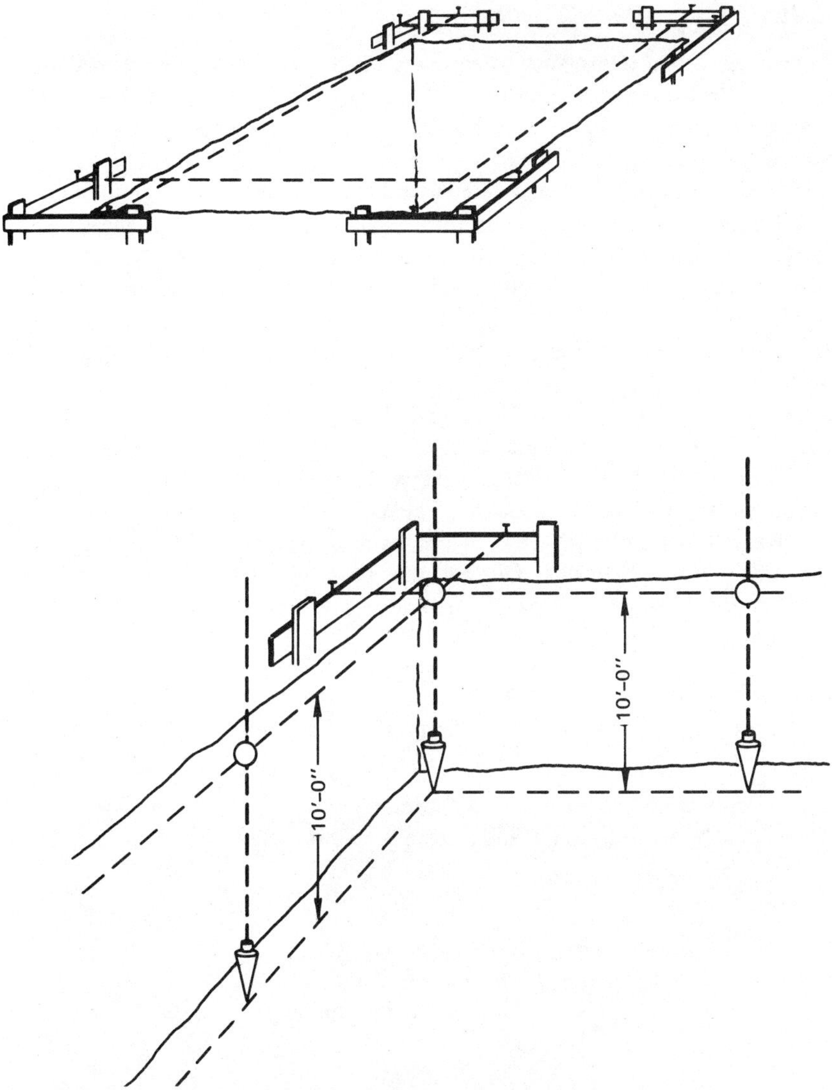


**Fig. 16-16 Batterboards**

process is repeated at higher levels. Points are needed to guide the placement of materials and for checking positions. Placement is controlled by direct sighting through one or more instruments. Reference points at these stages are crosses cut in concrete, nicks or punch marks in steel members, or nails driven into masonary or wooden structures.

Line and grade for pavements and curbs is given from hubs along offset lines parallel to their edges. Line and grade for railroad track is also given this way.

| | |
|---|---|
| CAUTION: | Steel tapes must not fall across railroad tracks or they can trip electrically controlled train signals. |

## VOLUMES

Volumes of certain items, such as earthwork and concrete, are determined at various stages of construction. Earthwork is classified in several categories, with each category being paid for at a different price. This necessitates keeping calculations of volumes separate for each category.

All materials used in fill are excavated as either cut or borrow. The volume is calculated only once tor payment purposes. It is normally based upon measurements of material in its natural undisturbed state. Differences between natural undisturbed ground levels and finished grade lines provide the vertical dimensions for calculations.

Highway earthwork volumes are calculated by multiplying the sums of adjacent cross-section areas by half the distance between them. That volume in cubic ft is then expressed in cubic yards by dividing it by 27.

In calculating cross-section areas, figure 16-17, cuts or fills at breaks are considered bases of triangles. Horizontal distances between adjacent breaks are considered altitudes of those triangles. This pattern is the same as for any irregular, straight-sided polygon. It is also the same as the pattern of triangles for irregular areas bordering streams. One-half the cut or fill is multiplied by the sum of both distances to adjacent breaks. The product is the area of a pair of triangles. Cross-section areas for station 9 + 50 are:

$$\text{Fill} = \frac{(3.1)(15.9)}{2} = 25 \text{ square feet}$$

$$\text{Cut} = \frac{(3.3)(21.6) + (8.2)(16.0) + (9.8)(17.1)}{2} = 185 \text{ square feet}$$

Assume the cross-section area in cut at station 9 is 210 and at station 10 is 152. The cut volume between station 9 and station 9 + 50 is then:

$$\frac{(210 + 185)(50)}{2 \times 27} = 366 \text{ cubic yards}$$

and cut volume between station 9 + 50 and station 10 is:

$$\frac{(185 + 152)(50)}{2 \times 27} = 312 \text{ cubic yards}$$

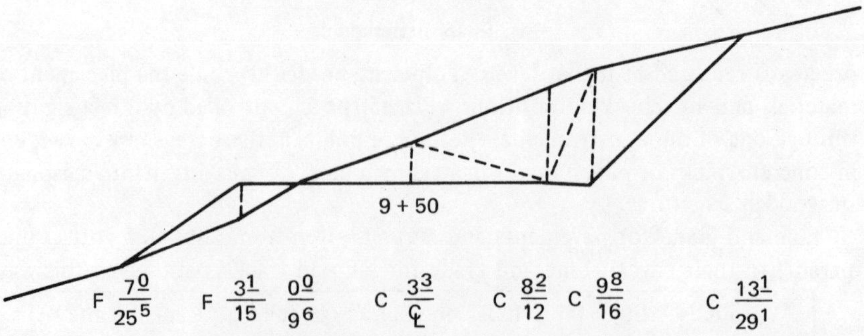


**Fig. 16-17 Pattern of triangles for cross section**

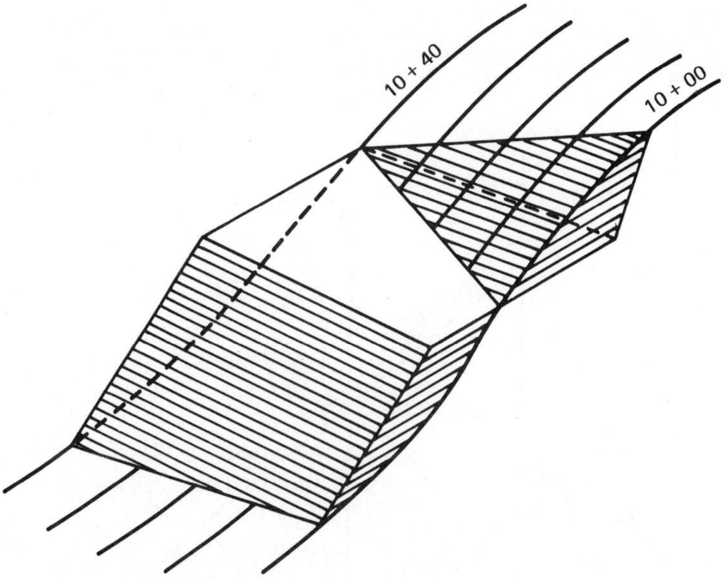


**Fig. 16-18 Pyramidal volume**

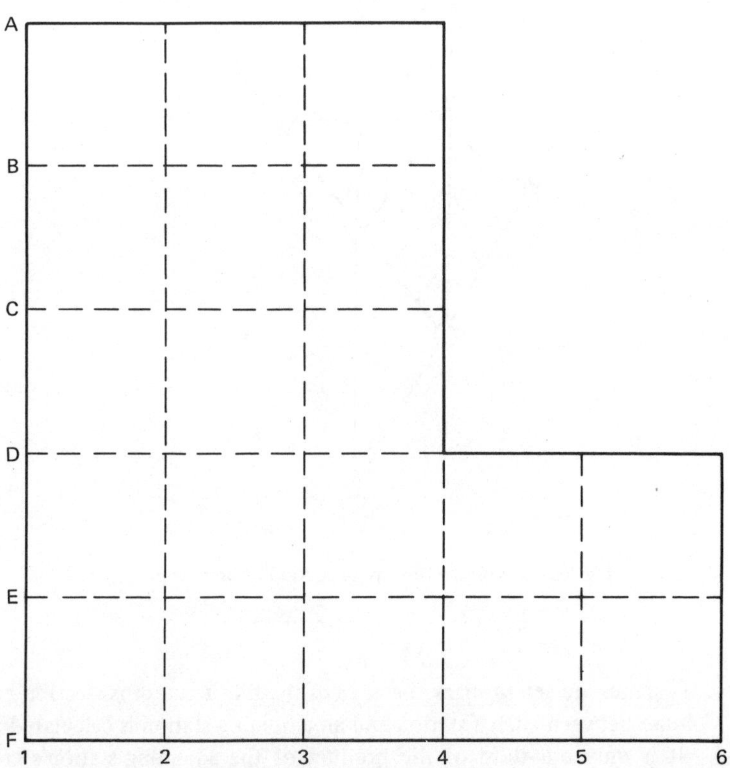


**Fig. 16-19 Grid for borrow pit**

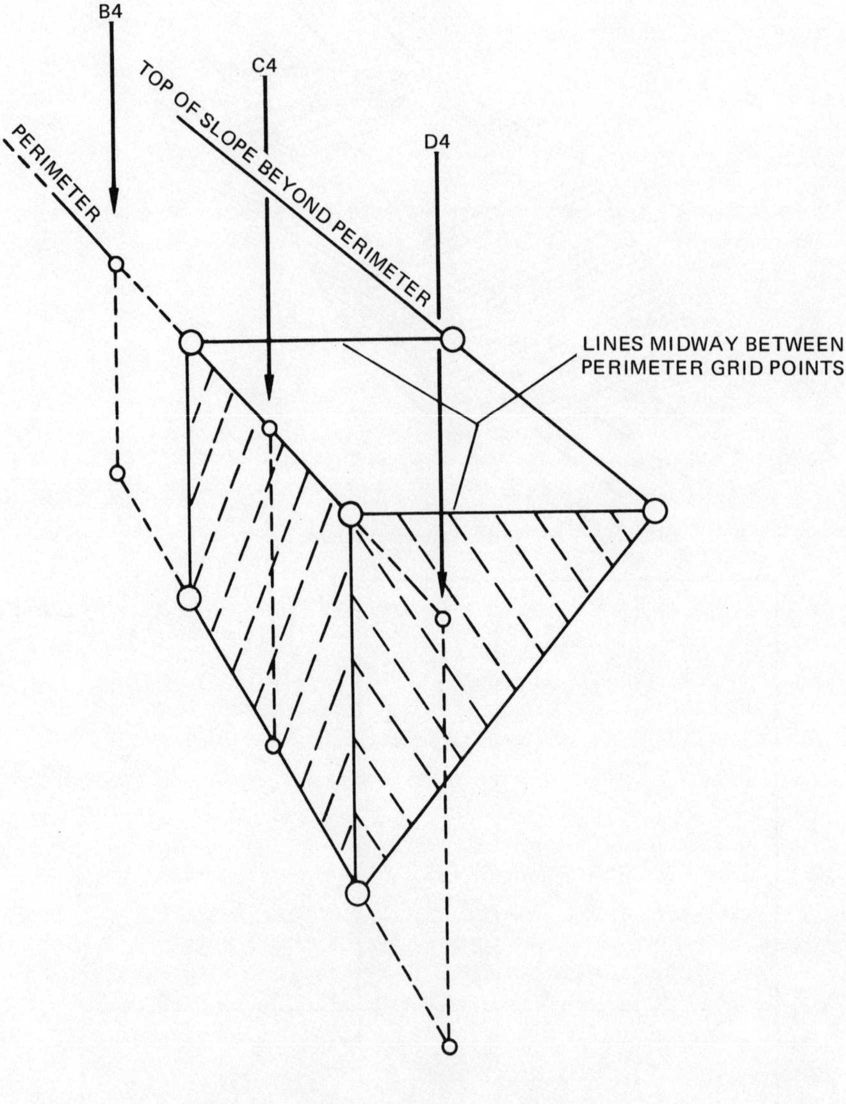


**Fig. 16-20 Volume of sloping portion of borrow pit**

Where 0$\underline{^{0}}$ stakes are set to mark the ends of the fill or cut, cross-section areas are 0. Volume between such a station and an adjoining station is calculated as a pyramid. It is only one-third of the product of the adjoining station's cross-section area and the distance between them. Assume the cross-section area in cut at station 10 + 40 is 0, figure 16-18. The cut volume between station 10 and station 10 + 40 is then:

$$\frac{(152)(40)}{3 \times 27} = 75 \text{ cubic yards}$$

Further refinements are made by some agencies to compensate for effects of the highway curvature on other characteristics. Such refinements are explained in manuals of instruction issued by those agencies.

In calculating volumes under the grid method, Cuts at grid points apply to areas of varying size. They apply only to as many quarters of one grid square's area as are in contact with the point. In figure 16-19, all interior points are contacted by four squares. Corner point D 4 is contacted by three squares. Corner points A 1, A 4, D 6, F 6, and F 1 are contacted by one square. All other perimeter points are contacted by two squares. Volumes sloping upward and outward from grid perimeters are based on one-half values of cuts at the perimeter grid points, figure 16-20. The half-values apply to areas bounded by the perimeter, top of the slope, and lines extended perpendicular to the perimeter midway between perimeter grid points. At corners, geometric shapes require study to determine the best method of computing volumes above sloping surfaces.

## MINE SURVEYING

Mine surveying is closely related to construction surveying. The primary objective is to chart horizontal alignment and elevations of all tunnels and shafts. These charts are needed by mine geologists and engineers to design ways of tapping ore bodies. They seek economic tunnel layouts, while maintaining a safe separation of all tunnels to preserve their structural integrity. Traversing and leveling are performed with instruments having electrically illuminated cross hairs. Backsights and foresights are illuminated the same as for night work outdoors. Plugs driven into tunnel roofs serve as transit points and bench marks. Leveling is performed by measuring downward from them. In deep shafts, plumb bobs are suspended by long thin wires. Their swinging movements are damped by submerging them in cans of light oil. Differences in elevation in deep shafts are measured by a surveyor's tape rather than by leveling equipment.

## SUMMARY

Major construction projects generate demands for many preliminary surveys and maps. Single major construction programs can involve land acquisition of thousands of parcels. Extension of geodetic control networks are sometimes required. The location of specific construction projects are tied in to control monuments and land boundaries. On-site measurements are made to aid architects and designers.

As the time of actual construction approaches, points are staked to give contractors alignment and grade. After construction begins, additional stakes are set for the control of successive stages. Periodic measurements of materials removed or put in place are made as a basis for payments due.

In previous chapters, survey techniques are explained which are applicable to surveys performed before construction begins. New combinations and applications of survey techniques are required during all construction phases. Some new survey techniques are also required.

Route surveying uses map study, photogrammetry, drafting, reconnaissance, curve layout, profiles, cross sections, boundary determinations, and tie-ins to control monuments.

Surveying connected with earthwork includes staking alignment, grades and slopes, referencing key points, and measuring quantities removed.

Slope stakes are set as part of a cross-section technique, but new techniques are also introduced. Catch points at the tops of cut slopes and toes of fill slopes are found by trial and error.

There are conventional ways for staking various kinds of construction, including marking methods.

Grid methods for obtaining topography are expanded when applied to staking out borrow pits along grid lines.

For fine grading, alignment and grades are restaked. This time points are set with greater precision than for rough grading. Stakes for structural excavation are set with comparable precision.

Points for line and grade to control the location of structures are set with still more precision. Points are marked by tacked hubs, batter boards, crosses cut in concrete, or nicks and punch marks in steel.

Earthwork is classified in several different categories. Volumes for each classification are measured separately. Volumes of both cut and fill are calculated for the purpose of balancing one against the other. However, volumes of material excavated and then used as fill are calculated only once for payment purposes. They are normally based upon measurement of material in its natural undisturbed state.

Highway earthwork volumes are calculated by multiplying sums of adjacent cross-section areas by half the distance between them. Refinements are sometimes made to compensate for highway curvature and prismoidal characteristics.

A cross-section area is easily calculated when it is recognized as being composed of pairs of triangles. The bases of the triangles are vertical, representing differences between the natural ground and the grade levels. The altitudes are horizontal distances between breaks. Their pattern is the same as for polygons considered in deriving DMD and coordinate formulas for area.

When calculating volumes by grid methods, cuts at grid points apply to horizontal areas of varying size. The size depends upon how many quarters of grid squares are in contact with the grid point.

Mine surveying is closely related to construction surveying. The principal difference is that traverse points and bench marks are overhead in tunnel roofs. In deep shafts, differences in elevation are measured by a tape. Interior lights in transits and levels keep cross hairs illuminated.

## ACTIVITIES

1. In the field, select an alignment of a 500- to 1000-foot section of tentative subdivision road in hilly terrain. Obtain centerline elevations, plot the centerline profile, set slope stakes, and compute the cut and fill quantities of earthwork.
2. Stake locations of buildings along a new street that you lay out across a rolling field. Take the field data obtained in activity 1 of Chapter 6 and stake out the front corners of the buildings in the same relative positions. Locate the centerline of the new street. Obtain elevations at the fronts of each building, and along the center line of the new street. Establish grade elevations for a sewer line running down the center line of the street on a 1% grade between manholes. Keep the sewer between 8 and 15 feet below ground level, allowing vertical drops at manholes. Keep it at a level which drains all the buildings. Set line and grade stakes for the sewer.

## REVIEW QUESTIONS

### A. Multiple Choice

1. The steepness to which cuts and fills are sloped are defined as a ratio. Construction designations for a slope falling 25 feet per 100 feet is:
   a. 1:4 c. 4:1
   b. ¼:1 d. 25:100

### B. Short Answer

2. A number of surveying activities are generated by major construction projects. One is land surveys of parcels that must be acquired. List three others.
3. Many kinds of organizations become engaged in staking out construction projects. List three.
4. Surveyors furnish many kinds of information for the preparation of site maps. One is the location and elevation of surrounding hills, tall buildings, and towers for airport sites. List three others.
5. A basic element of route surveying is the selection of control points between route terminals. List three features that constitute control points.
6. The following is a list of activities that are a part of surveying, or closely allied to it. Indicate whether it is performed in connection with route surveying, earthwork, or both.
   a. cross section
   b. reconnaissance
   c. measuring quantities removed

d. tie-ins to control monuments
e. staking a curved centerline
f. boundary determinations
g. locating catch points
h. photogrammetry

7. To surveyors, the term, grade, most often means the elevation to which land is brought. What are two other meanings it has in construction?
8. Points for line and grade to control the location of structures are marked by several means. List three.
9. From 4000 cubic yards of cut, 4300 cubic yards of fill are made because of the high percentage of rock. For how many cubic yards is the contractor due payment? In another case, 3000 cubic yards of cut produces only 2700 cubic yards of fill. For how many cubic yards is the contractor due payment?
10. List three differences between mine surveying and surveying above ground.

## C. Problems

11. A crew is seeking the location of a catch point for a 1:1 cut slope. Their first trial discloses a rise C of 10.9 feet at horizontal distance x of 10.2 feet. Should they move outward or inward for their next trial? At what distance is the catch point if the natural ground rises before them 0.3 feet per foot?
12. At station 20 + 50, the cross-section area is 211 in cut and 106 in fill. At station 20 + 75, the cross-section area is 195 in cut and 164 in fill. What are the cut and fill volumes in cubic yards between these two sections?
13. Continuing question 12, the cut pinches out to 0 at station 20 + 95. What is the cut volume in cubic yards between station 20 + 75 and 20 + 95?
14. Notes for a cross-section at station 11 + 50 area:

$$\frac{F\ 17^{\underline{2}}}{46^{\underline{3}}}\quad \frac{F\ 14^{\underline{2}}}{36^{\underline{3}}}\quad \frac{F\ 7^{\underline{5}}}{15^{\underline{0}}}\quad \frac{F\ 1^{\underline{2}}}{℄}\quad \frac{0^{\underline{0}}}{3^{\underline{0}}}\quad \frac{C\ 3^{\underline{2}}}{12^{\underline{0}}}\quad \frac{C\ 4^{\underline{5}}}{16^{\underline{0}}}\quad \frac{C\ 7^{\underline{4}}}{22^{\underline{4}}}\quad \frac{C\ 9^{\underline{6}}}{32^{\underline{4}}}$$

What are cross-section areas of fill and cut?

15. The volume of excavation for a square hole 20 feet on each side is calculated by the grid method. Sides of the holes are vertical. Grids are laid off in 10-foot squares. Cut at the intersection of the grid lines at the hole's center is 6.6 feet. Cut at the corners of the holes are 4.0, 5.1, 6.2 and 7.7 feet. Cut at the midpoints of the hole's sides are 4.8, 5.9, 7.4, and 6.9 feet. What is the volume in cubic feet? Recalculate by multiplying the average of all nine cut values by the 400-square foot area. How deficient is that approximation in cubic feet?

16. A topographic map shows that a proposed road has to rise at least 300 feet per mile to reach a mountain pass. Map study of feasible alternate lines is therefore based on a trial 6-percent grade. The map is at a scale of one inch equals 500 feet. Contour interval is 10 feet. At what spacing, in inches, are the dividers set to step off the distances between contours to reveal a 6-percent alignment?

# APPENDIX A

## TRIGONOMETRIC FORMULAS

### RIGHT TRIANGLES

Angles are A, B, and C with $C = 90^\circ$. Sides opposite are a, b, and c, respectively.

$$\sin A = \cos B = \frac{a}{c}$$

$$\sin B = \cos A = \frac{b}{c}$$

$$\tan A = \frac{a}{b}, \tan B = \frac{b}{a}$$

$$a = c \sin A = c \cos B = b \tan A$$

$$b = c \sin B = c \cos A = a \tan B$$

$$c = \frac{a}{\sin A} = \frac{b}{\sin B} = \frac{a}{\cos B} = \frac{b}{\cos A}$$

$$c^2 = a^2 + b^2$$

$$\text{Area} = \tfrac{1}{2} ab = \tfrac{1}{2} bc \sin A = \tfrac{1}{2} ac \sin B$$

### OBLIQUE TRIANGLES

Angles are A, B, and C. Sides opposite are a, b, and c, respectively.
Given two sides and one opposite angle, a, b, A

$$\sin B = \frac{\sin A}{a} b$$

$$C = 180^\circ - (A+B)$$

$$c = \frac{a}{\sin A} \sin C$$

Given one side and two angles a, B, C,

$$A = 180^\circ - (B+C)$$

$$c = \frac{a}{\sin A} \sin C$$

$$b = \frac{a}{\sin A} \sin B$$

Given two sides and included angle a, b, C

$$c = \sqrt{a^2 + b^2 - 2ab \cos C}$$

$$\sin A = \frac{\sin C}{c} a$$

$$B = 180^\circ - (A+C)$$

Given three sides a, b, c

$$\cos A = \frac{b^2 + c^2 - a^2}{2\,bc}$$

$$\cos B = \frac{a^2 + c^2 - b^2}{2\,ac} \qquad \cos C = \frac{a^2 + b^2 - c^2}{2\,ab}$$

$$c = 180^\circ - (A+B)$$

$$\text{Area} = \tfrac{1}{2}\,ab \sin C = \tfrac{1}{2}\,bc \sin A = \tfrac{1}{2}\,ac \sin B$$

or let $s = \frac{1}{2}(a + b + c)$

$$\text{Area} = \sqrt{s(s-a)(s-b)(s-c)}$$

# APPENDIX B

## TAPE CORRECTIONS

### TEMPERATURE

$C_t = 0.00000645\ L\ (T - T_o)$

where $C_t$ = Correction for distance L, in ft
$L$ = Measured length, in ft
$T$ = Temperature (Fahrenheit) when measurement is made
$T_o$ = Temperature (Fahrenheit) at which tape is standardized

### TENSION

$$C_p = \frac{(P - P_o)L}{AE}$$

where $C_p$ = Correction for distance L, in ft
$P$ = Applied tension, in pounds
$P_o$ = Tension for which tape is standardized, in pounds
$L$ = Measured length, in ft
$A$ = Cross-section area, in square inches
$E$ = Modulus of elasticity of steel, normally 28,000,000 to 30,000,000 pounds per sq in

### SAG

$$C_s = \frac{w^2\ L^3}{24\ P^2}$$

where $C_s$ = Correction for sag, in ft
$w$ = Weight of tape in pounds per ft of length
$L$ = Unsupported length, in ft
$P$ = Applied tension in pounds

$$P_n = \frac{0.204\ W\ \sqrt{AE}}{\sqrt{P_n - P_o}}$$

where $P_n$ = Tension required to offset sag, in pounds
$P_o$ = Tension for which tape is standardized
$W$ = Total weight of tape between supports in pounds

# APPENDIX C

## EARTH'S CURVATURE AND REFRACTION

### EARTH'S CURVATURE

$c = 0.66\ K^2$, or $0.024\ M^2$

where $c$ = Vertical distance, in ft, between horizontal plane tangent to earth's surface, and a level line through point of tangency

$K$ = Distance along level line from point of tangency, in miles

$M$ = Distance along level line from point of tangency, in thousands of ft

### REFRACTION

$r = 0.09K^2$, or $0.003M^2$

$r$ = Vertical distance, in ft, between horizontal plane tangent to earth's surface and a horizontal line of sight through point of tangency

### COMBINED EFFECTS OF CURVATURE AND REFRACTION

$c + r = 0.57K^2$, or $0.021M^2$

# APPENDIX D

## STANDARD 5 PLACE NATURAL SINES, COSINES, AND TANGENTS

### TABLE I

Natural Sines and Cosines

| ′ | 0° | | 1° | | 2° | | 3° | | 4° | | ′ |
|---|---|---|---|---|---|---|---|---|---|---|---|
| | Sine | Cosine | Sine | Cosine | Sine | Cosine | Sine | Cosine | Sine | Cosine | |
| 0 | .00000 | 1. | .01745 | .99985 | .03490 | .99939 | .05234 | .99863 | .06976 | .99756 | 60 |
| 1 | .00029 | 1. | .01774 | .99984 | .03519 | .99938 | .05263 | .99861 | .07005 | .99754 | 59 |
| 2 | .00058 | 1. | .01803 | .99984 | .03548 | .99937 | .05292 | .99860 | .07034 | .99752 | 58 |
| 3 | .00087 | 1. | .01832 | .99983 | .03577 | .99936 | .05321 | .99858 | .07063 | .99750 | 57 |
| 4 | .00116 | 1. | .01862 | .99983 | .03606 | .99935 | .05350 | .99857 | .07092 | .99748 | 56 |
| 5 | .00145 | 1. | .01891 | .99982 | .03635 | .99934 | .05379 | .99855 | .07121 | .99746 | 55 |
| 6 | .00175 | 1. | .01920 | .99982 | .03664 | .99933 | .05408 | .99854 | .07150 | .99744 | 54 |
| 7 | .00204 | 1. | .01949 | .99981 | .03693 | .99932 | .05437 | .99852 | .07179 | .99742 | 53 |
| 8 | .00233 | 1. | .01978 | .99980 | .03723 | .99931 | .05466 | .99851 | .07208 | .99740 | 52 |
| 9 | .00262 | 1. | .02007 | .99980 | .03752 | .99930 | .05495 | .99849 | .07237 | .99738 | 51 |
| 10 | .00291 | 1. | .02036 | .99979 | .03781 | .99929 | .05524 | .99847 | .07266 | .99736 | 50 |
| 11 | .00320 | .99999 | .02065 | .99979 | .03810 | .99927 | .05553 | .99846 | .07295 | .99734 | 49 |
| 12 | .00349 | .99999 | .02094 | .99978 | .03839 | .99926 | .05582 | .99844 | .07324 | .99731 | 48 |
| 13 | .00378 | .99999 | .02123 | .99977 | .03868 | .99925 | .05611 | .99842 | .07353 | .99729 | 47 |
| 14 | .00407 | .99999 | .02152 | .99977 | .03897 | .99924 | .05640 | .99841 | .07382 | .99727 | 46 |
| 15 | .00436 | .99999 | .02181 | .99976 | .03926 | .99923 | .05669 | .99839 | .07411 | .99725 | 45 |
| 16 | .00465 | .99999 | .02211 | .99976 | .03955 | .99922 | .05698 | .99838 | .07440 | .99723 | 44 |
| 17 | .00495 | .99999 | .02240 | .99975 | .03984 | .99921 | .05727 | .99836 | .07469 | .99721 | 43 |
| 18 | .00524 | .99999 | .02269 | .99974 | .04013 | .99919 | .05756 | .99834 | .07498 | .99719 | 42 |
| 19 | .00553 | .99998 | .02298 | .99974 | .04042 | .99918 | .05785 | .99833 | .07527 | .99716 | 41 |
| 20 | .00582 | .99998 | .02327 | .99973 | .04071 | .99917 | .05814 | .99831 | .07556 | .99714 | 40 |
| 21 | .00611 | .99998 | .02356 | .99972 | .04100 | .99916 | .05844 | .99829 | .07585 | .99712 | 39 |
| 22 | .00640 | .99998 | .02385 | .99972 | .04129 | .99915 | .05873 | .99827 | .07614 | .99710 | 38 |
| 23 | .00669 | .99998 | .02414 | .99971 | .04159 | .99913 | .05902 | .99826 | .07643 | .99708 | 37 |
| 24 | .00698 | .99998 | .02443 | .99970 | .04188 | .99912 | .05931 | .99824 | .07672 | .99705 | 36 |
| 25 | .00727 | .99997 | .02472 | .99969 | .04217 | .99911 | .05960 | .99822 | .07701 | .99703 | 35 |
| 26 | .00756 | .99997 | .02501 | .99969 | .04246 | .99910 | .05989 | .99821 | .07730 | .99701 | 34 |
| 27 | .00785 | .99997 | .02530 | .99968 | .04275 | .99909 | .06018 | .99819 | .07759 | .99699 | 33 |
| 28 | .00814 | .99997 | .02560 | .99967 | .04304 | .99907 | .06047 | .99817 | .07788 | .99696 | 32 |
| 29 | .00844 | .99996 | .02589 | .99966 | .04333 | .99906 | .06076 | .99815 | .07817 | .99694 | 31 |
| 30 | .00873 | .99996 | .02618 | .99966 | .04362 | .99905 | .06105 | .99813 | .07846 | .99692 | 30 |
| 31 | .00902 | .99996 | .02647 | .99965 | .04391 | .99904 | .06134 | .99812 | .07875 | .99689 | 29 |
| 32 | .00931 | .99996 | .02676 | .99964 | .04420 | .99902 | .06163 | .99810 | .07904 | .99687 | 28 |
| 33 | .00960 | .99995 | .02705 | .99963 | .04449 | .99901 | .06192 | .99808 | .07933 | .99685 | 27 |
| 34 | .00989 | .99995 | .02734 | .99963 | .04478 | .99900 | .06221 | .99806 | .07962 | .99683 | 26 |
| 35 | .01018 | .99995 | .02763 | .99962 | .04507 | .99898 | .06250 | .99804 | .07991 | .99680 | 25 |
| 36 | .01047 | .99995 | .02792 | .99961 | .04536 | .99897 | .06279 | .99803 | .08020 | .99678 | 24 |
| 37 | .01076 | .99994 | .02821 | .99960 | .04565 | .99896 | .06308 | .99801 | .08049 | .99676 | 23 |
| 38 | .01105 | .99994 | .02850 | .99959 | .04594 | .99894 | .06337 | .99799 | .08078 | .99673 | 22 |
| 39 | .01134 | .99994 | .02879 | .99959 | .04623 | .99893 | .06366 | .99797 | .08107 | .99671 | 21 |
| 40 | .01164 | .99993 | .02908 | .99958 | .04653 | .99892 | .06395 | .99795 | .08136 | .99668 | 20 |
| 41 | .01193 | .99993 | .02938 | .99957 | .04682 | .99890 | .06424 | .99793 | .08165 | .99666 | 19 |
| 42 | .01222 | .99993 | .02967 | .99956 | .04711 | .99889 | .06453 | .99792 | .08194 | .99664 | 18 |
| 43 | .01251 | .99992 | .02996 | .99955 | .04740 | .99888 | 06482 | .99790 | .08223 | .99661 | 17 |
| 44 | .01280 | .99992 | .03025 | .99954 | .04769 | .99886 | .06511 | .99788 | .08252 | .99659 | 16 |
| 45 | .01309 | .99991 | .03054 | .99953 | .04798 | .99885 | .06540 | .99786 | .08281 | .99657 | 15 |
| 46 | .01338 | .99991 | .03083 | .99952 | .04827 | .99883 | .06569 | .99784 | .08310 | .99654 | 14 |
| 47 | .01367 | .99991 | .03112 | .99952 | .04856 | .99882 | .06598 | .99782 | .08339 | .99652 | 13 |
| 48 | .01396 | .99990 | .03141 | .99951 | .04885 | .99881 | .06627 | .99780 | .08368 | .99649 | 12 |
| 49 | .01425 | .99990 | .03170 | .99950 | .04914 | .99879 | .06656 | .99778 | .08397 | .99647 | 11 |
| 50 | .01454 | .99989 | .03199 | .99949 | .04943 | .99878 | .06685 | .99776 | .08426 | .99644 | 10 |
| 51 | .01483 | .99989 | .03228 | .99948 | .04972 | .99876 | .06714 | .99774 | .08455 | .99642 | 9 |
| 52 | .01513 | .99989 | .03257 | .99947 | .05001 | .99875 | .06743 | .99772 | .08484 | .99639 | 8 |
| 53 | .01542 | .99988 | .03286 | .99946 | .05030 | .99873 | .06773 | .99770 | .08513 | .99637 | 7 |
| 54 | .01571 | .99988 | .03316 | .99945 | .05059 | .99872 | .06802 | .99768 | .08542 | .99635 | 6 |
| 55 | .01600 | .99987 | .03345 | .99944 | .05088 | .99870 | .06831 | .99766 | .08571 | .99632 | 5 |
| 56 | .01629 | .99987 | .03374 | .99943 | .05117 | .99869 | .06860 | .99764 | .08600 | .99630 | 4 |
| 57 | .01658 | .99986 | .03403 | .99942 | .05146 | .99867 | .06889 | .99762 | .08629 | .99627 | 3 |
| 58 | .01687 | .99986 | .03432 | .99941 | .05175 | .99866 | .06918 | .99760 | .08658 | .99625 | 2 |
| 59 | .01716 | .99985 | .03461 | .99940 | .05205 | .99864 | .06947 | .99758 | .08687 | .99622 | 1 |
| 60 | .01745 | .99985 | .03490 | .99939 | .05234 | .99863 | .06976 | .99756 | .08716 | .99619 | 0 |
| ′ | Cosine | Sine | Cosine | Sine | Cosine | Sine | Cosine | Sine | Cosine | Sine | ′ |
| | 89° | | 88° | | 87° | | 86° | | 85° | | |

**TABLE 1 cont.**
Natural Sines and Cosines

| ′ | 5° | | 6° | | 7° | | 8° | | 9° | | ′ |
|---|---|---|---|---|---|---|---|---|---|---|---|
| | Sine | Cosine | Sine | Cosine | Sine | Cosine | Sine | Cosine | Sine | Cosine | |
| 0 | .08716 | .99619 | .10453 | .99452 | .12187 | .99255 | .13917 | .99027 | .15643 | .98769 | 60 |
| 1 | .08745 | .99617 | .10482 | .99449 | .12216 | .99251 | .13946 | .99023 | .15672 | .98764 | 59 |
| 2 | .08774 | .99614 | .10511 | .99446 | .12245 | .99248 | .13975 | .99019 | .15701 | .98760 | 58 |
| 3 | .08803 | .99612 | .10540 | .99443 | .12274 | .99244 | .14004 | .99015 | .15730 | .98755 | 57 |
| 4 | .08831 | .99609 | .10569 | .99440 | .12302 | .99240 | .14033 | .99011 | .15758 | .98751 | 56 |
| 5 | .08860 | .99607 | .10597 | .99437 | .12331 | .99237 | .14061 | .99006 | .15787 | .98746 | 55 |
| 6 | .08889 | .99604 | .10626 | .99434 | .12360 | .99233 | .14090 | .99002 | .15816 | .98741 | 54 |
| 7 | .08918 | .99602 | .10655 | .99431 | .12389 | .99230 | .14119 | .98998 | .15845 | .98737 | 53 |
| 8 | .08947 | .99599 | .10684 | .99428 | .12418 | .99226 | .14148 | .98994 | .15873 | .98732 | 52 |
| 9 | .08976 | .99596 | .10713 | .99424 | .12447 | .99222 | .14177 | .98990 | .15902 | .98728 | 51 |
| 10 | .09005 | .99594 | .10742 | .99421 | .12476 | .99219 | .14205 | .98986 | .15931 | .98723 | 50 |
| 11 | .09034 | .99591 | .10771 | .99418 | .12504 | .99215 | .14234 | .98982 | .15959 | .98718 | 49 |
| 12 | .09063 | .99588 | .10800 | .99415 | .12533 | .99211 | .14263 | .98978 | .15988 | .98714 | 48 |
| 13 | .09092 | .99586 | .10829 | .99412 | .12562 | .99208 | .14292 | .98973 | .16017 | .98709 | 47 |
| 14 | .09121 | .99583 | .10858 | .99409 | .12591 | .99204 | .14320 | .98969 | .16046 | .98704 | 46 |
| 15 | .09150 | .99580 | .10887 | .99406 | .12620 | .99200 | .14349 | .98965 | .16074 | .98700 | 45 |
| 16 | .09179 | .99578 | .10916 | .99402 | .12649 | .99197 | .14378 | .98961 | .16103 | .98695 | 44 |
| 17 | .09208 | .99575 | .10945 | .99399 | .12678 | .99193 | .14407 | .98957 | .16132 | .98690 | 43 |
| 18 | .09237 | .99572 | .10973 | .99396 | .12706 | .99189 | .14436 | .98953 | .16160 | .98686 | 42 |
| 19 | .09266 | .99570 | .11002 | .99393 | .12735 | .99186 | .14464 | .98948 | .16189 | .98681 | 41 |
| 20 | .09295 | .99567 | .11031 | .99390 | .12764 | .99182 | .14493 | .98944 | .16218 | .98676 | 40 |
| 21 | .09324 | .99564 | .11060 | .99386 | .12793 | .99178 | .14522 | .98940 | .16246 | .98671 | 39 |
| 22 | .09353 | .99562 | .11089 | .99383 | .12822 | .99175 | .14551 | .98936 | .16275 | .98667 | 38 |
| 23 | .09382 | .99559 | .11118 | .99380 | .12851 | .99171 | .14580 | .98931 | .16304 | .98662 | 37 |
| 24 | .09411 | .99556 | .11147 | .99377 | .12880 | .99167 | .14608 | .98927 | .16333 | .98657 | 36 |
| 25 | .09440 | .99553 | .11176 | .99374 | .12908 | .99163 | .14637 | .98923 | .16361 | .98652 | 35 |
| 26 | .09469 | .99551 | .11205 | .99370 | .12937 | .99160 | .14666 | .98919 | .16390 | .98648 | 34 |
| 27 | .09498 | .99548 | .11234 | .99367 | .12966 | .99156 | .14695 | .98914 | .16419 | .98643 | 33 |
| 28 | .09527 | .99545 | .11263 | .99364 | .12995 | .99152 | .14723 | .98910 | .16447 | .98638 | 32 |
| 29 | .09556 | .99542 | .11291 | .99360 | .13024 | .99148 | .14752 | .98906 | .16476 | .98633 | 31 |
| 30 | .09585 | .99540 | .11320 | .99357 | .13053 | .99144 | .14781 | .98902 | .16505 | .98629 | 30 |
| 31 | .09614 | .99537 | .11349 | .99354 | .13081 | .99141 | .14810 | .98897 | .16533 | .98624 | 29 |
| 32 | .09642 | .99534 | .11378 | .99351 | .13110 | 99137 | .14838 | .98893 | .16562 | .98619 | 28 |
| 33 | 09671 | 99531 | .11407 | .99347 | .13139 | .99133 | .14867 | .98889 | .16591 | .98614 | 27 |
| 34 | .09700 | .99528 | .11436 | .99344 | .13168 | .99129 | .14896 | .98884 | .16620 | .98609 | 26 |
| 35 | .09729 | .99526 | .11465 | .99341 | .13197 | .99125 | .14925 | .98880 | .16648 | .98604 | 25 |
| 36 | .09758 | .99523 | .11494 | .99337 | .13226 | .99122 | .14954 | .98876 | .16677 | .98600 | 24 |
| 37 | .09787 | .99520 | .11523 | .99334 | .13254 | .99118 | .14982 | .98871 | .16706 | .98595 | 23 |
| 38 | .09816 | .99517 | .11552 | .99331 | .13283 | .99114 | .15011 | .98867 | .16734 | .98590 | 22 |
| 39 | .09845 | .99514 | .11580 | .99327 | .13312 | .99110 | .15040 | .98863 | .16763 | .98585 | 21 |
| 40 | .09874 | .99511 | .11609 | .99324 | .13341 | .99106 | .15069 | .98858 | .16792 | .98580 | 20 |
| 41 | .09903 | .99508 | .11638 | .99320 | .13370 | .99102 | .15097 | .98854 | .16820 | .98575 | 19 |
| 42 | .09932 | .99506 | .11667 | .99317 | .13399 | .99098 | .15126 | .98849 | .16849 | .98570 | 18 |
| 43 | .09961 | .99503 | .11696 | .99314 | .13427 | .99094 | .15155 | .98845 | .16878 | .98565 | 17 |
| 44 | .09990 | .99500 | .11725 | .99310 | .13456 | .99091 | .15184 | .98841 | .16906 | .98561 | 16 |
| 45 | .10019 | .99497 | .11754 | .99307 | .13485 | .99087 | .15212 | .98836 | .16935 | .98556 | 15 |
| 46 | .10048 | .99494 | .11783 | .99303 | .13514 | .99083 | .15241 | .98832 | .16964 | .98551 | 14 |
| 47 | .10077 | .99491 | .11812 | .99300 | .13543 | .99079 | .15270 | .98827 | .16992 | .98546 | 13 |
| 48 | .10106 | .99488 | .11840 | .99297 | .13572 | .99075 | .15299 | .98823 | .17021 | .98541 | 12 |
| 49 | .10135 | .99485 | .11869 | .99293 | .13600 | .99071 | .15327 | .98818 | .17050 | .98536 | 11 |
| 50 | .10164 | .99482 | .11898 | .99290 | .13629 | .99067 | .15356 | .98814 | .17078 | .98531 | 10 |
| 51 | .10192 | .99479 | .11927 | .99286 | .13658 | .99063 | .15385 | .98809 | .17107 | .98526 | 9 |
| 52 | .10221 | .99476 | .11956 | .99283 | .13687 | .99059 | .15414 | .98805 | .17136 | .98521 | 8 |
| 53 | .10250 | .99473 | .11985 | .99279 | .13716 | .99055 | .15442 | .98800 | .17164 | .98516 | 7 |
| 54 | .10279 | .99470 | .12014 | .99276 | .13744 | .99051 | .15471 | .98796 | .17193 | .98511 | 6 |
| 55 | .10308 | .99467 | .12043 | .99272 | .13773 | .99047 | .15500 | .98791 | .17222 | .98506 | 5 |
| 56 | .10337 | .99464 | .12071 | .99269 | .13802 | .99043 | .15529 | .98787 | .17250 | .98501 | 4 |
| 57 | .10366 | .99461 | .12100 | .99265 | .13831 | .99039 | .15557 | .98782 | .17279 | .98496 | 3 |
| 58 | .10395 | .99458 | .12129 | .99262 | .13860 | .99035 | .15586 | .98778 | .17308 | .98491 | 2 |
| 59 | .10424 | .99455 | .12158 | .99258 | .13889 | .99031 | .15615 | .98773 | .17336 | .98486 | 1 |
| 60 | .10453 | .99452 | .12187 | .99255 | .13917 | .99027 | .15643 | .98769 | .17365 | .98481 | 0 |
| ′ | Cosine | Sine | Cosine | Sine | Cosine | Sine | Cosine | Sine | Cosine | Sine | ′ |
| | 84° | | 83° | | 82° | | 81° | | 80° | | |

**TABLE 1 cont.**

Natural Sines and Cosines

| ′ | 10° | | 11° | | 12° | | 13° | | 14° | | ′ |
|---|---|---|---|---|---|---|---|---|---|---|---|
| | Sine | Cosine | Sine | Cosine | Sine | Cosine | Sine | Cosine | Sine | Cosine | |
| 0 | .17365 | .98481 | .19081 | .98163 | .20791 | .97815 | .22495 | .97437 | .24192 | .97030 | 60 |
| 1 | .17393 | .98476 | .19109 | .98157 | .20820 | .97809 | .22523 | .97430 | .24220 | .97023 | 59 |
| 2 | .17422 | .98471 | .19138 | .98152 | .20848 | .97803 | .22552 | .97424 | .24249 | .97015 | 58 |
| 3 | .17451 | .98466 | .19167 | .98146 | .20877 | .97797 | .22580 | .97417 | .24277 | .97008 | 57 |
| 4 | .17479 | .98461 | .19195 | .98140 | .20905 | .97791 | .22608 | .97411 | .24305 | .97001 | 56 |
| 5 | .17508 | .98455 | .19224 | .98135 | .20933 | .97784 | .22637 | .97404 | .24333 | .96994 | 55 |
| 6 | .17537 | .98450 | .19252 | .98129 | .20962 | .97778 | .22665 | .97398 | .24362 | .96987 | 54 |
| 7 | .17565 | .98445 | .19281 | .98124 | .20990 | .97772 | .22693 | .97391 | .24390 | .96980 | 53 |
| 8 | .17594 | .98440 | .19309 | .98118 | .21019 | .97766 | .22722 | .97384 | .24418 | .96973 | 52 |
| 9 | .17623 | .98435 | .19338 | .98112 | .21047 | .97760 | .22750 | .97378 | .24446 | .96966 | 51 |
| 10 | .17651 | .98430 | .19366 | .98107 | .21076 | .97754 | .22778 | .97371 | .24474 | .96959 | 50 |
| 11 | .17680 | .98425 | .19395 | .98101 | .21104 | .97748 | .22807 | .97365 | .24503 | .96952 | 49 |
| 12 | .17708 | .98420 | .19423 | .98096 | .21132 | .97742 | .22835 | .97358 | .24531 | .96945 | 48 |
| 13 | .17737 | .98414 | .19452 | .98090 | .21161 | .97735 | .22863 | .97351 | .24559 | .96937 | 47 |
| 14 | .17766 | .98409 | .19481 | .98084 | .21189 | .97729 | .22892 | .97345 | .24587 | .96930 | 46 |
| 15 | .17794 | .98404 | .19509 | .98079 | .21218 | .97723 | .22920 | .97338 | .24615 | .96923 | 45 |
| 16 | .17823 | .98399 | .19538 | .98073 | .21246 | .97717 | .22948 | .97331 | .24644 | .96916 | 44 |
| 17 | .17852 | .98394 | .19566 | .98067 | .21275 | .97711 | .22977 | .97325 | .24672 | .96909 | 43 |
| 18 | .17880 | .98389 | .19595 | .98061 | .21303 | .97705 | .23005 | .97318 | .24700 | .96902 | 42 |
| 19 | .17909 | .98383 | .19623 | .98056 | .21331 | .97698 | .23033 | .97311 | .24728 | .96894 | 41 |
| 20 | .17937 | .98378 | .19652 | .98050 | .21360 | .97692 | .23062 | .97304 | .24756 | .96887 | 40 |
| 21 | .17966 | .98373 | .19680 | .98044 | .21388 | .97686 | .23090 | .97298 | .24784 | .96880 | 39 |
| 22 | .17995 | .98368 | .19709 | .98039 | .21417 | .97680 | .23118 | .97291 | .24813 | .96873 | 38 |
| 23 | .18023 | .98362 | .19737 | .98033 | .21445 | .97673 | .23146 | .97284 | .24841 | .96866 | 37 |
| 24 | .18052 | .98357 | .19766 | .98027 | .21474 | .97667 | .23175 | .97278 | .24869 | .96858 | 36 |
| 25 | .18081 | .98352 | .19794 | .98021 | .21502 | .97661 | .23203 | .97271 | .24897 | .96851 | 35 |
| 26 | .18109 | .98347 | .19823 | .98016 | .21530 | .97655 | .23231 | .97264 | .24925 | .96844 | 34 |
| 27 | .18138 | .98341 | .19851 | .98010 | .21559 | .97648 | .23260 | .97257 | .24954 | .96837 | 33 |
| 28 | .18166 | .98336 | .19880 | .98004 | .21587 | .97642 | .23288 | .97251 | .24982 | .96829 | 32 |
| 29 | .18195 | .98331 | .19908 | .97998 | .21616 | .97636 | .23316 | .97244 | .25010 | .96822 | 31 |
| 30 | .18224 | .98325 | .19937 | .97992 | .21644 | .97630 | .23345 | .97237 | .25038 | .96815 | 30 |
| 31 | .18252 | .98320 | .19965 | .97987 | .21672 | .97623 | .23373 | .97230 | .25066 | .96807 | 29 |
| 32 | .18281 | .98315 | .19994 | .97981 | .21701 | .97617 | .23401 | .97223 | .25094 | .96800 | 28 |
| 33 | .18309 | .98310 | .20022 | .97975 | .21729 | .97611 | .23429 | .97217 | .25122 | .96793 | 27 |
| 34 | .18338 | .98304 | .20051 | .97969 | .21758 | .97604 | .23458 | .97210 | .25151 | .96786 | 26 |
| 35 | .18367 | .98299 | .20079 | .97963 | .21786 | .97598 | .23486 | .97203 | .25179 | .96778 | 25 |
| 36 | .18395 | .98294 | .20108 | .97958 | .21814 | .97592 | .23514 | .97196 | .25207 | .96771 | 24 |
| 37 | .18424 | .98288 | .20136 | .97952 | .21843 | .97585 | .23542 | .97189 | .25235 | .96764 | 23 |
| 38 | .18452 | .98283 | .20165 | .97946 | .21871 | .97579 | .23571 | .97182 | .25263 | .96756 | 22 |
| 39 | .18481 | .98277 | .20193 | .97940 | .21899 | .97573 | .23599 | .97176 | .25291 | .96749 | 21 |
| 40 | .18509 | .98272 | .20222 | .97934 | .21928 | .97566 | .23627 | .97169 | .25320 | .96742 | 20 |
| 41 | .18538 | .98267 | .20250 | .97928 | .21956 | .97560 | .23656 | .97162 | .25348 | .96734 | 19 |
| 42 | .18567 | .98261 | .20279 | .97922 | .21985 | .97553 | .23684 | .97155 | .25376 | .96727 | 18 |
| 43 | .18595 | .98256 | .20307 | .97916 | .22013 | .97547 | .23712 | .97148 | .25404 | .96719 | 17 |
| 44 | .18624 | .98250 | .20336 | .97910 | .22041 | .97541 | .23740 | .97141 | .25432 | .96712 | 16 |
| 45 | .18652 | .98245 | .20364 | .97905 | .22070 | .97534 | .23769 | .97134 | .25460 | .96705 | 15 |
| 46 | .18681 | .98240 | .20393 | .97899 | .22098 | .97528 | .23797 | .97127 | .25488 | .96697 | 14 |
| 47 | .18710 | .98234 | .20421 | .97893 | .22126 | .97521 | .23825 | .97120 | .25516 | .96690 | 13 |
| 48 | .18738 | .98229 | .20450 | .97887 | .22155 | .97515 | .23853 | .97113 | .25545 | .96682 | 12 |
| 49 | .18767 | .98223 | .20478 | .97881 | .22183 | .97508 | .23832 | .97106 | .25573 | .96675 | 11 |
| 50 | .18795 | .98218 | .20507 | .97875 | .22212 | .97502 | .23910 | .97100 | .25601 | .96667 | 10 |
| 51 | .18824 | .98212 | .20535 | .97869 | .22240 | .97496 | .23938 | .97093 | .25629 | .96660 | 9 |
| 52 | .18852 | .98207 | .20563 | .97863 | .22268 | .97489 | .23966 | .97086 | .25657 | .96653 | 8 |
| 53 | .18881 | .98201 | .20592 | .97857 | .22297 | .97483 | .23995 | .97079 | .25685 | .96645 | 7 |
| 54 | .18910 | .98196 | .20620 | .97851 | .22325 | .97476 | .24023 | .97072 | .25713 | .96638 | 6 |
| 55 | .18938 | .98190 | .20649 | .97845 | .22353 | .97470 | .24051 | .97065 | .25741 | .96630 | 5 |
| 56 | .18967 | .98185 | .20677 | .97839 | .22382 | .97463 | .24079 | .97058 | .25769 | .96623 | 4 |
| 57 | .18995 | .98179 | .20706 | .97833 | .22410 | .97457 | .24108 | .97051 | .25798 | .96615 | 3 |
| 58 | .19024 | .98174 | .20734 | .97827 | .22438 | .97450 | .24136 | .97044 | .25826 | .96608 | 2 |
| 59 | .19052 | .98168 | .20763 | .97821 | .22467 | .97444 | .24164 | .97037 | .25854 | .96600 | 1 |
| 60 | .19081 | .98163 | .20791 | .97815 | .22495 | .97437 | .24192 | .97030 | .25882 | .96593 | 0 |
| ′ | Cosine | Sine | Cosine | Sine | Cosine | Sine | Cosine | Sine | Cosine | Sine | ′ |
| | 79° | | 78° | | 77° | | 76° | | 75° | | |

**TABLE 1 cont.**

Natural Sines and Cosines

| ′ | 15° | | 16° | | 17° | | 18° | | 19° | | ′ |
|---|---|---|---|---|---|---|---|---|---|---|---|
| | Sine | Cosine | Sine | Cosine | Sine | Cosine | Sine | Cosine | Sine | Cosine | |
| 0 | .25882 | .96593 | .27564 | .96126 | .29237 | .95630 | .30902 | .95106 | .32557 | .94552 | 60 |
| 1 | .25910 | .96585 | .27592 | .96118 | .29265 | .95622 | .30929 | .95097 | .32584 | .94542 | 59 |
| 2 | .25938 | .96578 | .27620 | .96110 | .29293 | .95613 | .30957 | .95088 | .32612 | .94533 | 58 |
| 3 | .25966 | .96570 | .27648 | .96102 | .29321 | .95605 | .30985 | .95079 | .32639 | .94523 | 57 |
| 4 | .25994 | .96562 | .27676 | .96094 | .29348 | .95596 | .31012 | .95070 | .32667 | .94514 | 56 |
| 5 | .26022 | .96555 | .27704 | .96086 | .29376 | .95588 | .31040 | .95061 | .32694 | .94504 | 55 |
| 6 | .26050 | .96547 | .27731 | .96078 | .29404 | .95579 | .31068 | .95052 | .32722 | .94495 | 54 |
| 7 | .26079 | .96540 | .27759 | .96070 | .29432 | .95571 | .31095 | .95043 | .32749 | .94485 | 53 |
| 8 | .26107 | .96532 | .27787 | .96062 | .29460 | .95562 | .31123 | .95033 | .32777 | .94476 | 52 |
| 9 | .26135 | .96524 | .27815 | .96054 | .29487 | .95554 | .31151 | .95024 | .32804 | .94466 | 51 |
| 10 | .26163 | .96517 | .27843 | .96046 | .29515 | .95545 | .31178 | .95015 | .32832 | .94457 | 50 |
| 11 | .26191 | .96509 | .27871 | .96037 | .29543 | .95536 | .31206 | .95006 | .32859 | .94447 | 49 |
| 12 | .26219 | .96502 | .27899 | .96029 | .29571 | .95528 | .31233 | .94997 | .32887 | .94438 | 48 |
| 13 | .26247 | .96494 | .27927 | .96021 | .29599 | .95519 | .31261 | .94988 | .32914 | .94428 | 47 |
| 14 | .26275 | .96486 | .27955 | .96013 | .29626 | .95511 | .31289 | .94979 | .32942 | .94418 | 46 |
| 15 | .26303 | .96479 | .27983 | .96005 | .29654 | .95502 | .31316 | .94970 | .32969 | .94409 | 45 |
| 16 | .26331 | .96471 | .28011 | .95997 | .29682 | .95493 | .31344 | .94961 | .32997 | .94399 | 44 |
| 17 | .26359 | .96463 | .28039 | .95989 | .29710 | .95485 | .31372 | .94952 | .33024 | .94390 | 43 |
| 18 | .26387 | .96456 | .28067 | .95981 | .29737 | .95476 | .31399 | .94943 | .33051 | .94380 | 42 |
| 19 | .26415 | .96448 | .28095 | .95972 | .29765 | .95467 | .31427 | .94933 | .33079 | .94370 | 41 |
| 20 | .26443 | .96440 | .28123 | .95964 | .29793 | .95459 | .31454 | .94924 | .33106 | .94361 | 40 |
| 21 | .26471 | .96433 | .28150 | .95956 | .29821 | .95450 | .31482 | .94915 | .33134 | .94351 | 39 |
| 22 | .26500 | .96425 | .28178 | .95948 | .29849 | .95441 | .31510 | .94906 | .33161 | .94342 | 38 |
| 23 | .26528 | .96417 | .28206 | .95940 | .29876 | .95433 | .31537 | .94897 | .33189 | .94332 | 37 |
| 24 | .26556 | .96410 | .28234 | .95931 | .29904 | .95424 | .31565 | .94888 | .33216 | .94322 | 36 |
| 25 | .26584 | .96402 | .28262 | .95923 | .29932 | .95415 | .31593 | .94878 | .33244 | .94313 | 35 |
| 26 | .26612 | .96394 | .28290 | .95915 | .29960 | .95407 | .31620 | .94869 | .33271 | .94303 | 34 |
| 27 | .26640 | .96386 | .28318 | .95907 | .29987 | .95398 | .31648 | .94860 | .33298 | .94293 | 33 |
| 28 | .26668 | .96379 | .28346 | .95898 | .30015 | .95389 | .31675 | .94851 | .33326 | .94284 | 32 |
| 29 | .26696 | .96371 | .28374 | .95890 | .30043 | .95380 | .31703 | .94842 | .33353 | .94274 | 31 |
| 30 | .26724 | .96363 | .28402 | .95882 | .30071 | .95372 | .31730 | .94832 | .33381 | .94264 | 30 |
| 31 | .26752 | .96355 | .28429 | .95874 | .30098 | .95363 | .31758 | .94823 | .33408 | .94254 | 29 |
| 32 | .26780 | .96347 | .28457 | .95865 | .30126 | .95354 | .31786 | .94814 | .33436 | .94245 | 28 |
| 33 | .26808 | .96340 | .28485 | .95857 | .30154 | .95345 | .31813 | .94805 | .33463 | .94235 | 27 |
| 34 | .26836 | .96332 | .28513 | .95849 | .30182 | .95337 | .31841 | .94795 | .33490 | .94225 | 26 |
| 35 | .26864 | .96324 | .28541 | .95841 | .30209 | .95328 | .31868 | .94786 | .33518 | .94215 | 25 |
| 36 | .26892 | .96316 | .28569 | .95832 | .30237 | .95319 | .31896 | .94777 | .33545 | .94206 | 24 |
| 37 | .26920 | .96308 | .28597 | .95824 | .30265 | .95310 | .31923 | .94768 | .33573 | .94196 | 23 |
| 38 | .26948 | .96301 | .28625 | .95816 | .30292 | .95301 | .31951 | .94758 | .33600 | .94186 | 22 |
| 39 | .26976 | .96293 | .28652 | .95807 | .30320 | .95293 | .31979 | .94749 | .33627 | .94176 | 21 |
| 40 | .27004 | .96285 | .28680 | .95799 | .30348 | .95284 | .32006 | .94740 | .33655 | .94167 | 20 |
| 41 | .27032 | .96277 | .28708 | .95791 | .30376 | .95275 | .32034 | .94730 | .33682 | .94157 | 19 |
| 42 | .27060 | .96269 | .28736 | .95782 | .30403 | .95266 | .32061 | .94721 | .33710 | .94147 | 18 |
| 43 | .27088 | .96261 | .28764 | .95774 | .30431 | .95257 | .32089 | .94712 | .33737 | .94137 | 17 |
| 44 | .27116 | .96253 | .28792 | .95766 | .30459 | .95248 | .32116 | .94702 | .33764 | .94127 | 16 |
| 45 | .27144 | .96246 | .28820 | .95757 | .30486 | .95240 | .32144 | .94693 | .33792 | .94118 | 15 |
| 46 | .27172 | .96238 | .28847 | .95749 | .30514 | .95231 | .32171 | .94684 | .33819 | .94108 | 14 |
| 47 | .27200 | .96230 | .28875 | .95740 | .30542 | .95222 | .32199 | .94674 | .33846 | .94098 | 13 |
| 48 | .27228 | .96222 | .28903 | .95732 | .30570 | .95213 | .32227 | .94665 | .33874 | .94088 | 12 |
| 49 | .27256 | .96214 | .28931 | .95724 | .30597 | .95204 | .32254 | .94656 | .33901 | .94078 | 11 |
| 50 | .27284 | .96206 | .28959 | .95715 | .30625 | .95195 | .32282 | .94646 | .33929 | .94068 | 10 |
| 51 | .27312 | .96198 | .28987 | .95707 | .30653 | .95186 | .32309 | .94637 | .33956 | .94058 | 9 |
| 52 | .27340 | .96190 | .29015 | .95698 | .30680 | .95177 | .32337 | .94627 | .33983 | .94049 | 8 |
| 53 | .27368 | .96182 | .29042 | .95690 | .30708 | .95168 | .32364 | .94618 | .34011 | .94039 | 7 |
| 54 | .27396 | .96174 | .29070 | .95681 | .30736 | .95159 | .32392 | .94609 | .34038 | .94029 | 6 |
| 55 | .27424 | .96166 | .29098 | .95673 | .30763 | .95150 | .32419 | .94599 | .34065 | .94019 | 5 |
| 56 | .27452 | .96158 | .29126 | .95664 | .30791 | .95142 | .32447 | .94590 | .34093 | .94009 | 4 |
| 57 | .27480 | .96150 | .29154 | .95656 | .30819 | .95133 | .32474 | .94580 | .34120 | .93999 | 3 |
| 58 | .27508 | .96142 | .29182 | .95647 | .30846 | .95124 | .32502 | .94571 | .34147 | .93989 | 2 |
| 59 | .27536 | .96134 | .29209 | .95639 | .30874 | .95115 | .32529 | .94561 | .34175 | .93979 | 1 |
| 60 | .27564 | .96126 | .29237 | .95630 | .30902 | .95106 | .32557 | .94552 | .34202 | .93969 | 0 |
| | Cosine | Sine | Cosine | Sine | Cosine | Sine | Cosine | Sine | Cosine | Sine | |
| ′ | 74° | | 73° | | 72° | | 71° | | 70° | | ′ |

## TABLE 1 cont.

Natural Sines and Cosines

| ′ | 20° | | 21° | | 22° | | 23° | | 24° | | ′ |
|---|---|---|---|---|---|---|---|---|---|---|---|
| | Sine | Cosine | Sine | Cosine | Sine | Cosine | Sine | Cosine | Sine | Cosine | |
| 0 | .34202 | .93969 | .35837 | .93358 | .37461 | .92718 | .39073 | .92050 | .40674 | .91355 | 60 |
| 1 | .34229 | .93959 | .35864 | .93348 | .37488 | .92707 | .39100 | .92039 | .40700 | .91343 | 59 |
| 2 | .34257 | .93949 | .35891 | .93337 | .37515 | .92697 | .39127 | .92028 | .40727 | .91331 | 58 |
| 3 | .34284 | .93939 | .35918 | .93327 | .37542 | .92686 | .39153 | .92016 | .40753 | .91319 | 57 |
| 4 | .34311 | .93929 | .35945 | .93316 | .37569 | .92675 | .39180 | .92005 | .40780 | .91307 | 56 |
| 5 | .34339 | .93919 | .35973 | .93306 | .37595 | .92664 | .39207 | .91994 | .40806 | .91295 | 55 |
| 6 | .34366 | .93909 | .36000 | .93295 | .37622 | .92653 | .39234 | .91982 | .40833 | .91283 | 54 |
| 7 | .34393 | .93899 | .36027 | .93285 | .37649 | .92642 | .39260 | .91971 | .40860 | .91272 | 53 |
| 8 | .34421 | .93889 | .36054 | .93274 | .37676 | .92631 | .39287 | .91959 | .40886 | .91260 | 52 |
| 9 | .34448 | .93879 | .36081 | .93264 | .37703 | .92620 | .39314 | .91948 | .40913 | .91248 | 51 |
| 10 | .34475 | .93869 | .36108 | .93253 | .37730 | .92609 | .39341 | .91936 | .40939 | .91236 | 50 |
| 11 | .34503 | .93859 | .36135 | .93243 | .37757 | .92598 | .39367 | .91925 | .40966 | .91224 | 49 |
| 12 | .34530 | .93849 | .36162 | .93232 | .37784 | .92587 | .39394 | .91914 | .40992 | .91212 | 48 |
| 13 | .34557 | .93839 | .36190 | .93222 | .37811 | .92576 | .39421 | .91902 | .41019 | .91200 | 47 |
| 14 | .34584 | .93829 | .36217 | .93211 | .37838 | .92565 | .39448 | .91891 | .41045 | .91188 | 46 |
| 15 | .34612 | .93819 | .36244 | .93201 | .37865 | .92554 | .39474 | .91879 | .41072 | .91176 | 45 |
| 16 | .34639 | .93809 | .36271 | .93190 | .37892 | .92543 | .39501 | .91868 | .41098 | .91164 | 44 |
| 17 | .34666 | .93799 | .36298 | .93180 | .37919 | .92532 | .39528 | .91856 | .41125 | .91152 | 43 |
| 18 | .34694 | .93789 | .36325 | .93169 | .37946 | .92521 | .39555 | .91845 | .41151 | .91140 | 42 |
| 19 | .34721 | .93779 | .36352 | .93159 | .37973 | .92510 | .39581 | .91833 | .41178 | .91128 | 41 |
| 20 | .34748 | .93769 | .36379 | .93148 | .37999 | .92499 | .39608 | .91822 | .41204 | .91116 | 40 |
| 21 | .34775 | .93759 | .36406 | .93137 | .38026 | .92488 | .39635 | .91810 | .41231 | .91104 | 39 |
| 22 | .34803 | .93748 | .36434 | .93127 | .38053 | .92477 | .39661 | .91799 | .41257 | .91092 | 38 |
| 23 | .34830 | .93738 | .36461 | .93116 | .38080 | .92466 | .39688 | .91787 | .41284 | .91080 | 37 |
| 24 | .34857 | .93728 | .36488 | .93106 | .38107 | .92455 | .39715 | .91775 | .41310 | .91068 | 36 |
| 25 | .34884 | .93718 | .36515 | .93095 | .38134 | .92444 | .39741 | .91764 | .41337 | .91056 | 35 |
| 26 | .34912 | .93708 | .36542 | .93084 | .38161 | .92432 | .39768 | .91752 | .41363 | .91044 | 34 |
| 27 | .34939 | .93698 | .36569 | .93074 | .38188 | .92421 | .39795 | .91741 | .41390 | .91032 | 33 |
| 28 | .34966 | .93688 | .36596 | .93063 | .38215 | .92410 | .39822 | .91729 | .41416 | .91020 | 32 |
| 29 | .34993 | .93677 | .36623 | .93052 | .38241 | .92399 | .39848 | .91718 | .41443 | .91008 | 31 |
| 30 | .35021 | .93667 | .36650 | .93042 | .38268 | .92388 | .39875 | .91706 | .41469 | .90996 | 30 |
| 31 | .35048 | .93657 | .36677 | .93031 | .38295 | .92377 | .39902 | .91694 | .41496 | .90984 | 29 |
| 32 | .35075 | .93647 | .36704 | .93020 | .38322 | .92366 | .39928 | .91683 | .41522 | .90972 | 28 |
| 33 | .35102 | .93637 | .36731 | .93010 | .38349 | .92355 | .39955 | .91671 | .41549 | .90960 | 27 |
| 34 | .35130 | .93626 | .36758 | .92999 | .38376 | .92343 | .39982 | .91660 | .41575 | .90948 | 26 |
| 35 | .35157 | .93616 | .36785 | .92988 | .38403 | .92332 | .40008 | .91648 | .41602 | .90936 | 25 |
| 36 | .35184 | .93606 | .36812 | .92978 | .38430 | .92321 | .40035 | .91636 | .41628 | .90924 | 24 |
| 37 | .35211 | .93596 | .36839 | .92967 | .38456 | .92310 | .40062 | .91625 | .41655 | .90911 | 23 |
| 38 | .35239 | .93585 | .36867 | .92956 | .38483 | .92299 | .40088 | .91613 | .41681 | .90899 | 22 |
| 39 | .35266 | .93575 | .36894 | .92945 | .38510 | .92287 | .40115 | .91601 | .41707 | .90887 | 21 |
| 40 | .35293 | .93565 | .36921 | .92935 | .38537 | .92276 | .40141 | .91590 | .41734 | .90875 | 20 |
| 41 | .35320 | .93555 | .36948 | .92924 | .38564 | .92265 | .40168 | .91578 | .41760 | .90863 | 19 |
| 42 | .35347 | .93544 | .36975 | .92913 | .38591 | .92254 | .40195 | .91566 | .41787 | .90851 | 18 |
| 43 | .35375 | .93534 | .37002 | .92902 | .38617 | .92243 | .40221 | .91555 | .41813 | .90839 | 17 |
| 44 | .35402 | .93524 | .37029 | .92892 | .38644 | .92231 | .40248 | .91543 | .41840 | .90826 | 16 |
| 45 | .35429 | .93514 | .37056 | .92881 | .38671 | .92220 | .40275 | .91531 | .41866 | .90814 | 15 |
| 46 | .35456 | .93503 | .37083 | .92870 | .38698 | .92209 | .40301 | .91519 | .41892 | .90802 | 14 |
| 47 | .35484 | .93493 | .37110 | .92859 | .38725 | .92198 | .40328 | .91508 | .41919 | .90790 | 13 |
| 48 | .35511 | .93483 | .37137 | .92849 | .38752 | .92186 | .40355 | .91496 | .41945 | .90778 | 12 |
| 49 | .35538 | .93472 | .37164 | .92838 | .38778 | .92175 | .40381 | .91484 | .41972 | .90766 | 11 |
| 50 | .35565 | .93462 | .37191 | .92827 | .38805 | .92164 | .40408 | .91472 | .41998 | .90753 | 10 |
| 51 | .35592 | .93452 | .37218 | .92816 | .38832 | .92152 | .40434 | .91461 | .42024 | .90741 | 9 |
| 52 | .35619 | .93441 | .37245 | .92805 | .38859 | .92141 | .40461 | .91449 | .42051 | .90729 | 8 |
| 53 | .35647 | .93431 | .37272 | .92794 | .38886 | .92130 | .40488 | .91437 | .42077 | .90717 | 7 |
| 54 | .35674 | .93420 | .37299 | .92784 | .38912 | .92119 | .40514 | .91425 | .42104 | .90704 | 6 |
| 55 | .35701 | .93410 | .37326 | .92773 | .38939 | .92107 | .40541 | .91414 | .42130 | .90692 | 5 |
| 56 | .35728 | .93400 | .37353 | .92762 | .38966 | .92096 | .40567 | .91402 | .42156 | .90680 | 4 |
| 57 | .35755 | .93389 | .37380 | .92751 | .38993 | .92085 | .40594 | .91390 | .42183 | .90668 | 3 |
| 58 | .35782 | .93379 | .37407 | .92740 | .39020 | .92073 | .40621 | .91378 | .42209 | .90655 | 2 |
| 59 | .35810 | .93368 | .37434 | .92729 | .39046 | .92062 | .40647 | .91366 | .42235 | .90643 | 1 |
| 60 | .35837 | .93358 | .37461 | .92718 | .39073 | .92050 | .40674 | .91355 | .42262 | .90631 | 0 |
| ′ | Cosine | Sine | Cosine | Sine | Cosine | Sine | Cosine | Sine | Cosine | Sine | ′ |
| | 69° | | 68° | | 67° | | 66° | | 65° | | |

**TABLE 1 cont.**

Natural Sines and Cosines

| ′ | 25° Sine | 25° Cosine | 26° Sine | 26° Cosine | 27° Sine | 27° Cosine | 28° Sine | 28° Cosine | 29° Sine | 29° Cosine | ′ |
|---|---|---|---|---|---|---|---|---|---|---|---|
| 0 | .42262 | .90631 | .43837 | .89879 | .45399 | .89101 | .46947 | .88295 | .48481 | .87462 | 60 |
| 1 | .42288 | .90618 | .43863 | .89867 | .45425 | .89087 | .46973 | .88281 | .48506 | .87448 | 59 |
| 2 | .42315 | .90606 | .43889 | .89854 | .45451 | .89074 | .46999 | .88267 | .48532 | .87434 | 58 |
| 3 | .42341 | .90594 | .43916 | .89841 | .45477 | .89061 | .47024 | .88254 | .48557 | .87420 | 57 |
| 4 | .42367 | .90582 | .43942 | .89828 | .45503 | .89048 | .47050 | .88240 | .48583 | .87406 | 56 |
| 5 | .42394 | .90569 | .43968 | .89816 | .45529 | .89035 | .47076 | .88226 | .48608 | .87391 | 55 |
| 6 | .42420 | .90557 | .43994 | .89803 | .45554 | .89021 | .47101 | .88213 | .48634 | .87377 | 54 |
| 7 | .42446 | .90545 | .44020 | .89790 | .45580 | .89008 | .47127 | .88199 | .48659 | .87363 | 53 |
| 8 | .42473 | .90532 | .44046 | .89777 | .45606 | .88995 | .47153 | .88185 | .48684 | .87349 | 52 |
| 9 | .42499 | .90520 | .44072 | .89764 | .45632 | .88981 | .47178 | .88172 | .48710 | .87335 | 51 |
| 10 | .42525 | .90507 | .44098 | .89752 | .45658 | .88968 | .47204 | .88158 | .48735 | .87321 | 50 |
| 11 | .42552 | .90495 | .44124 | .89739 | .45684 | .88955 | .47229 | .88144 | .48761 | .87306 | 49 |
| 12 | .42578 | .90483 | .44151 | .89726 | .45710 | .88942 | .47255 | .88130 | .48786 | .87292 | 48 |
| 13 | .42604 | .90470 | .44177 | .89713 | .45736 | .88928 | .47281 | .88117 | .48811 | .87278 | 47 |
| 14 | .42631 | .90458 | .44203 | .89700 | .45762 | .88915 | .47306 | .88103 | .48837 | .87264 | 46 |
| 15 | .42657 | .90446 | .44229 | .89687 | .45787 | .88902 | .47332 | .88089 | .48862 | .87250 | 45 |
| 16 | .42683 | .90433 | .44255 | .89674 | .45813 | .88888 | .47358 | .88075 | .48888 | .87235 | 44 |
| 17 | .42709 | .90421 | .44281 | .89662 | .45839 | .88875 | .47383 | .88062 | .48913 | .87221 | 43 |
| 18 | .42736 | .90408 | .44307 | .89649 | .45865 | .88862 | .47409 | .88048 | .48938 | .87207 | 42 |
| 19 | .42762 | .90396 | .44333 | .89636 | .45891 | .88848 | .47434 | .88034 | .48964 | .87193 | 41 |
| 20 | .42788 | .90383 | .44359 | .89623 | .45917 | .88835 | .47460 | .88020 | .48989 | .87178 | 40 |
| 21 | .42815 | .90371 | .44385 | .89610 | .45942 | .88822 | .47486 | .88006 | .49014 | .87164 | 39 |
| 22 | .42841 | .90358 | .44411 | .89597 | .45968 | .88808 | .47511 | .87993 | .49040 | .87150 | 38 |
| 23 | .42867 | .90346 | .44437 | .89584 | .45994 | .88795 | .47537 | .87979 | .49065 | .87136 | 37 |
| 24 | .42894 | .90334 | .44464 | .89571 | .46020 | .88782 | .47562 | .87965 | .49090 | .87121 | 36 |
| 25 | .42920 | .90321 | .44490 | .89558 | .46046 | .88768 | .47588 | .87951 | .49116 | .87107 | 35 |
| 26 | .42946 | .90309 | .44516 | .89545 | .46072 | .88755 | .47614 | .87937 | .49141 | .87093 | 34 |
| 27 | .42972 | .90296 | .44542 | .89532 | .46097 | .88741 | .47639 | .87923 | .49166 | .87079 | 33 |
| 28 | .42999 | .90284 | .44568 | .89519 | .46123 | .88728 | .47665 | .87909 | .49192 | .87064 | 32 |
| 29 | .43025 | .90271 | .44594 | .89506 | .46149 | .88715 | .47690 | .87896 | .49217 | .87050 | 31 |
| 30 | .43051 | .90259 | .44620 | .89493 | .46175 | .88701 | .47716 | .87882 | .49242 | .87036 | 30 |
| 31 | .43077 | .90246 | .44646 | .89480 | .46201 | .88688 | .47741 | .87868 | .49268 | .87021 | 29 |
| 32 | .43104 | .90233 | .44672 | .89467 | .46226 | .88674 | .47767 | .87854 | .49293 | .87007 | 28 |
| 33 | .43130 | .90221 | .44698 | .89454 | .46252 | .88661 | .47793 | .87840 | .49318 | .86993 | 27 |
| 34 | .43156 | .90208 | .44724 | .89441 | .46278 | .88647 | .47818 | .87826 | .49344 | .86978 | 26 |
| 35 | .43182 | .90196 | .44750 | .89428 | .46304 | .88634 | .47844 | .87812 | .49369 | .86964 | 25 |
| 36 | .43209 | .90183 | .44776 | .89415 | .46330 | .88620 | .47869 | .87798 | .49394 | .86949 | 24 |
| 37 | .43235 | .90171 | .44802 | .89402 | .46355 | .88607 | .47895 | .87784 | .49419 | .86935 | 23 |
| 38 | .43261 | .90158 | .44828 | .89389 | .46381 | .88593 | .47920 | .87770 | .49445 | .86921 | 22 |
| 39 | .43287 | .90146 | .44854 | .89376 | .46407 | .88580 | .47946 | .87756 | .49470 | .86906 | 21 |
| 40 | .43313 | .90133 | .44880 | .89363 | .46433 | .88566 | .47971 | .87743 | .49495 | .86892 | 20 |
| 41 | .43340 | .90120 | .44906 | .89350 | .46458 | .88553 | .47997 | .87729 | .49521 | .86878 | 19 |
| 42 | .43366 | .90108 | .44932 | .89337 | .46484 | .88539 | .48022 | .87715 | .49546 | .86863 | 18 |
| 43 | .43392 | .90095 | .44958 | .89324 | .46510 | .88526 | .48048 | .87701 | .49571 | .86849 | 17 |
| 44 | .43418 | .90082 | .44984 | .89311 | .46536 | .88512 | .48073 | .87687 | .49596 | .86834 | 16 |
| 45 | .43445 | .90070 | .45010 | .89298 | .46561 | .88499 | .48099 | .87673 | .49622 | .86820 | 15 |
| 46 | .43471 | .90057 | .45036 | .89285 | .46587 | .88485 | .48124 | .87659 | .49647 | .86805 | 14 |
| 47 | .43497 | .90045 | .45062 | .89272 | .46613 | .88472 | .48150 | .87645 | .49672 | .86791 | 13 |
| 48 | .43523 | .90032 | .45088 | .89259 | .46639 | .88458 | .48175 | .87631 | .49697 | .86777 | 12 |
| 49 | .43549 | .90019 | .45114 | .89245 | .46664 | .88445 | .48201 | .87617 | .49723 | .86762 | 11 |
| 50 | .43575 | .90007 | .45140 | .89232 | .46690 | .88431 | .48226 | .87603 | .49748 | .86748 | 10 |
| 51 | .43602 | .89994 | .45166 | .89219 | .46716 | .88417 | .48252 | .87589 | .49773 | .86733 | 9 |
| 52 | .43628 | .89981 | .45192 | .89206 | .46742 | .88404 | .48277 | .87575 | .49798 | .86719 | 8 |
| 53 | .43654 | .89968 | .45218 | .89193 | .46767 | .88390 | .48303 | .87561 | .49824 | .86704 | 7 |
| 54 | .43680 | .89956 | .45243 | .89180 | .46793 | .88377 | .48328 | .87546 | .49849 | .86690 | 6 |
| 55 | .43706 | .89943 | .45269 | .89167 | .46819 | .88363 | .48354 | .87532 | .49874 | .86675 | 5 |
| 56 | .43733 | .89930 | .45295 | .89153 | .46844 | .88349 | .48379 | .87518 | .49899 | .86661 | 4 |
| 57 | .43759 | .89918 | .45321 | .89140 | .46870 | .88336 | .48405 | .87504 | .49924 | .86646 | 3 |
| 58 | .43785 | .89905 | .45347 | .89127 | .46896 | .88322 | .48430 | .87490 | .49950 | .86632 | 2 |
| 59 | .43811 | .89892 | .45373 | .89114 | .46921 | .88308 | .48456 | .87476 | .49975 | .86617 | 1 |
| 60 | .43837 | .89879 | .45399 | .89101 | .46947 | .88295 | .48481 | .87462 | .50000 | .86603 | 0 |
| ′ | Cosine 64° | Sine 64° | Cosine 63° | Sine 63° | Cosine 62° | Sine 62° | Cosine 61° | Sine 61° | Cosine 60° | Sine 60° | ′ |

**TABLE 1 cont.**
Natural Sines and Cosines

| ′ | 30° | | 31° | | 32° | | 33° | | 34° | | ′ |
|---|---|---|---|---|---|---|---|---|---|---|---|
| | Sine | Cosine | Sine | Cosine | Sine | Cosine | Sine | Cosine | Sine | Cosine | |
| 0 | .50000 | .86603 | .51504 | .85717 | .52992 | .84805 | .54464 | .83867 | .55919 | .82904 | 60 |
| 1 | .50025 | .86588 | .51529 | .85702 | .53017 | .84789 | .54488 | .83851 | .55943 | .82887 | 59 |
| 2 | .50050 | .86573 | .51554 | .85687 | .53041 | .84774 | .54513 | .83835 | .55968 | .82871 | 58 |
| 3 | .50076 | .86559 | .51579 | .85672 | .53066 | .84759 | .54537 | .83819 | .55992 | .82855 | 57 |
| 4 | .50101 | .86544 | .51604 | .85657 | .53091 | .84743 | .54561 | .83804 | .56016 | .82839 | 56 |
| 5 | .50126 | .86530 | .51628 | .85642 | .53115 | .84728 | .54586 | .83788 | .56040 | .82822 | 55 |
| 6 | .50151 | .86515 | .51653 | .85627 | .53140 | .84712 | .54610 | .83772 | .56064 | .82806 | 54 |
| 7 | .50176 | .86501 | .51678 | .85612 | .53164 | .84697 | .54635 | .83756 | .56088 | .82790 | 53 |
| 8 | .50201 | .86486 | .51703 | .85597 | .53189 | .84681 | .54659 | .83740 | .56112 | .82773 | 52 |
| 9 | .50227 | .86471 | .51728 | .85582 | .53214 | .84666 | .54683 | .83724 | .56136 | .82757 | 51 |
| 10 | .50252 | .86457 | .51753 | .85567 | .53238 | .84650 | .54708 | .83708 | .56160 | .82741 | 50 |
| 11 | .50277 | .86442 | .51778 | .85551 | .53263 | .84635 | .54732 | .83692 | .56184 | .82724 | 49 |
| 12 | .50302 | .86427 | .51803 | .85536 | .53288 | .84619 | .54756 | .83676 | .56208 | .82708 | 48 |
| 13 | .50327 | .86413 | .51828 | .85521 | .53312 | .84604 | .54781 | .83660 | .56232 | .82692 | 47 |
| 14 | .50352 | .86398 | .51852 | .85506 | .53337 | .84588 | .54805 | .83645 | .56256 | .82675 | 46 |
| 15 | .50377 | .86384 | .51877 | .85491 | .53361 | .84573 | .54829 | .83629 | .56280 | .82659 | 45 |
| 16 | .50403 | .86369 | .51902 | .85476 | .53386 | .84557 | .54854 | .83613 | .56305 | .82643 | 44 |
| 17 | .50428 | .86354 | .51927 | .85461 | .53411 | .84542 | .54878 | .83597 | .56329 | .82626 | 43 |
| 18 | .50453 | .86340 | .51952 | .85446 | .53435 | .84526 | .54902 | .83581 | .56353 | .82610 | 42 |
| 19 | .50478 | .86325 | .51977 | .85431 | .53460 | .84511 | .54927 | .83565 | .56377 | .82593 | 41 |
| 20 | .50503 | .86310 | .52002 | .85416 | .53484 | .84495 | .54951 | .83549 | .56401 | .82577 | 40 |
| 21 | .50528 | .86295 | .52026 | .85401 | .53509 | .84480 | .54975 | .83533 | .56425 | .82561 | 39 |
| 22 | .50553 | .86281 | .52051 | .85385 | .53534 | .84464 | .54999 | .83517 | .56449 | .82544 | 38 |
| 23 | .50578 | .86266 | .52076 | .85370 | .53558 | .84448 | .55024 | .83501 | .56473 | .82528 | 37 |
| 24 | .50603 | .86251 | .52101 | .85355 | .53583 | .84433 | .55048 | .83485 | .56497 | .82511 | 36 |
| 25 | .50628 | .86237 | .52126 | .85340 | .53607 | .84417 | .55072 | .83469 | .56521 | .82495 | 35 |
| 26 | .50654 | .86222 | .52151 | .85325 | .53632 | .84402 | .55097 | .83453 | .56545 | .82478 | 34 |
| 27 | .50679 | .86207 | .52175 | .85310 | .53656 | .84386 | .55121 | .83437 | .56569 | .82462 | 33 |
| 28 | .50704 | .86192 | .52200 | .85294 | .53681 | .84370 | .55145 | .83421 | .56593 | .82446 | 32 |
| 29 | .50729 | .86178 | .52225 | .85279 | .53705 | .84355 | .55169 | .83405 | .56617 | .82429 | 31 |
| 30 | .50754 | .86163 | .52250 | .85264 | .53730 | .84339 | .55194 | .83389 | .56641 | .82413 | 30 |
| 31 | .50779 | .86148 | .52275 | .85249 | .53754 | .84324 | .55218 | .83373 | .56665 | .82396 | 29 |
| 32 | .50804 | .86133 | .52299 | .85234 | .53779 | .84308 | .55242 | .83356 | .56689 | .82380 | 28 |
| 33 | .50829 | .86119 | .52324 | .85218 | .53804 | .84292 | .55266 | .83340 | .56713 | .82363 | 27 |
| 34 | .50854 | .86104 | .52349 | .85203 | .53828 | .84277 | .55291 | .83324 | .56736 | .82347 | 26 |
| 35 | .50879 | .86089 | .52374 | .85188 | .53853 | .84261 | .55315 | .83308 | .56760 | .82330 | 25 |
| 36 | .50904 | .86074 | .52399 | .85173 | .53877 | .84245 | .55339 | .83292 | .56784 | .82314 | 24 |
| 37 | .50929 | .86059 | .52423 | .85157 | .53902 | .84230 | .55363 | .83276 | .56808 | .82297 | 23 |
| 38 | .50954 | .86045 | .52448 | .85142 | .53926 | .84214 | .55388 | .83260 | .56832 | .82281 | 22 |
| 39 | .50979 | .86030 | .52473 | .85127 | .53951 | .84198 | .55412 | .83244 | .56856 | .82264 | 21 |
| 40 | .51004 | .86015 | .52498 | .85112 | .53975 | .84182 | .55436 | .83228 | .56880 | .82248 | 20 |
| 41 | .51029 | .86000 | .52522 | .85096 | .54000 | .84167 | .55460 | .83212 | .56904 | .82231 | 19 |
| 42 | .51054 | .85985 | .52547 | .85081 | .54024 | .84151 | .55484 | .83195 | .56928 | .82214 | 18 |
| 43 | .51079 | .85970 | .52572 | .85066 | .54049 | .84135 | .55509 | .83179 | .56952 | .82198 | 17 |
| 44 | .51104 | .85956 | .52597 | .85051 | .54073 | .84120 | .55533 | .83163 | .56976 | .82181 | 16 |
| 45 | .51129 | .85941 | .52621 | .85035 | .54097 | .84104 | .55557 | .83147 | .57000 | .82165 | 15 |
| 46 | .51154 | .85926 | .52646 | .85020 | .54122 | .84088 | .55581 | .83131 | .57024 | .82148 | 14 |
| 47 | .51179 | .85911 | .52671 | .85005 | .54146 | .84072 | .55605 | .83115 | .57047 | .82132 | 13 |
| 48 | .51204 | .85896 | .52696 | .84989 | .54171 | .84057 | .55630 | .83098 | .57071 | .82115 | 12 |
| 49 | .51229 | .85881 | .52720 | .84974 | .54195 | .84041 | .55654 | .83082 | .57095 | .82098 | 11 |
| 50 | .51254 | .85866 | .52745 | .84959 | .54220 | .84025 | .55678 | .83066 | .57119 | .82082 | 10 |
| 51 | .51279 | .85851 | .52770 | .84943 | .54244 | .84009 | .55702 | .83050 | .57143 | .82065 | 9 |
| 52 | .51304 | .85836 | .52794 | .84928 | .54269 | .83994 | .55726 | .83034 | .57167 | .82048 | 8 |
| 53 | .51329 | .85821 | .52819 | .84913 | .54293 | .83978 | .55750 | .83017 | .57191 | .82032 | 7 |
| 54 | .51354 | .85806 | .52844 | .84897 | .54317 | .83962 | .55775 | .83001 | .57215 | .82015 | 6 |
| 55 | .51379 | .85792 | .52869 | .84882 | .54342 | .83946 | .55799 | .82985 | .57238 | .81999 | 5 |
| 56 | .51404 | .85777 | .52893 | .84866 | .54366 | .83930 | .55823 | .82969 | .57262 | .81982 | 4 |
| 57 | .51429 | .85762 | .52918 | .84851 | .54391 | .83915 | .55847 | .82953 | .57286 | .81965 | 3 |
| 58 | .51454 | .85747 | .52943 | .84836 | .54415 | .83899 | .55871 | .82936 | .57310 | .81949 | 2 |
| 59 | .51479 | .85732 | .52967 | .84820 | .54440 | .83883 | .55895 | .82920 | .57334 | .81932 | 1 |
| 60 | .51504 | .85717 | .52992 | .84805 | .54464 | .83867 | .55919 | .82904 | .57358 | .81915 | 0 |
| ′ | Cosine | Sine | Cosine | Sine | Cosine | Sine | Cosine | Sine | Cosine | Sine | ′ |
| | 59° | | 58° | | 57° | | 56° | | 55° | | |

**TABLE 1 cont.**
Natural Sines and Cosines

| ′ | 35° Sine | 35° Cosine | 36° Sine | 36° Cosine | 37° Sine | 37° Cosine | 38° Sine | 38° Cosine | 39° Sine | 39° Cosine | ′ |
|---|---|---|---|---|---|---|---|---|---|---|---|
| 0 | .57358 | .81915 | .58779 | .80902 | .60182 | .79864 | .61566 | .78801 | .62932 | .77715 | 60 |
| 1 | .57381 | .81899 | .58802 | .80885 | .60205 | .79846 | .61589 | .78783 | .62955 | .77696 | 59 |
| 2 | .57405 | .81882 | .58826 | .80867 | .60228 | .79829 | .61612 | .78765 | .62977 | .77678 | 58 |
| 3 | .57429 | .81865 | .58849 | .80850 | .60251 | .79811 | .61635 | .78747 | .63000 | .77660 | 57 |
| 4 | .57453 | .81848 | .58873 | .80833 | .60274 | .79793 | .61658 | .78729 | .63022 | .77641 | 56 |
| 5 | .57477 | .81832 | .58896 | .80816 | .60298 | .79776 | .61681 | .78711 | .63045 | .77623 | 55 |
| 6 | .57501 | .81815 | .58920 | .80799 | .60321 | .79758 | .61704 | .78694 | .63068 | .77605 | 54 |
| 7 | .57524 | .81798 | .58943 | .80782 | .60344 | .79741 | .61726 | .78676 | .63090 | .77586 | 53 |
| 8 | .57548 | .81782 | .58967 | .80765 | .60367 | .79723 | .61749 | .78658 | .63113 | .77568 | 52 |
| 9 | .57572 | .81765 | .58990 | .80748 | .60390 | .79706 | .61772 | .78640 | .63135 | .77550 | 51 |
| 10 | .57596 | .81748 | .59014 | .80730 | .60414 | .79688 | .61795 | .78622 | .63158 | .77531 | 50 |
| 11 | .57619 | .81731 | .59037 | .80713 | .60437 | .79671 | .61818 | .78604 | .63180 | .77513 | 49 |
| 12 | .57643 | .81714 | .59061 | .80696 | .60460 | .79653 | .61841 | .78586 | .63203 | .77494 | 48 |
| 13 | .57667 | .81698 | .59084 | .80679 | .60483 | .79635 | .61864 | .78568 | .63225 | .77476 | 47 |
| 14 | .57691 | .81681 | .59108 | .80662 | .60506 | .79618 | .61887 | .78550 | .63248 | .77458 | 46 |
| 15 | .57715 | .81664 | .59131 | .80644 | .60529 | .79600 | .61909 | .78532 | .63271 | .77439 | 45 |
| 16 | .57738 | .81647 | .59154 | .80627 | .60553 | .79583 | .61932 | .78514 | .63293 | .77421 | 44 |
| 17 | .57762 | .81631 | .59178 | .80610 | .60576 | .79565 | .61955 | .78496 | .63316 | .77402 | 43 |
| 18 | .57786 | .81614 | .59201 | .80593 | .60599 | .79547 | .61978 | .78478 | .63338 | .77384 | 42 |
| 19 | .57810 | .81597 | .59225 | .80576 | .60622 | .79530 | .62001 | .78460 | .63361 | .77366 | 41 |
| 20 | .57833 | .81580 | .59248 | .80558 | .60645 | .79512 | .62024 | .78442 | .63383 | .77347 | 40 |
| 21 | .57857 | .81563 | .59272 | .80541 | .60668 | .79494 | .62046 | .78424 | .63406 | .77329 | 39 |
| 22 | .57881 | .81546 | .59295 | .80524 | .60691 | .79477 | .62069 | .78405 | .63428 | .77310 | 38 |
| 23 | .57904 | .81530 | .59318 | .80507 | .60714 | .79459 | .62092 | .78387 | .63451 | .77292 | 37 |
| 24 | .57928 | .81513 | .59342 | .80489 | .60738 | .79441 | .62115 | .78369 | .63473 | .77273 | 36 |
| 25 | .57952 | .81496 | .59365 | .80472 | .60761 | .79424 | .62138 | .78351 | .63496 | .77255 | 35 |
| 26 | .57976 | .81479 | .59389 | .80455 | .60784 | .79406 | .62160 | .78333 | .63518 | .77236 | 34 |
| 27 | .57999 | .81462 | .59412 | .80438 | .60807 | .79388 | .62183 | .78315 | .63540 | .77218 | 33 |
| 28 | .58023 | .81445 | .59436 | .80420 | .60830 | .79371 | .62206 | .78297 | .63563 | .77199 | 32 |
| 29 | .58047 | .81428 | .59459 | .80403 | .60853 | .79353 | .62229 | .78279 | .63585 | .77181 | 31 |
| 30 | .58070 | .81412 | .59482 | .80386 | .60876 | .79335 | .62251 | .78261 | .63608 | .77162 | 30 |
| 31 | .58094 | .81395 | .59506 | .80368 | .60899 | .79318 | .62274 | .78243 | .63630 | .77144 | 29 |
| 32 | .58118 | .81378 | .59529 | .80351 | .60922 | .79300 | .62297 | .78225 | .63653 | .77125 | 28 |
| 33 | .58141 | .81361 | .59552 | .80334 | .60945 | .79282 | .62320 | .78206 | .63675 | .77107 | 27 |
| 34 | .58165 | .81344 | .59576 | .80316 | .60968 | .79264 | .62342 | .78188 | .63698 | .77088 | 26 |
| 35 | .58189 | .81327 | .59599 | .80299 | .60991 | .79247 | .62365 | .78170 | .63720 | .77070 | 25 |
| 36 | .58212 | .81310 | .59622 | .80282 | .61015 | .79229 | .62388 | .78152 | .63742 | .77051 | 24 |
| 37 | .58236 | .81293 | .59646 | .80264 | .61038 | .79211 | .62411 | .78134 | .63765 | .77033 | 23 |
| 38 | .58260 | .81276 | .59669 | .80247 | .61061 | .79193 | .62433 | .78116 | .63787 | .77014 | 22 |
| 39 | .58283 | .81259 | .59693 | .80230 | .61084 | .79176 | .62456 | .78098 | .63810 | .76996 | 21 |
| 40 | .58307 | .81242 | .59716 | .80212 | .61107 | .79158 | .62479 | .78079 | .63832 | .76977 | 20 |
| 41 | .58330 | .81225 | .59739 | .80195 | .61130 | .79140 | .62502 | .78061 | .63854 | .76959 | 19 |
| 42 | .58354 | .81208 | .59763 | .80178 | .61153 | .79122 | .62524 | .78043 | .63877 | .76940 | 18 |
| 43 | .58378 | .81191 | .59786 | .80160 | .61176 | .79105 | .62547 | .78025 | .63899 | .76921 | 17 |
| 44 | .58401 | .81174 | .59809 | .80143 | .61199 | .79087 | .62570 | .78007 | .63922 | .76903 | 16 |
| 45 | .58425 | .81157 | .59832 | .80125 | .61222 | .79069 | .62592 | .77988 | .63944 | .76884 | 15 |
| 46 | .58449 | .81140 | .59856 | .80108 | .61245 | .79051 | .62615 | .77970 | .63966 | .76866 | 14 |
| 47 | .58472 | .81123 | .59879 | .80091 | .61268 | .79033 | .62638 | .77952 | .63989 | .76847 | 13 |
| 48 | .58496 | .81106 | .59902 | .80073 | .61291 | .79016 | .62660 | .77934 | .64011 | .76828 | 12 |
| 49 | .58519 | .81089 | .59926 | .80056 | .61314 | .78998 | .62683 | .77916 | .64033 | .76810 | 11 |
| 50 | .58543 | .81072 | .59949 | .80038 | .61337 | .78980 | .62706 | .77897 | .64056 | .76791 | 10 |
| 51 | .58567 | .81055 | .59972 | .80021 | .61360 | .78962 | .62728 | .77879 | .64078 | .76772 | 9 |
| 52 | .58590 | .81038 | .59995 | .80003 | .61383 | .78944 | .62751 | .77861 | .64100 | .76754 | 8 |
| 53 | .58614 | .81021 | .60019 | .79986 | .61406 | .78926 | .62774 | .77843 | .64123 | .76735 | 7 |
| 54 | .58637 | .81004 | .60042 | .79968 | .61429 | .78908 | .62796 | .77824 | .64145 | .76717 | 6 |
| 55 | .58661 | .80987 | .60065 | .79951 | .61451 | .78891 | .62819 | .77806 | .64167 | .76698 | 5 |
| 56 | .58684 | .80970 | .60089 | .79934 | .61474 | .78873 | .62842 | .77788 | .64190 | .76679 | 4 |
| 57 | .58708 | .80953 | .60112 | .79916 | .61497 | .78855 | .62864 | .77769 | .64212 | .76661 | 3 |
| 58 | .58731 | .80936 | .60135 | .79899 | .61520 | .78837 | .62887 | .77751 | .64234 | .76642 | 2 |
| 59 | .58755 | .80919 | .60158 | .79881 | .61543 | .78819 | .62909 | .77733 | .64256 | .76623 | 1 |
| 60 | .58779 | .80902 | .60182 | .79864 | .61566 | .78801 | .62932 | .77715 | .64279 | .76604 | 0 |
| ′ | 54° Cosine | 54° Sine | 53° Cosine | 53° Sine | 52° Cosine | 52° Sine | 51° Cosine | 51° Sine | 50° Cosine | 50° Sine | ′ |

## TABLE 1 cont.

Natural Sines and Cosines

| ′ | 40° | | 41° | | 42° | | 43° | | 44° | | ′ |
|---|---|---|---|---|---|---|---|---|---|---|---|
| | Sine | Cosine | Sine | Cosine | Sine | Cosine | Sine | Cosine | Sine | Cosine | |
| 0 | .64279 | .76604 | .65606 | .75471 | .66913 | .74314 | .68200 | .73135 | .69466 | .71934 | 60 |
| 1 | .64301 | .76586 | .65628 | .75452 | .66935 | .74295 | .68221 | .73116 | .69487 | .71914 | 59 |
| 2 | .64323 | .76567 | .65650 | .75433 | .66956 | .74276 | .68242 | .73096 | .69508 | .71894 | 58 |
| 3 | .64346 | .76548 | .65672 | .75414 | .66978 | .74256 | .68264 | .73076 | .69529 | .71873 | 57 |
| 4 | .64368 | .76530 | .65694 | .75395 | .66999 | .74237 | .68285 | .73056 | .69549 | .71853 | 56 |
| 5 | .64390 | .76511 | .65716 | .75375 | .67021 | .74217 | .68306 | .73036 | .69570 | .71833 | 55 |
| 6 | .64412 | .76492 | .65738 | .75356 | .67043 | .74198 | .68327 | .73016 | .69591 | .71813 | 54 |
| 7 | .64435 | .76473 | .65759 | .75337 | .67064 | .74178 | .68349 | .72996 | .69612 | .71792 | 53 |
| 8 | .64457 | .76455 | .65781 | .75318 | .67086 | .74159 | .68370 | .72976 | .69633 | .71772 | 52 |
| 9 | .64479 | .76436 | .65803 | .75299 | .67107 | .74139 | .68391 | .72957 | .69654 | .71752 | 51 |
| 10 | .64501 | .76417 | .65825 | .75280 | .67129 | .74120 | .68412 | .72937 | .69675 | .71732 | 50 |
| 11 | .64524 | .76398 | .65847 | .75261 | .67151 | .74100 | .68434 | .72917 | .69696 | .71711 | 49 |
| 12 | .64546 | .76380 | .65869 | .75241 | .67172 | .74080 | .68455 | .72897 | .69717 | .71691 | 48 |
| 13 | .64568 | .76361 | .65891 | .75222 | .67194 | .74061 | .68476 | .72877 | .69737 | .71671 | 47 |
| 14 | .64590 | .76342 | .65913 | .75203 | .67215 | .74041 | .68497 | .72857 | .69758 | .71650 | 46 |
| 15 | .64612 | .76323 | .65935 | .75184 | .67237 | .74022 | .68518 | .72837 | .69779 | .71630 | 45 |
| 16 | .64635 | .76304 | .65956 | .75165 | .67258 | .74002 | .68539 | .72817 | .69800 | .71610 | 44 |
| 17 | .64657 | .76286 | .65978 | .75146 | .67280 | .73983 | .68561 | .72797 | .69821 | .71590 | 43 |
| 18 | .64679 | .76267 | .66000 | .75126 | .67301 | .73963 | .68582 | .72777 | .69842 | .71569 | 42 |
| 19 | .64701 | .76248 | .66022 | .75107 | .67323 | .73944 | .68603 | .72757 | .69862 | .71549 | 41 |
| 20 | .64723 | .76229 | .66044 | .75088 | .67344 | .73924 | .68624 | .72737 | .69883 | .71529 | 40 |
| 21 | .64746 | .76210 | .66066 | .75069 | .67366 | .73904 | .68645 | .72717 | .69904 | .71508 | 39 |
| 22 | .64768 | .76192 | .66088 | .75050 | .67387 | .73885 | .68666 | .72697 | .69925 | .71488 | 38 |
| 23 | .64790 | .76173 | .66109 | .75030 | .67409 | .73865 | .68688 | .72677 | .69946 | .71468 | 37 |
| 24 | .64812 | .76154 | .66131 | .75011 | .67430 | .73846 | .68709 | .72657 | .69966 | .71447 | 36 |
| 25 | .64834 | .76135 | .66153 | .74992 | .67452 | .73826 | .68730 | .72637 | .69987 | .71427 | 35 |
| 26 | .64856 | .76116 | .66175 | .74973 | .67473 | .73806 | .68751 | .72617 | .70008 | .71407 | 34 |
| 27 | .64878 | .76097 | .66197 | .74953 | .67495 | .73787 | .68772 | .72597 | .70029 | .71386 | 33 |
| 28 | .64901 | .76078 | .66218 | .74934 | .67516 | .73767 | .68793 | .72577 | .70049 | .71366 | 32 |
| 29 | .64923 | .76059 | .66240 | .74915 | .67538 | .73747 | .68814 | .72557 | .70070 | .71345 | 31 |
| 30 | .64945 | .76041 | .66262 | .74896 | .67559 | .73728 | .68835 | .72537 | .70091 | .71325 | 30 |
| 31 | .64967 | .76022 | .66284 | .74876 | .67580 | .73708 | .68857 | .72517 | .70112 | .71305 | 29 |
| 32 | .64989 | .76003 | .66306 | .74857 | .67602 | .73688 | .68878 | .72497 | .70132 | .71284 | 28 |
| 33 | .65011 | .75984 | .66327 | .74838 | .67623 | .73669 | .68899 | .72477 | .70153 | .71264 | 27 |
| 34 | .65033 | .75965 | .66349 | .74818 | .67645 | .73649 | .68920 | .72457 | .70174 | .71243 | 26 |
| 35 | .65055 | .75946 | .66371 | .74799 | .67666 | .73629 | .68941 | .72437 | .70195 | .71223 | 25 |
| 36 | .65077 | .75927 | .66393 | .74780 | .67688 | .73610 | .68962 | .72417 | .70215 | .71203 | 24 |
| 37 | .65100 | .75908 | .66414 | .74760 | .67709 | .73590 | .68983 | .72397 | .70236 | .71182 | 23 |
| 38 | .65122 | .75889 | .66436 | .74741 | .67730 | .73570 | .69004 | .72377 | .70257 | .71162 | 22 |
| 39 | .65144 | .75870 | .66458 | .74722 | .67752 | .73551 | .69025 | .72357 | .70277 | .71141 | 21 |
| 40 | .65166 | .75851 | .66480 | .74703 | .67773 | .73531 | .69046 | .72337 | .70298 | .71121 | 20 |
| 41 | .65188 | .75832 | .66501 | .74683 | .67795 | .73511 | .69067 | .72317 | .70319 | .71100 | 19 |
| 42 | .65210 | .75813 | .66523 | .74664 | .67816 | .73491 | .69088 | .72297 | .70339 | .71080 | 18 |
| 43 | .65232 | .75794 | .66545 | .74644 | .67837 | .73472 | .69109 | .72277 | .70360 | .71059 | 17 |
| 44 | .65254 | .75775 | .66566 | .74625 | .67859 | .73452 | .69130 | .72257 | .70381 | .71039 | 16 |
| 45 | .65276 | .75756 | .66588 | .74606 | .67880 | .73432 | .69151 | .72236 | .70401 | .71019 | 15 |
| 46 | .65298 | .75738 | 66610 | .74586 | .67901 | .73413 | .69172 | .72216 | .70422 | .70998 | 14 |
| 47 | .65320 | .75719 | .66632 | .74567 | .67923 | .73393 | .69193 | .72196 | .70443 | .70978 | 13 |
| 48 | .65342 | .75700 | .66653 | .74548 | .67944 | .73373 | .69214 | .72176 | .70463 | .70957 | 12 |
| 49 | .65364 | .75680 | .66675 | .74528 | .67965 | .73353 | .69235 | .72156 | .70484 | .70937 | 11 |
| 50 | .65386 | .75661 | .66697 | .74509 | .67987 | .73333 | .69256 | .72136 | .70505 | .70916 | 10 |
| 51 | .65408 | .75642 | .66718 | .74489 | .68008 | .73314 | .69277 | .72116 | .70525 | .70896 | 9 |
| 52 | .65430 | .75623 | .66740 | .74470 | .68029 | .73294 | .69298 | .72095 | .70546 | .70875 | 8 |
| 53 | .65452 | .75604 | .66762 | .74451 | .68051 | .73274 | .69319 | .72075 | .70567 | .70855 | 7 |
| 54 | .65474 | .75585 | .66783 | .74431 | .68072 | .73254 | .69340 | .72055 | .70587 | .70834 | 6 |
| 55 | .65496 | .75566 | .66805 | .74412 | .68093 | .73234 | .69361 | .72035 | .70608 | .70813 | 5 |
| 56 | .65518 | .75547 | .66827 | .74392 | .68115 | .73215 | .69382 | .72015 | .70628 | .70793 | 4 |
| 57 | .65540 | .75528 | .66848 | .74373 | .68136 | .73195 | .69403 | .71995 | .70649 | .70772 | 3 |
| 58 | .65562 | .75509 | .66870 | .74353 | .68157 | .73175 | .69424 | .71974 | .70670 | .70752 | 2 |
| 59 | .65584 | .75490 | .66891 | .74334 | .68179 | .73155 | .69445 | .71954 | .70690 | .70731 | 1 |
| 60 | .65606 | .75471 | .66913 | .74314 | .68200 | .73135 | .69466 | .71934 | .70711 | .70711 | 0 |
| ′ | Cosine | Sine | Cosine | Sine | Cosine | Sine | Cosine | Sine | Cosine | Sine | ′ |
| | 49° | | 48° | | 47° | | 46° | | 45° | | |

**TABLE 2**

Natural Tangents and Cotangents

| ′ | 0° | | 1° | | 2° | | 3° | | 4° | | ′ |
|---|---|---|---|---|---|---|---|---|---|---|---|
| | Tang | Cotang | Tang | Cotang | Tang | Cotang | Tang | Cotang | Tang | Cotang | |
| 0 | .00000 | Infinite | .01746 | 57.2900 | .03492 | 28.6363 | .05241 | 19.0811 | .06993 | 14.3007 | 60 |
| 1 | .00029 | 3437.75 | .01775 | 56.3506 | .03521 | 28.3994 | .05270 | 18.9755 | .07022 | 14.2411 | 59 |
| 2 | .00058 | 1718.87 | .01804 | 55.4415 | .03550 | 28.1664 | .05299 | 18.8711 | .07051 | 14.1821 | 58 |
| 3 | .00087 | 1145.92 | .01833 | 54.5613 | .03579 | 27.9372 | .05328 | 18.7678 | .07080 | 14.1235 | 57 |
| 4 | .00116 | 859.436 | .01862 | 53.7086 | .03609 | 27.7117 | .05357 | 18.6656 | .07110 | 14.0655 | 56 |
| 5 | .00145 | 687.549 | .01891 | 52.8821 | .03638 | 27.4899 | .05387 | 18.5645 | .07139 | 14.0079 | 55 |
| 6 | .00175 | 572.957 | .01920 | 52.0807 | .03667 | 27.2715 | .05416 | 18.4645 | .07168 | 13.9507 | 54 |
| 7 | .00204 | 491.106 | .01949 | 51.3032 | .03696 | 27.0566 | .05445 | 18.3655 | .07197 | 13.8940 | 53 |
| 8 | .00233 | 429.718 | .01978 | 50.5485 | .03725 | 26.8450 | .05474 | 18.2677 | .07227 | 13.8378 | 52 |
| 9 | .00262 | 381.971 | .02007 | 49.8157 | .03754 | 26.6367 | .05503 | 18.1708 | .07256 | 13.7821 | 51 |
| 10 | .00291 | 343.774 | .02036 | 49.1039 | .03783 | 26.4316 | .05533 | 18.0750 | .07285 | 13.7267 | 50 |
| 11 | .00320 | 312.521 | .02066 | 48.4121 | .03812 | 26.2296 | .05562 | 17.9802 | .07314 | 13.6719 | 49 |
| 12 | .00349 | 286.478 | .02095 | 47.7395 | .03842 | 26.0307 | .05591 | 17.8863 | .07344 | 13.6174 | 48 |
| 13 | .00378 | 264.441 | .02124 | 47.0853 | .03871 | 25.8348 | .05620 | 17.7934 | .07373 | 13.5634 | 47 |
| 14 | .00407 | 245.552 | .02153 | 46.4489 | .03900 | 25.6418 | .05649 | 17.7015 | .07402 | 13.5098 | 46 |
| 15 | .00436 | 229.182 | .02182 | 45.8294 | .03929 | 25.4517 | .05678 | 17.6106 | .07431 | 13.4566 | 45 |
| 16 | .00465 | 214.858 | .02211 | 45.2261 | .03958 | 25.2644 | .05708 | 17.5205 | .07461 | 13.4039 | 44 |
| 17 | .00495 | 202.219 | .02240 | 44.6386 | .03987 | 25.0798 | .05737 | 17.4314 | .07490 | 13.3515 | 43 |
| 18 | .00524 | 190.984 | .02269 | 44.0661 | .04016 | 24.8978 | .05766 | 17.3432 | .07519 | 13.2996 | 42 |
| 19 | .00553 | 180.932 | .02298 | 43.5081 | .04046 | 24.7185 | .05795 | 17.2558 | .07548 | 13.2480 | 41 |
| 20 | .00582 | 171.885 | .02328 | 42.9641 | .04075 | 24.5418 | .05824 | 17.1693 | .07578 | 13.1969 | 40 |
| 21 | .00611 | 163.700 | .02357 | 42.4335 | .04104 | 24.3675 | .05854 | 17.0837 | .07607 | 13.1461 | 39 |
| 22 | .00640 | 156.259 | .02386 | 41.9158 | .04133 | 24.1957 | .05883 | 16.9990 | .07636 | 13.0958 | 38 |
| 23 | .00669 | 149.465 | .02415 | 41.4106 | .04162 | 24.0263 | .05912 | 16.9150 | .07665 | 13.0458 | 37 |
| 24 | .00698 | 143.237 | .02444 | 40.9174 | .04191 | 23.8593 | .05941 | 16.8319 | .07695 | 12.9962 | 36 |
| 25 | .00727 | 137.507 | .02473 | 40.4358 | .04220 | 23.6945 | .05970 | 16.7496 | .07724 | 12.9469 | 35 |
| 26 | .00756 | 132.219 | .02502 | 39.9655 | .04250 | 23.5321 | .05999 | 16.6681 | .07753 | 12.8981 | 34 |
| 27 | .00785 | 127.321 | .02531 | 39.5059 | .04279 | 23.3718 | .06029 | 16.5874 | .07782 | 12.8496 | 33 |
| 28 | .00815 | 122.774 | .02560 | 39.0568 | .04308 | 23.2137 | .06058 | 16.5075 | .07812 | 12.8014 | 32 |
| 29 | .00844 | 118.540 | .02589 | 38.6177 | .04337 | 23.0577 | .06087 | 16.4283 | .07841 | 12.7536 | 31 |
| 30 | .00873 | 114.589 | .02619 | 38.1885 | .04366 | 22.9038 | .06116 | 16.3499 | .07870 | 12.7062 | 30 |
| 31 | .00902 | 110.892 | .02648 | 37.7686 | .04395 | 22.7519 | .06145 | 16.2722 | .07899 | 12.6591 | 29 |
| 32 | .00931 | 107.426 | .02677 | 37.3579 | .04424 | 22.6020 | .06175 | 16.1952 | .07929 | 12.6124 | 28 |
| 33 | .00960 | 104.171 | .02706 | 36.9560 | .04454 | 22.4541 | .06204 | 16.1190 | .07958 | 12.5660 | 27 |
| 34 | .00989 | 101.107 | .02735 | 36.5627 | .04483 | 22.3081 | .06233 | 16.0435 | .07987 | 12.5199 | 26 |
| 35 | .01018 | 98.2179 | .02764 | 36.1776 | .04512 | 22.1640 | .06262 | 15.9687 | .08017 | 12.4742 | 25 |
| 36 | .01047 | 95.4895 | .02793 | 35.8006 | .04541 | 22.0217 | .06291 | 15.8945 | .08046 | 12.4288 | 24 |
| 37 | .01076 | 92.9085 | .02822 | 35.4313 | .04570 | 21.8813 | .06321 | 15.8211 | .08075 | 12.3838 | 23 |
| 38 | .01105 | 90.4633 | .02851 | 35.0695 | .04599 | 21.7426 | .06350 | 15.7483 | .08104 | 12.3390 | 22 |
| 39 | .01135 | 88.1436 | .02881 | 34.7151 | .04628 | 21.6056 | .06379 | 15.6762 | .08134 | 12.2946 | 21 |
| 40 | .01164 | 85.9398 | .02910 | 34.3678 | .04658 | 21.4704 | .06408 | 15.6048 | .08163 | 12.2505 | 20 |
| 41 | .01193 | 83.8435 | .02939 | 34.0273 | .04687 | 21.3369 | .06437 | 15.5340 | .08192 | 12.2067 | 19 |
| 42 | .01222 | 81.8470 | .02968 | 33.6935 | .04716 | 21.2049 | .06467 | 15.4638 | .08221 | 12.1632 | 18 |
| 43 | .01251 | 79.9434 | .02997 | 33.3662 | .04745 | 21.0747 | .06496 | 15.3943 | .08251 | 12.1201 | 17 |
| 44 | .01280 | 78.1263 | .03026 | 33.0452 | .04774 | 20.9460 | .06525 | 15.3254 | .08280 | 12.0772 | 16 |
| 45 | .01309 | 76.3900 | .03055 | 32.7303 | .04803 | 20.8188 | .06554 | 15.2571 | .08309 | 12.0346 | 15 |
| 46 | .01338 | 74.7292 | .03084 | 32.4213 | .04833 | 20.6932 | .06584 | 15.1893 | .08339 | 11.9923 | 14 |
| 47 | .01367 | 73.1390 | .03114 | 32.1181 | .04862 | 20.5691 | .06613 | 15.1222 | .08368 | 11.9504 | 13 |
| 48 | .01396 | 71.6151 | .03143 | 31.8205 | .04891 | 20.4465 | .06642 | 15.0557 | .08397 | 11.9087 | 12 |
| 49 | .01425 | 70.1533 | .03172 | 31.5284 | .04920 | 20.3253 | .06671 | 14.9898 | .08427 | 11.8673 | 11 |
| 50 | .01455 | 68.7501 | .03201 | 31.2416 | .04949 | 20.2056 | .06700 | 14.9244 | .08456 | 11.8262 | 10 |
| 51 | .01484 | 67.4019 | .03230 | 30.9599 | .04978 | 20.0872 | .06730 | 14.8596 | .08485 | 11.7853 | 9 |
| 52 | .01513 | 66.1055 | .03259 | 30.6833 | .05007 | 19.9702 | .06759 | 14.7954 | .08514 | 11.7448 | 8 |
| 53 | .01542 | 64.8580 | .03288 | 30.4116 | .05037 | 19.8546 | .06788 | 14.7317 | .08544 | 11.7045 | 7 |
| 54 | .01571 | 63.6567 | .03317 | 30.1446 | .05066 | 19.7403 | .06817 | 14.6685 | .08573 | 11.6645 | 6 |
| 55 | .01600 | 62.4992 | .03346 | 29.8823 | .05095 | 19.6273 | .06847 | 14.6059 | .08602 | 11.6248 | 5 |
| 56 | .01629 | 61.3829 | .03376 | 29.6245 | .05124 | 19.5156 | .06876 | 14.5438 | .08632 | 11.5853 | 4 |
| 57 | .01658 | 60.3058 | .03405 | 29.3711 | .05153 | 19.4051 | .06905 | 14.4823 | .08661 | 11.5461 | 3 |
| 58 | .01687 | 59.2659 | .03434 | 29.1220 | .05182 | 19.2959 | .06934 | 14.4212 | .08690 | 11.5072 | 2 |
| 59 | .01716 | 58.2612 | .03463 | 28.8771 | .05212 | 19.1879 | .06963 | 14.3607 | .08720 | 11.4685 | 1 |
| 60 | .01746 | 57.2900 | .03492 | 28.6363 | .05241 | 19.0811 | .06993 | 14.3007 | .08749 | 11.4301 | 0 |
| ′ | Cotang | Tang | Cotang | Tang | Cotang | Tang | Cotang | Tang | Cotang | Tang | ′ |
| | 89° | | 88° | | 87° | | 86° | | 85° | | |

## TABLE 2 cont.

Natural Tangents and Cotangents

| ′ | 5° Tang | 5° Cotang | 6° Tang | 6° Cotang | 7° Tang | 7° Cotang | 8° Tang | 8° Cotang | 9° Tang | 9° Cotang | ′ |
|---|---|---|---|---|---|---|---|---|---|---|---|
| 0 | .08749 | 11.4301 | .10510 | 9.51436 | .12278 | 8.14435 | .14054 | 7.11537 | .15838 | 6.31375 | 60 |
| 1 | .08778 | 11.3919 | .10540 | 9.48781 | .12308 | 8.12481 | .14084 | 7.10038 | .15868 | 6.30189 | 59 |
| 2 | .08807 | 11.3540 | .10569 | 9.46141 | .12338 | 8.10536 | .14113 | 7.08546 | .15898 | 6.29007 | 58 |
| 3 | .08837 | 11.3163 | .10599 | 9.43515 | .12367 | 8.08600 | .14143 | 7.07059 | .15928 | 6.27829 | 57 |
| 4 | .08866 | 11.2789 | .10628 | 9.40904 | .12397 | 8.06674 | .14173 | 7.05579 | .15958 | 6.26655 | 56 |
| 5 | .08895 | 11.2417 | .10657 | 9.38307 | .12426 | 8.04756 | .14202 | 7.04105 | .15988 | 6.25486 | 55 |
| 6 | .08925 | 11.2048 | .10687 | 9.35724 | .12456 | 8.02848 | .14232 | 7.02637 | .16017 | 6.24321 | 54 |
| 7 | .08954 | 11.1681 | .10716 | 9.33155 | .12485 | 8.00948 | .14262 | 7.01174 | .16047 | 6.23160 | 53 |
| 8 | .08983 | 11.1316 | .10746 | 9.30599 | .12515 | 7.99058 | .14291 | 6.99718 | .16077 | 6.22003 | 52 |
| 9 | .09013 | 11.0954 | .10775 | 9.28058 | .12544 | 7.97176 | .14321 | 6.98268 | .16107 | 6.20851 | 51 |
| 10 | .09042 | 11.0594 | .10805 | 9.25530 | .12574 | 7.95302 | .14351 | 6.96823 | .16137 | 6.19703 | 50 |
| 11 | .09071 | 11.0237 | .10834 | 9.23016 | .12603 | 7.93438 | .14381 | 6.95385 | .16167 | 6.18559 | 49 |
| 12 | .09101 | 10.9882 | .10863 | 9.20516 | .12633 | 7.91582 | .14410 | 6.93952 | .16196 | 6.17419 | 48 |
| 13 | .09130 | 10.9529 | .10893 | 9.18028 | .12662 | 7.89734 | .14440 | 6.92525 | .16226 | 6.16283 | 47 |
| 14 | .09159 | 10.9178 | .10922 | 9.15554 | .12692 | 7.87895 | .14470 | 6.91104 | .16256 | 6.15151 | 46 |
| 15 | .09189 | 10.8829 | .10952 | 9.13093 | .12722 | 7.86064 | .14499 | 6.89688 | .16286 | 6.14023 | 45 |
| 16 | .09218 | 10.8483 | .10981 | 9.10646 | .12751 | 7.84242 | .14529 | 6.88278 | .16316 | 6.12899 | 44 |
| 17 | .09247 | 10.8139 | .11011 | 9.08211 | .12781 | 7.82428 | .14559 | 6.86874 | .16346 | 6.11779 | 43 |
| 18 | .09277 | 10.7797 | .11040 | 9.05789 | .12810 | 7.80622 | .14588 | 6.85475 | .16376 | 6.10664 | 42 |
| 19 | .09306 | 10.7457 | .11070 | 9.03379 | .12840 | 7.78825 | .14618 | 6.84082 | .16405 | 6.09552 | 41 |
| 20 | .09335 | 10.7119 | .11099 | 9.00983 | .12869 | 7.77035 | .14648 | 6.82694 | .16435 | 6.08444 | 40 |
| 21 | .09365 | 10.6783 | .11128 | 8.98598 | .12899 | 7.75254 | .14678 | 6.81312 | .16465 | 6.07340 | 39 |
| 22 | .09394 | 10.6450 | .11158 | 8.96227 | .12929 | 7.73480 | .14707 | 6.79936 | .16495 | 6.06240 | 38 |
| 23 | .09423 | 10.6118 | .11187 | 8.93867 | .12958 | 7.71715 | .14737 | 6.78564 | .16525 | 6.05143 | 37 |
| 24 | .09453 | 10.5789 | .11217 | 8.91520 | .12988 | 7.69957 | .14767 | 6.77199 | .16555 | 6.04051 | 36 |
| 25 | .09482 | 10.5462 | .11246 | 8.89185 | .13017 | 7.68208 | .14796 | 6.75838 | .16585 | 6.02962 | 35 |
| 26 | .09511 | 10.5136 | .11276 | 8.86862 | .13047 | 7.66466 | .14826 | 6.74483 | .16615 | 6.01878 | 34 |
| 27 | .09541 | 10.4813 | .11305 | 8.84551 | .13076 | 7.64732 | .14856 | 6.73133 | .16645 | 6.00797 | 33 |
| 28 | .09570 | 10.4491 | .11335 | 8.82252 | .13106 | 7.63005 | .14886 | 6.71789 | .16674 | 5.99720 | 32 |
| 29 | .09600 | 10.4172 | .11364 | 8.79964 | .13136 | 7.61287 | .14915 | 6.70450 | .16704 | 5.98646 | 31 |
| 30 | .09629 | 10.3854 | .11394 | 8.77689 | .13165 | 7.59575 | .14945 | 6.69116 | .16734 | 5.97576 | 30 |
| 31 | .09658 | 10.3538 | .11423 | 8.75425 | .13195 | 7.57872 | .14975 | 6.67787 | .16764 | 5.96510 | 29 |
| 32 | .09688 | 10.3224 | .11452 | 8.73172 | .13224 | 7.56176 | .15005 | 6.66463 | .16794 | 5.95448 | 28 |
| 33 | .09717 | 10.2913 | .11482 | 8.70931 | .13254 | 7.54487 | .15034 | 6.65144 | .16824 | 5.94390 | 27 |
| 34 | .09746 | 10.2602 | .11511 | 8.68701 | .13284 | 7.52806 | .15064 | 6.63831 | .16854 | 5.93335 | 26 |
| 35 | .09776 | 10.2294 | .11541 | 8.66482 | .13313 | 7.51132 | .15094 | 6.62523 | .16884 | 5.92283 | 25 |
| 36 | .09805 | 10.1988 | .11570 | 8.64275 | .13343 | 7.49465 | .15124 | 6.61219 | .16914 | 5.91236 | 24 |
| 37 | .09834 | 10.1683 | .11600 | 8.62078 | .13372 | 7.47806 | .15153 | 6.59921 | .16944 | 5.90191 | 23 |
| 38 | .09864 | 10.1381 | .11629 | 8.59893 | .13402 | 7.46154 | .15183 | 6.58627 | .16974 | 5.89151 | 22 |
| 39 | .09893 | 10.1080 | .11659 | 8.57718 | .13432 | 7.44509 | .15213 | 6.57339 | .17004 | 5.88114 | 21 |
| 40 | .09923 | 10.0780 | .11688 | 8.55555 | .13461 | 7.42871 | .15243 | 6.56055 | .17033 | 5.87080 | 20 |
| 41 | .09952 | 10.0483 | .11718 | 8.53402 | .13491 | 7.41240 | .15272 | 6.54777 | .17063 | 5.86051 | 19 |
| 42 | .09981 | 10.0187 | .11747 | 8.51259 | .13521 | 7.39616 | .15302 | 6.53503 | .17093 | 5.85024 | 18 |
| 43 | .10011 | 9.98931 | .11777 | 8.49128 | .13550 | 7.37999 | .15332 | 6.52234 | .17123 | 5.84001 | 17 |
| 44 | .10040 | 9.96007 | .11806 | 8.47007 | .13580 | 7.36389 | .15362 | 6.50970 | .17153 | 5.82982 | 16 |
| 45 | .10069 | 9.93101 | .11836 | 8.44896 | .13609 | 7.34786 | .15391 | 6.49710 | .17183 | 5.81966 | 15 |
| 46 | .10099 | 9.90211 | .11865 | 8.42795 | .13639 | 7.33190 | .15421 | 6.48456 | .17213 | 5.80953 | 14 |
| 47 | .10128 | 9.87338 | .11895 | 8.40705 | .13669 | 7.31600 | .15451 | 6.47206 | .17243 | 5.79944 | 13 |
| 48 | .10158 | 9.84482 | .11924 | 8.38625 | .13698 | 7.30018 | .15481 | 6.45961 | .17273 | 5.78938 | 12 |
| 49 | .10187 | 9.81641 | .11954 | 8.36555 | .13728 | 7.28442 | .15511 | 6.44720 | .17303 | 5.77936 | 11 |
| 50 | .10216 | 9.78817 | .11983 | 8.34496 | .13758 | 7.26873 | .15540 | 6.43484 | .17333 | 5.76937 | 10 |
| 51 | .10246 | 9.76009 | .12013 | 8.32446 | .13787 | 7.25310 | .15570 | 6.42253 | .17363 | 5.75941 | 9 |
| 52 | .10275 | 9.73217 | .12042 | 8.30406 | .13817 | 7.23754 | .15600 | 6.41026 | .17393 | 5.74949 | 8 |
| 53 | .10305 | 9.70441 | .12072 | 8.28376 | .13846 | 7.22204 | .15630 | 6.39804 | .17423 | 5.73960 | 7 |
| 54 | .10334 | 9.67680 | .12101 | 8.26355 | .13876 | 7.20661 | .15660 | 6.38587 | .17453 | 5.72974 | 6 |
| 55 | .10363 | 9.64935 | .12131 | 8.24345 | .13906 | 7.19125 | .15689 | 6.37374 | .17483 | 5.71992 | 5 |
| 56 | .10393 | 9.62205 | .12160 | 8.22344 | .13935 | 7.17594 | .15719 | 6.36165 | .17513 | 5.71013 | 4 |
| 57 | .10422 | 9.59490 | .12190 | 8.20352 | .13965 | 7.16071 | .15749 | 6.34961 | .17543 | 5.70037 | 3 |
| 58 | .10452 | 9.56791 | .12219 | 8.18370 | .13995 | 7.14553 | .15779 | 6.33761 | .17573 | 5.69064 | 2 |
| 59 | .10481 | 9.54106 | .12249 | 8.16398 | .14024 | 7.13042 | .15809 | 6.32566 | .17603 | 5.68094 | 1 |
| 60 | .10510 | 9.51436 | .12278 | 8.14435 | .14054 | 7.11537 | .15838 | 6.31375 | .17633 | 5.67128 | 0 |
| ′ | Cotang | Tang | Cotang | Tang | Cotang | Tang | Cotang | Tang | Cotang | Tang | ′ |
| | 84° | | 83° | | 82° | | 81° | | 80° | | |

**TABLE 2 cont.**
Natural Tangents and Cotangents

| ′ | 10° | | 11° | | 12° | | 13° | | 14° | | ′ |
|---|---|---|---|---|---|---|---|---|---|---|---|
| | Tang | Cotang | Tang | Cotang | Tang | Cotang | Tang | Cotang | Tang | Cotang | |
| 0 | 17633 | 5.67128 | .19438 | 5.14455 | .21256 | 4.70463 | .23087 | 4.33148 | .24933 | 4.01078 | 60 |
| 1 | .17663 | 5.66165 | .19468 | 5.13658 | .21286 | 4.69791 | .23117 | 4.32573 | .24964 | 4.00582 | 59 |
| 2 | .17693 | 5.65205 | .19498 | 5.12862 | .21316 | 4.69121 | .23148 | 4.32001 | .24995 | 4.00086 | 58 |
| 3 | .17723 | 5.64248 | .19529 | 5.12069 | .21347 | 4.68452 | .23179 | 4.31430 | .25026 | 3.99592 | 57 |
| 4 | .17753 | 5.63295 | .19559 | 5.11279 | .21377 | 4.67786 | .23209 | 4.30860 | .25056 | 3.99099 | 56 |
| 5 | .17783 | 5.62344 | .19589 | 5.10490 | .21408 | 4.67121 | .23240 | 4.30291 | .25087 | 3.98607 | 55 |
| 6 | .17813 | 5.61397 | .19619 | 5.09704 | .21438 | 4.66458 | .23271 | 4.29724 | .25118 | 3.98117 | 54 |
| 7 | .17843 | 5.60452 | .19649 | 5.08921 | .21469 | 4.65797 | .23301 | 4.29159 | .25149 | 3.97627 | 53 |
| 8 | .17873 | 5.59511 | .19680 | 5.08139 | .21499 | 4.65138 | .23332 | 4.28595 | .25180 | 3.97139 | 52 |
| 9 | .17903 | 5.58573 | .19710 | 5.07360 | .21529 | 4.64480 | .23363 | 4.28032 | .25211 | 3.96651 | 51 |
| 10 | .17933 | 5.57638 | .19740 | 5.06584 | .21560 | 4.63825 | .23393 | 4.27471 | .25242 | 3.96165 | 50 |
| 11 | .17963 | 5.56706 | .19770 | 5.05809 | .21590 | 4.63171 | .23424 | 4.26911 | .25273 | 3.95680 | 49 |
| 12 | .17993 | 5.55777 | .19801 | 5.05037 | .21621 | 4.62518 | .23455 | 4.26352 | .25304 | 3.95196 | 48 |
| 13 | .18023 | 5.54851 | .19831 | 5.04267 | .21651 | 4.61868 | .23485 | 4.25795 | .25335 | 3.94713 | 47 |
| 14 | .18053 | 5.53927 | .19861 | 5.03499 | .21682 | 4.61219 | .23516 | 4.25239 | .25366 | 3.94232 | 46 |
| 15 | .18083 | 5.53007 | .19891 | 5.02734 | .21712 | 4.60572 | .23547 | 4.24685 | .25397 | 3.93751 | 45 |
| 16 | .18113 | 5.52090 | .19921 | 5.01971 | .21743 | 4.59927 | .23578 | 4.24132 | .25428 | 3.93271 | 44 |
| 17 | .18143 | 5.51176 | .19952 | 5.01210 | .21773 | 4.59283 | .23608 | 4.23580 | .25459 | 3.92793 | 43 |
| 18 | .18173 | 5.50264 | .19982 | 5.00451 | .21804 | 4.58641 | .23639 | 4.23030 | .25490 | 3.92316 | 42 |
| 19 | .18203 | 5.49356 | .20012 | 4.99695 | .21834 | 4.58001 | .23670 | 4.22481 | .25521 | 3.91839 | 41 |
| 20 | .18233 | 5.48451 | .20042 | 4.98940 | .21864 | 4.57363 | .23700 | 4.21933 | .25552 | 3.91364 | 40 |
| 21 | .18263 | 5.47548 | .20073 | 4.98188 | .21895 | 4.56726 | .23731 | 4.21387 | .25583 | 3.90890 | 39 |
| 22 | .18293 | 5.46648 | .20103 | 4.97438 | .21925 | 4.56091 | .23762 | 4.20842 | .25614 | 3.90417 | 38 |
| 23 | .18323 | 5.45751 | .20133 | 4.96690 | .21956 | 4.55458 | .23793 | 4.20298 | .25645 | 3.89945 | 37 |
| 24 | .18353 | 5.44857 | .20164 | 4.95945 | .21986 | 4.54826 | .23823 | 4.19756 | .25676 | 3.89474 | 36 |
| 25 | .18384 | 5.43966 | .20194 | 4.95201 | .22017 | 4.54196 | .23854 | 4.19215 | .25707 | 3.89004 | 35 |
| 26 | .18414 | 5.43077 | .20224 | 4.94460 | .22047 | 4.53568 | .23885 | 4.18675 | .25738 | 3.88536 | 34 |
| 27 | .18444 | 5.42192 | .20254 | 4.93721 | .22078 | 4.52941 | .23916 | 4.18137 | .25769 | 3.88068 | 33 |
| 28 | .18474 | 5.41309 | .20285 | 4.92984 | .22108 | 4.52316 | .23946 | 4.17600 | .25800 | 3.87601 | 32 |
| 29 | .18504 | 5.40429 | .20315 | 4.92249 | .22139 | 4.51693 | .23977 | 4.17064 | .25831 | 3.87136 | 31 |
| 30 | .18534 | 5.39552 | .20345 | 4.91516 | .22169 | 4.51071 | .24008 | 4.16530 | .25862 | 3.86671 | 30 |
| 31 | .18564 | 5.38677 | .20376 | 4.90785 | .22200 | 4.50451 | .24039 | 4.15997 | .25893 | 3.86208 | 29 |
| 32 | .18594 | 5.37805 | .20406 | 4.90056 | .22231 | 4.49832 | .24069 | 4.15465 | .25924 | 3.85745 | 28 |
| 33 | .18624 | 5.36936 | .20436 | 4.89330 | .22261 | 4.49215 | .24100 | 4.14934 | .25955 | 3.85284 | 27 |
| 34 | .18654 | 5.36070 | .20466 | 4.88605 | .22292 | 4.48600 | .24131 | 4.14405 | .25986 | 3.84824 | 26 |
| 35 | .18684 | 5.35206 | .20497 | 4.87882 | .22322 | 4.47986 | .24162 | 4.13877 | .26017 | 3.84364 | 25 |
| 36 | .18714 | 5.34345 | .20527 | 4.87162 | .22353 | 4.47374 | .24193 | 4.13350 | .26048 | 3.83906 | 24 |
| 37 | .18745 | 5.33487 | .20557 | 4.86444 | .22383 | 4.46764 | .24223 | 4.12825 | .26079 | 3.83449 | 23 |
| 38 | .18775 | 5.32631 | .20588 | 4.85727 | .22414 | 4.46155 | .24254 | 4.12301 | .26110 | 3.82992 | 22 |
| 39 | .18805 | 5.31778 | .20618 | 4.85013 | .22444 | 4.45543 | .24285 | 4.11778 | .26141 | 3.82537 | 21 |
| 40 | .18835 | 5.30928 | .20648 | 4.84300 | .22475 | 4.44942 | .24316 | 4.11256 | .26172 | 3.82083 | 20 |
| 41 | .18865 | 5.30080 | .20679 | 4.83590 | .22505 | 4.44338 | .24347 | 4.10736 | .26203 | 3.81630 | 19 |
| 42 | .18895 | 5.29235 | .20709 | 4.82882 | .22536 | 4.43735 | .24377 | 4.10216 | .26235 | 3.81177 | 18 |
| 43 | .18925 | 5.28393 | .20739 | 4.82175 | .22567 | 4.43134 | .24408 | 4.09699 | .26266 | 3.80726 | 17 |
| 44 | .18955 | 5.27553 | .20770 | 4.81471 | .22597 | 4.42534 | .24439 | 4.09182 | .26297 | 3.80276 | 16 |
| 45 | .18986 | 5.26715 | .20800 | 4.80769 | .22628 | 4.41936 | .24470 | 4.08666 | .26328 | 3.79827 | 15 |
| 46 | .19016 | 5.25880 | .20830 | 4.80068 | .22658 | 4.41340 | .24501 | 4.08152 | .26359 | 3.79378 | 14 |
| 47 | .19046 | 5.25048 | .20861 | 4.79370 | .22689 | 4.40745 | .24532 | 4.07639 | .26390 | 3.78931 | 13 |
| 48 | .19076 | 5.24218 | .20891 | 4.78673 | .22719 | 4.40152 | .24562 | 4.07127 | .26421 | 3.78485 | 12 |
| 49 | .19106 | 5.23391 | .20921 | 4.77978 | .22750 | 4.39560 | .24593 | 4.06616 | .26452 | 3.78040 | 11 |
| 50 | .19136 | 5.22566 | .20952 | 4.77286 | .22781 | 4.38969 | .24624 | 4.06107 | .26483 | 3.77595 | 10 |
| 51 | .19166 | 5.21744 | .20982 | 4.76595 | .22811 | 4.38381 | .24655 | 4.05599 | .26515 | 3.77152 | 9 |
| 52 | .19197 | 5.20925 | .21013 | 4.75906 | .22842 | 4.37793 | .24686 | 4.05092 | .26546 | 3.76709 | 8 |
| 53 | .19227 | 5.20107 | .21043 | 4.75219 | .22872 | 4.37207 | .24717 | 4.04586 | .26577 | 3.76268 | 7 |
| 54 | .19257 | 5.19293 | .21073 | 4.74534 | .22903 | 4.36623 | .24747 | 4.04081 | .26608 | 3.75828 | 6 |
| 55 | .19287 | 5.18480 | .21104 | 4.73851 | .22934 | 4.36040 | .24778 | 4.03578 | .26639 | 3.75388 | 5 |
| 56 | .19317 | 5.17671 | .21134 | 4.73170 | .22964 | 4.35459 | .24809 | 4.03076 | .26670 | 3.74950 | 4 |
| 57 | .19347 | 5.16863 | .21164 | 4.72490 | .22995 | 4.34879 | .24840 | 4.02574 | .26701 | 3.74512 | 3 |
| 58 | .19378 | 5.16058 | .21195 | 4.71813 | .23026 | 4.34300 | .24871 | 4.02074 | .26733 | 3.74075 | 2 |
| 59 | .19408 | 5.15256 | .21225 | 4.71137 | .23056 | 4.33723 | .24902 | 4.01576 | .26764 | 3.73640 | 1 |
| 60 | .19438 | 5.14455 | .21256 | 4.70463 | .23087 | 4.33148 | .24933 | 4.01078 | .26795 | 3.73205 | 0 |
| ′ | Cotang | Tang | Cotang | Tang | Cotang | Tang | Cotang | Tang | Cotang | Tang | ′ |
| | 79° | | 78° | | 77° | | 76° | | 75° | | |

## TABLE 2 cont.
Natural Tangents and Cotangents

| ′ | 15° | | 16° | | 17° | | 18° | | 19° | | ′ |
|---|---|---|---|---|---|---|---|---|---|---|---|
| | Tang | Cotang | Tang | Cotang | Tang | Cotang | Tang | Cotang | Tang | Cotang | |
| 0 | .26795 | 3.73205 | .28675 | 3.48741 | .30573 | 3.27085 | .32492 | 3.07768 | .34433 | 2.90421 | 60 |
| 1 | .26826 | 3.72771 | .28706 | 3.48359 | .30605 | 3.26745 | .32524 | 3.07464 | .34465 | 2.90147 | 59 |
| 2 | .26857 | 3.72338 | .28738 | 3.47977 | .30637 | 3.26406 | .32556 | 3.07160 | .34498 | 2.89873 | 58 |
| 3 | .26888 | 3.71907 | .28769 | 3.47596 | .30669 | 3.26067 | .32588 | 3.06857 | .34530 | 2.89600 | 57 |
| 4 | .26920 | 3.71476 | .28800 | 3.47216 | .30700 | 3.25729 | .32621 | 3.06554 | .34563 | 2.89327 | 56 |
| 5 | .26951 | 3.71046 | .28832 | 3.46837 | .30732 | 3.25392 | .32653 | 3.06252 | .34596 | 2.89055 | 55 |
| 6 | .26982 | 3.70616 | .28864 | 3.46458 | .30764 | 3.25055 | .32685 | 3.05950 | .34628 | 2.88783 | 54 |
| 7 | .27013 | 3.70188 | .28895 | 3.46080 | .30796 | 3.24719 | .32717 | 3.05649 | .34661 | 2.88511 | 53 |
| 8 | .27044 | 3.69761 | .28927 | 3.45703 | .30828 | 3.24383 | .32749 | 3.05349 | .34693 | 2.88240 | 52 |
| 9 | .27076 | 3.69335 | .28958 | 3.45327 | .30860 | 3.24049 | .32782 | 3.05049 | .34726 | 2.87970 | 51 |
| 10 | .27107 | 3.68909 | .28990 | 3.44951 | .30891 | 3.23714 | .32814 | 3.04749 | .34758 | 2.87700 | 50 |
| 11 | .27138 | 3.68485 | .29021 | 3.44576 | .30923 | 3.23381 | .32846 | 3.04450 | .34791 | 2.87430 | 49 |
| 12 | .27169 | 3.68061 | .29053 | 3.44202 | .30955 | 3.23048 | .32878 | 3.04152 | .34824 | 2.87161 | 48 |
| 13 | .27201 | 3.67638 | .29084 | 3.43829 | .30987 | 3.22715 | .32911 | 3.03854 | .34856 | 2.86892 | 47 |
| 14 | .27232 | 3.67217 | .29116 | 3.43456 | .31019 | 3.22384 | .32943 | 3.03556 | .34889 | 2.86624 | 46 |
| 15 | .27263 | 3.66796 | .29147 | 3.43084 | .31051 | 3.22053 | .32975 | 3.03260 | .34922 | 2.86356 | 45 |
| 16 | .27294 | 3.66376 | .29179 | 3.42713 | .31083 | 3.21722 | .33007 | 3.02963 | .34954 | 2.86089 | 44 |
| 17 | .27326 | 3.65957 | .29210 | 3.42343 | .31115 | 3.21392 | .33040 | 3.02667 | .34987 | 2.85822 | 43 |
| 18 | .27357 | 3.65538 | .29242 | 3.41973 | .31147 | 3.21063 | .33072 | 3.02372 | .35020 | 2.85555 | 42 |
| 19 | .27388 | 3.65121 | .29274 | 3.41604 | .31178 | 3.20734 | .33104 | 3.02077 | .35052 | 2.85289 | 41 |
| 20 | .27419 | 3.64705 | .29305 | 3.41236 | .31210 | 3.20406 | .33136 | 3.01783 | .35085 | 2.85023 | 40 |
| 21 | .27451 | 3.64289 | .29337 | 3.40869 | .31242 | 3.20079 | .33169 | 3.01489 | .35118 | 2.84758 | 39 |
| 22 | .27482 | 3.63874 | .29368 | 3.40502 | .31274 | 3.19752 | .33201 | 3.01196 | .35150 | 2.84494 | 38 |
| 23 | .27513 | 3.63461 | .29400 | 3.40136 | .31306 | 3.19426 | .33233 | 3.00903 | .35183 | 2.84229 | 37 |
| 24 | .27545 | 3.63048 | .29432 | 3.39771 | .31338 | 3.19100 | .33266 | 3.00611 | .35216 | 2.83965 | 36 |
| 25 | .27576 | 3.62636 | .29463 | 3.39406 | .31370 | 3.18775 | .33298 | 3.00319 | .35248 | 2.83702 | 35 |
| 26 | .27607 | 3.62224 | .29495 | 3.39042 | .31402 | 3.18451 | .33330 | 3.00028 | .35281 | 2.83439 | 34 |
| 27 | .27638 | 3.61814 | .29526 | 3.38679 | .31434 | 3.18127 | .33363 | 2.99738 | .35314 | 2.83176 | 33 |
| 28 | .27670 | 3.61405 | .29558 | 3.38317 | .31466 | 3.17804 | .33395 | 2.99447 | .35346 | 2.82914 | 32 |
| 29 | .27701 | 3.60996 | .29590 | 3.37955 | .31498 | 3.17481 | .33427 | 2.99158 | .25379 | 2.82653 | 31 |
| 30 | .27732 | 3.60588 | .29621 | 3.37594 | .31530 | 3.17159 | .33460 | 2.98868 | .35412 | 2.82391 | 30 |
| 31 | .27764 | 3.60181 | .29653 | 3.37234 | .31562 | 3.16838 | .33492 | 2.98580 | .35445 | 2.82130 | 29 |
| 32 | .27795 | 3.59775 | .29685 | 3.36875 | .31594 | 3.16517 | .33524 | 2.98292 | .35477 | 2.81870 | 28 |
| 33 | .27826 | 3.59370 | .29716 | 3.36516 | .31626 | 3.16197 | .33557 | 2.98004 | .35510 | 2.81610 | 27 |
| 34 | .27858 | 3.58966 | .29748 | 3.36158 | .31658 | 3.15877 | .33589 | 2.97717 | .35543 | 2.81350 | 26 |
| 35 | .27889 | 3.58562 | .29780 | 3.35800 | .31690 | 3.15558 | .33621 | 2.97430 | .35576 | 2.81091 | 25 |
| 36 | .27921 | 3.58160 | .29811 | 3.35443 | .31722 | 3.15240 | .33654 | 2.97144 | .35608 | 2.80833 | 24 |
| 37 | .27952 | 3.57758 | .29843 | 3.35087 | .31754 | 3.14922 | .33686 | 2.96858 | .35641 | 2.80574 | 23 |
| 38 | .27983 | 3.57357 | .29875 | 3.34732 | .31786 | 3.14605 | .33718 | 2.96573 | .35674 | 2.80316 | 22 |
| 39 | .28015 | 3.56957 | .29906 | 3.34377 | .31818 | 3.14288 | .33751 | 2.96288 | .35707 | 2.80059 | 21 |
| 40 | .28046 | 3.56557 | .29938 | 3.34023 | .31850 | 3.13972 | .33783 | 2.96004 | .35740 | 2.79802 | 20 |
| 41 | .28077 | 3.56159 | .29970 | 3.33670 | .31882 | 3.13656 | .33816 | 2.95721 | .35772 | 2.79545 | 19 |
| 42 | .28109 | 3.55761 | .30001 | 3.33317 | .31914 | 3.13341 | .33848 | 2.95437 | .35805 | 2.79289 | 18 |
| 43 | .28140 | 3.55364 | .30033 | 3.32965 | .31946 | 3.13027 | .33881 | 2.95155 | .35838 | 2.79033 | 17 |
| 44 | .28172 | 3.54968 | .30065 | 3.32614 | .31978 | 3.12713 | .33913 | 2.94872 | .35871 | 2.78778 | 16 |
| 45 | .28203 | 3.54573 | .30097 | 3.32264 | .32010 | 3.12400 | .33945 | 2.94591 | .35904 | 2.78523 | 15 |
| 46 | .28234 | 3.54179 | .30128 | 3.31914 | .32042 | 3.12087 | .33978 | 2.94309 | .35937 | 2.78269 | 14 |
| 47 | .28266 | 3.53785 | .30160 | 3.31565 | .32074 | 3.11775 | .34010 | 2.94028 | .35969 | 2.78014 | 13 |
| 48 | .28297 | 3.53393 | .30192 | 3.31216 | .32106 | 3.11464 | .34043 | 2.93748 | .36002 | 2.77761 | 12 |
| 49 | .28329 | 3.53001 | .30224 | 3.30868 | .32139 | 3.11153 | .34075 | 2.93468 | .36035 | 2.77507 | 11 |
| 50 | .28360 | 3.52609 | .30255 | 3.30521 | .32171 | 3.10842 | .34108 | 2.93189 | .36068 | 2.77254 | 10 |
| 51 | .28391 | 3.52219 | .30287 | 3.30174 | .32203 | 3.10532 | .34140 | 2.92910 | .36101 | 2.77002 | 9 |
| 52 | .28423 | 3.51829 | .30319 | 3.29829 | .32235 | 3.10223 | .34173 | 2.92632 | .36134 | 2.76750 | 8 |
| 53 | .28454 | 3.51441 | .30351 | 3.29483 | .32267 | 3.09914 | .34205 | 2.92354 | .36167 | 2.76498 | 7 |
| 54 | .28486 | 3.51053 | .30382 | 3.29139 | .32299 | 3.09606 | .34238 | 2.92076 | .36199 | 2.76247 | 6 |
| 55 | .28517 | 3.50666 | .30414 | 3.28795 | .32331 | 3.09298 | .34270 | 2.91799 | .36232 | 2.75996 | 5 |
| 56 | .28549 | 3.50279 | .30446 | 3.28452 | .32363 | 3.08991 | .34303 | 2.91523 | .36265 | 2.75746 | 4 |
| 57 | .28580 | 3.49894 | .30478 | 3.28109 | .32396 | 3.08685 | .34335 | 2.91246 | .36298 | 2.75496 | 3 |
| 58 | .28612 | 3.49509 | .30509 | 3.27767 | .32428 | 3.08379 | .34368 | 2.90971 | .36331 | 2.75246 | 2 |
| 59 | .28643 | 3.49125 | .30541 | 3.27426 | .32460 | 3.08073 | .34400 | 2.90696 | .36364 | 2.74997 | 1 |
| 60 | .28675 | 3.48741 | .30573 | 3.27085 | .32492 | 3.07768 | .34433 | 2.90421 | .36397 | 2.74748 | 0 |
| ′ | Cotang | Tang | Cotang | Tang | Cotang | Tang | Cotang | Tang | Cotang | Tang | ′ |
| | 74° | | 73° | | 72° | | 71° | | 70° | | |

## TABLE 2 cont.

Natural Tangents and Cotangents

| ′ | 20° | | 21° | | 22° | | 23° | | 24° | | ′ |
|---|---|---|---|---|---|---|---|---|---|---|---|
| | Tang | Cotang | Tang | Cotang | Tang | Cotang | Tang | Cotang | Tang | Cotang | |
| 0 | .36397 | 2.74748 | .38386 | 2.60509 | .40403 | 2.47509 | .42447 | 2.35585 | .44523 | 2.24604 | 60 |
| 1 | .36430 | 2.74499 | .38420 | 2.60283 | .40436 | 2.47302 | .42482 | 2.35395 | .44558 | 2.24428 | 59 |
| 2 | .36463 | 2.74251 | .38453 | 2.60057 | .40470 | 2.47095 | .42516 | 2.35205 | .44593 | 2.24252 | 58 |
| 3 | .36496 | 2.74004 | .38487 | 2.59831 | .40504 | 2.46888 | .42551 | 2.35015 | .44627 | 2.24077 | 57 |
| 4 | .36529 | 2.73756 | .38520 | 2.59606 | .40538 | 2.46682 | .42585 | 2.34825 | .44662 | 2.23902 | 56 |
| 5 | .36562 | 2.73509 | .38553 | 2.59381 | .40572 | 2.46476 | .42619 | 2.34636 | .44697 | 2.23727 | 55 |
| 6 | .36595 | 2.73263 | .38587 | 2.59156 | .40606 | 2.46270 | .42654 | 2.34447 | .44732 | 2.23553 | 54 |
| 7 | .36628 | 2.73017 | .38620 | 2.58932 | .40640 | 2.46065 | .42688 | 2.34258 | .44767 | 2.23378 | 53 |
| 8 | .36661 | 2.72771 | .38654 | 2.58708 | .40674 | 2.45860 | .42722 | 2.34069 | .44802 | 2.23204 | 52 |
| 9 | .36694 | 2.72526 | .38687 | 2.58484 | .40707 | 2.45655 | .42757 | 2.33881 | .44837 | 2.23030 | 51 |
| 10 | .36727 | 2.72281 | .38721 | 2.58261 | .40741 | 2.45451 | .42791 | 2.33693 | .44872 | 2.22857 | 50 |
| 11 | .36760 | 2.72036 | .38754 | 2.58038 | .40775 | 2.45246 | .42826 | 2.33505 | .44907 | 2.22683 | 49 |
| 12 | .36793 | 2.71792 | .38787 | 2.57815 | .40809 | 2.45043 | .42860 | 2.33317 | .44942 | 2.22510 | 48 |
| 13 | .36826 | 2.71548 | .38821 | 2.57593 | .40843 | 2.44839 | .42894 | 2.33130 | .44977 | 2.22337 | 47 |
| 14 | .36859 | 2.71305 | .38854 | 2.57371 | .40877 | 2.44636 | .42929 | 2.32943 | .45012 | 2.22164 | 46 |
| 15 | .36892 | 2.71062 | .38888 | 2.57150 | .40911 | 2.44433 | .42963 | 2.32756 | .45047 | 2.21992 | 45 |
| 16 | .36925 | 2.70819 | .38921 | 2.56928 | .40945 | 2.44230 | .42998 | 2.32570 | .45082 | 2.21819 | 44 |
| 17 | .36958 | 2.70577 | .38955 | 2.56707 | .40979 | 2.44027 | .43032 | 2.32383 | .45117 | 2.21647 | 43 |
| 18 | .36991 | 2.70335 | .38988 | 2.56487 | .41013 | 2.43825 | .43067 | 2.32197 | .45152 | 2.21475 | 42 |
| 19 | .37024 | 2.70094 | .39022 | 2.56266 | .41047 | 2.43623 | .43101 | 2.32012 | .45187 | 2.21304 | 41 |
| 20 | .37057 | 2.69853 | .39055 | 2.56046 | .41081 | 2.43422 | .43136 | 2.31826 | .45222 | 2.21132 | 40 |
| 21 | .37090 | 2.69612 | .39089 | 2.55827 | .41115 | 2.43220 | .43170 | 2.31641 | .45257 | 2.20961 | 39 |
| 22 | .37123 | 2.69371 | .39122 | 2.55608 | .41149 | 2.43019 | .43205 | 2.31456 | .45292 | 2.20790 | 38 |
| 23 | .37157 | 2.69131 | .39156 | 2.55389 | .41183 | 2.42819 | .43230 | 2.31271 | .45327 | 2.20619 | 37 |
| 24 | .37190 | 2.68892 | .39190 | 2.55170 | .41217 | 2.42618 | .43274 | 2.31086 | .45362 | 2.20449 | 36 |
| 25 | .37223 | 2.68653 | .39223 | 2.54952 | .41251 | 2.42418 | .43308 | 2.30902 | .45397 | 2.20278 | 35 |
| 26 | .37256 | 2.68414 | .39257 | 2.54734 | .41285 | 2.42218 | .43343 | 2.30718 | .45432 | 2.20108 | 34 |
| 27 | .37289 | 2.68175 | .39290 | 2.54516 | .41319 | 2.42019 | .43378 | 2.30534 | .45467 | 2.19938 | 33 |
| 28 | .37322 | 2.67937 | .39324 | 2.54299 | .41353 | 2.41819 | .43412 | 2.30351 | .45502 | 2.19769 | 32 |
| 29 | .37355 | 2.67700 | .39357 | 2.54082 | .41387 | 2.41620 | .43447 | 2.30167 | .45538 | 2.19599 | 31 |
| 30 | .37388 | 2.67462 | .39391 | 2.53865 | .41421 | 2.41421 | .43481 | 2.29984 | .45573 | 2.19430 | 30 |
| 31 | .37422 | 2.67225 | .39425 | 2.53648 | .41455 | 2.41223 | .43516 | 2.29801 | .45608 | 2.19261 | 29 |
| 32 | .37455 | 2.66989 | .39458 | 2.53432 | .41490 | 2.41025 | .43550 | 2.29619 | .45643 | 2.19092 | 28 |
| 33 | .37488 | 2.66752 | .39492 | 2.53217 | .41524 | 2.40827 | .43585 | 2.29437 | .45678 | 2.18923 | 27 |
| 34 | .37521 | 2.66516 | .39526 | 2.53001 | .41558 | 2.40629 | .43620 | 2.29254 | .45713 | 2.18755 | 26 |
| 35 | .37554 | 2.66281 | .39559 | 2.52786 | .41592 | 2.40432 | .43654 | 2.29073 | .45748 | 2.18587 | 25 |
| 36 | .37588 | 2.66046 | .39593 | 2.52571 | .41626 | 2.40235 | .43689 | 2.28891 | .45784 | 2.18419 | 24 |
| 37 | .37621 | 2.65811 | .39626 | 2.52357 | .41660 | 2.40038 | .43724 | 2.28710 | .45819 | 2.18251 | 23 |
| 38 | .37654 | 2.65576 | .39660 | 2.52142 | .41694 | 2.39841 | .43758 | 2.28528 | .45854 | 2.18084 | 22 |
| 39 | .37687 | 2.65342 | .39694 | 2.51929 | .41728 | 2.39645 | .43793 | 2.28348 | .45889 | 2.17916 | 21 |
| 40 | .37720 | 2.65109 | .39727 | 2.51715 | .41763 | 2.39449 | .43828 | 2.28167 | .45924 | 2.17749 | 20 |
| 41 | .37754 | 2.64875 | .39761 | 2.51502 | .41797 | 2.39253 | .43862 | 2.27987 | .45960 | 2.17582 | 19 |
| 42 | .37787 | 2.64642 | .39795 | 2.51289 | .41831 | 2.39058 | .43897 | 2.27806 | .45995 | 2.17416 | 18 |
| 43 | .37820 | 2.64410 | .39829 | 2.51076 | .41865 | 2.38863 | .43932 | 2.27626 | .46030 | 2.17249 | 17 |
| 44 | .37853 | 2.64177 | .39862 | 2.50864 | .41899 | 2.38668 | .43966 | 2.27447 | .46065 | 2.17083 | 16 |
| 45 | .37887 | 2.63945 | .39896 | 2.50652 | .41933 | 2.38473 | .44001 | 2.27267 | .46101 | 2.16917 | 15 |
| 46 | .37920 | 2.63714 | .39930 | 2.50440 | .41968 | 2.38279 | .44036 | 2.27088 | .46136 | 2.16751 | 14 |
| 47 | .37953 | 2.63483 | .39963 | 2.50229 | .42002 | 2.38084 | .44071 | 2.26909 | .46171 | 2.16585 | 13 |
| 48 | .37986 | 2.63252 | .39997 | 2.50018 | .42036 | 2.37891 | .44105 | 2.26730 | .46206 | 2.16420 | 12 |
| 49 | .38020 | 2.63021 | .40031 | 2.49807 | .42070 | 2.37697 | .44140 | 2.26552 | .46242 | 2.16255 | 11 |
| 50 | .38053 | 2.62791 | .40065 | 2.49597 | .42105 | 2.37504 | .44175 | 2.26374 | .46277 | 2.16090 | 10 |
| 51 | .38086 | 2.62561 | .40098 | 2.49386 | .42139 | 2.37311 | .44210 | 2.26196 | .46312 | 2.15925 | 9 |
| 52 | .38120 | 2.62332 | .40132 | 2.49177 | .42173 | 2.37118 | .44244 | 2.26018 | .46348 | 2.15760 | 8 |
| 53 | .38153 | 2.62103 | .40166 | 2.48967 | .42207 | 2.36925 | .44279 | 2.25840 | .46383 | 2.15596 | 7 |
| 54 | .38186 | 2.61874 | .40200 | 2.48758 | .42242 | 2.36733 | .44314 | 2.25663 | .46418 | 2.15432 | 6 |
| 55 | .38220 | 2.61646 | .40234 | 2.48549 | .42276 | 2.36541 | .44349 | 2.25486 | .46454 | 2.15268 | 5 |
| 56 | .38253 | 2.61418 | .40267 | 2.48340 | .42310 | 2.36349 | .44384 | 2.25309 | .46489 | 2.15104 | 4 |
| 57 | .38286 | 2.61190 | .40301 | 2.48132 | .42345 | 2.36158 | .44418 | 2.25132 | .46525 | 2.14940 | 3 |
| 58 | .38320 | 2.60963 | .40335 | 2.47924 | .42379 | 2.35967 | .44453 | 2.24956 | .46560 | 2.14777 | 2 |
| 59 | .38353 | 2.60736 | .40369 | 2.47716 | .42413 | 2.35776 | .44488 | 2.24780 | .46595 | 2.14614 | 1 |
| 60 | .38386 | 2.60509 | .40403 | 2.47509 | .42447 | 2.35585 | .44523 | 2.24604 | .46631 | 2.14451 | 0 |
| | Cotang | Tang | Cotang | Tang | Cotang | Tang | Cotang | Tang | Cotang | Tang | |
| ′ | 69° | | 68° | | 67° | | 66° | | 65° | | ′ |

## TABLE 2 cont.
Natural Tangents and Cotangents

| ′ | 25° | | 26° | | 27° | | 28° | | 29° | | ′ |
|---|---|---|---|---|---|---|---|---|---|---|---|
| | Tang | Cotang | Tang | Cotang | Tang | Cotang | Tang | Cotang | Tang | Cotang | |
| 0 | .46631 | 2.14451 | .48773 | 2.05030 | .50953 | 1.96261 | .53171 | 1.88073 | .55431 | 1.80405 | 60 |
| 1 | .46666 | 2.14288 | .48809 | 2.04879 | .50989 | 1.96120 | .53208 | 1.87941 | .55469 | 1.80281 | 59 |
| 2 | .46702 | 2.14125 | .48845 | 2.04728 | .51026 | 1.95979 | .53246 | 1.87809 | .55507 | 1.80158 | 58 |
| 3 | .46737 | 2.13963 | .48881 | 2.04577 | .51063 | 1.95838 | .53283 | 1.87677 | .55545 | 1.80034 | 57 |
| 4 | .46772 | 2.13801 | .48917 | 2.04426 | .51099 | 1.95698 | .53320 | 1.87546 | .55583 | 1.79911 | 56 |
| 5 | .46808 | 2.13639 | .48953 | 2.04276 | .51136 | 1.95557 | .53358 | 1.87415 | .55621 | 1.79788 | 55 |
| 6 | .46843 | 2.13477 | .48989 | 2.04125 | .51173 | 1.95417 | .53395 | 1.87283 | .55659 | 1.79665 | 54 |
| 7 | .46879 | 2.13316 | .49026 | 2.03975 | .51209 | 1.95277 | .53432 | 1.87152 | .55697 | 1.79542 | 53 |
| 8 | .46914 | 2.13154 | .49062 | 2.03825 | .51246 | 1.95137 | .53470 | 1.87021 | .55736 | 1.79419 | 52 |
| 9 | .46950 | 2.12993 | .49098 | 2.03675 | .51283 | 1.94997 | .53507 | 1.86891 | .55774 | 1.79296 | 51 |
| 10 | .46985 | 2.12832 | .49134 | 2.03526 | .51319 | 1.94858 | .53545 | 1.86760 | .55812 | 1.79174 | 50 |
| 11 | .47021 | 2.12671 | .49170 | 2.03376 | .51356 | 1.94718 | .53582 | 1.86630 | .55850 | 1.79051 | 49 |
| 12 | .47056 | 2.12511 | .49206 | 2.03227 | .51393 | 1.94579 | .53620 | 1.86499 | .55888 | 1.78929 | 48 |
| 13 | .47092 | 2.12350 | .49242 | 2.03078 | .51430 | 1.94440 | .53657 | 1.86369 | .55926 | 1.78807 | 47 |
| 14 | .47128 | 2.12190 | .49278 | 2.02929 | .51467 | 1.94301 | .53694 | 1.86239 | .55964 | 1.78685 | 46 |
| 15 | .47163 | 2.12030 | .49315 | 2.02780 | .51503 | 1.94162 | .53732 | 1.86109 | .56003 | 1.78563 | 45 |
| 16 | .47199 | 2.11871 | .49351 | 2.02631 | .51540 | 1.94023 | .53769 | 1.85979 | .56041 | 1.78441 | 44 |
| 17 | .47234 | 2.11711 | .49387 | 2.02483 | .51577 | 1.93885 | .53807 | 1.85850 | .56079 | 1.78319 | 43 |
| 18 | .47270 | 2.11552 | .49423 | 2.02335 | .51614 | 1.93746 | .53844 | 1.85720 | .56117 | 1.78198 | 42 |
| 19 | .47305 | 2.11392 | .49459 | 2.02187 | .51651 | 1.93608 | .53882 | 1.85591 | .56156 | 1.78077 | 41 |
| 20 | .47341 | 2.11233 | .49495 | 2.02039 | .51688 | 1.93470 | .53920 | 1.85462 | .56194 | 1.77955 | 40 |
| 21 | .47377 | 2.11075 | .49532 | 2.01891 | .51724 | 1.93332 | .53957 | 1.85333 | .56232 | 1.77834 | 39 |
| 22 | .47412 | 2.10916 | .49568 | 2.01743 | .51761 | 1.93195 | .53995 | 1.85204 | .56270 | 1.77713 | 38 |
| 23 | .47448 | 2.10758 | .49604 | 2.01596 | .51798 | 1.93057 | .54032 | 1.85075 | .56309 | 1.77592 | 37 |
| 24 | .47483 | 2.10600 | .49640 | 2.01449 | .51835 | 1.92920 | .54070 | 1.84946 | .56347 | 1.77471 | 36 |
| 25 | .47519 | 2.10442 | .49677 | 2.01302 | .51872 | 1.92782 | .54107 | 1.84818 | .56385 | 1.77351 | 35 |
| 26 | .47555 | 2.10284 | .49713 | 2.01155 | .51909 | 1.92645 | .54145 | 1.84689 | .56424 | 1.77230 | 34 |
| 27 | .47590 | 2.10126 | .49749 | 2.01008 | .51946 | 1.92508 | .54183 | 1.84561 | .56462 | 1.77110 | 33 |
| 28 | .47626 | 2.09969 | .49786 | 2.00862 | .51983 | 1.92371 | .54220 | 1.84433 | .56501 | 1.76990 | 32 |
| 29 | .47662 | 2.09811 | .49822 | 2.00715 | .52020 | 1.92235 | .54258 | 1.84305 | .56539 | 1.76869 | 31 |
| 30 | .47698 | 2.09654 | .49858 | 2.00569 | .52057 | 1.92098 | .54296 | 1.84177 | .56577 | 1.76749 | 30 |
| 31 | .47733 | 2.09498 | .49894 | 2.00423 | .52094 | 1.91962 | .54333 | 1.84049 | .56616 | 1.76629 | 29 |
| 32 | .47769 | 2.09341 | .49931 | 2.00277 | .52131 | 1.91826 | .54371 | 1.83922 | .56654 | 1.76510 | 28 |
| 33 | .47805 | 2.09184 | .49967 | 2.00131 | .52168 | 1.91690 | .54409 | 1.83794 | .56693 | 1.76390 | 27 |
| 34 | .47840 | 2.09028 | .50004 | 1.99986 | .52205 | 1.91554 | .54446 | 1.83667 | .56731 | 1.76271 | 26 |
| 35 | .47876 | 2.08872 | .50040 | 1.99841 | .52242 | 1.91418 | .54484 | 1.83540 | .56769 | 1.76151 | 25 |
| 36 | .47912 | 2.08716 | .50076 | 1.99695 | .52279 | 1.91282 | .54522 | 1.83413 | .56808 | 1.76032 | 24 |
| 37 | .47948 | 2.08560 | .50113 | 1.99550 | .52316 | 1.91147 | .54560 | 1.83286 | .56846 | 1.75913 | 23 |
| 38 | .47984 | 2.08405 | .50149 | 1.99406 | .52353 | 1.91012 | .54597 | 1.83159 | .56885 | 1.75794 | 22 |
| 39 | .48019 | 2.08250 | .50185 | 1.99261 | .52390 | 1.90876 | .54635 | 1.83033 | .56923 | 1.75675 | 21 |
| 40 | .48055 | 2.08094 | .50222 | 1.99116 | .52427 | 1.90741 | .54673 | 1.82906 | .56962 | 1.75556 | 20 |
| 41 | .48091 | 2.07939 | .50258 | 1.98972 | .52464 | 1.90607 | .54711 | 1.82780 | .57000 | 1.75437 | 19 |
| 42 | .48127 | 2.07785 | .50295 | 1.98828 | .52501 | 1.90472 | .54748 | 1.82654 | .57039 | 1.75319 | 18 |
| 43 | .48163 | 2.07630 | .50331 | 1.98684 | .52538 | 1.90337 | .54786 | 1.82528 | .57078 | 1.75200 | 17 |
| 44 | .48198 | 2.07476 | .50368 | 1.98540 | .52575 | 1.90203 | .54824 | 1.82402 | .57116 | 1.75082 | 16 |
| 45 | .48234 | 2.07321 | .50404 | 1.98396 | .52613 | 1.90069 | .54862 | 1.82276 | .57155 | 1.74964 | 15 |
| 46 | .48270 | 2.07167 | .50441 | 1.98253 | .52650 | 1.89935 | .54900 | 1.82150 | .57193 | 1.74846 | 14 |
| 47 | .48306 | 2.07014 | .50477 | 1.98110 | .52687 | 1.89801 | .54938 | 1.82025 | .57232 | 1.74728 | 13 |
| 48 | .48342 | 2.06860 | .50514 | 1.97966 | .52724 | 1.89667 | .54975 | 1.81899 | .57271 | 1.74610 | 12 |
| 49 | .48378 | 2.06706 | .50550 | 1.97823 | .52761 | 1.89533 | .55013 | 1.81774 | .57309 | 1.74492 | 11 |
| 50 | .48414 | 2.06553 | .50587 | 1.97681 | .52798 | 1.89400 | .55051 | 1.81649 | .57348 | 1.74375 | 10 |
| 51 | .48450 | 2.06400 | .50623 | 1.97538 | .52836 | 1.89266 | .55089 | 1.81524 | .57386 | 1.74257 | 9 |
| 52 | .48486 | 2.06247 | .50660 | 1.97395 | .52873 | 1.89133 | .55127 | 1.81399 | .57425 | 1.74140 | 8 |
| 53 | .48521 | 2.06094 | .50696 | 1.97253 | .52910 | 1.89000 | .55165 | 1.81274 | .57464 | 1.74022 | 7 |
| 54 | .48557 | 2.05942 | .50733 | 1.97111 | .52947 | 1.88867 | .55203 | 1.81150 | .57503 | 1.73905 | 6 |
| 55 | .48593 | 2.05790 | .50769 | 1.96969 | .52985 | 1.88734 | .55241 | 1.81025 | .57541 | 1.73788 | 5 |
| 56 | .48629 | 2.05637 | .50806 | 1.96827 | .53022 | 1.88602 | .55279 | 1.80901 | .57580 | 1.73671 | 4 |
| 57 | .48665 | 2.05485 | .50843 | 1.96685 | .53059 | 1.88469 | .55317 | 1.80777 | .57619 | 1.73555 | 3 |
| 58 | .48701 | 2.05333 | .50879 | 1.96544 | .53096 | 1.88337 | .55355 | 1.80653 | .57657 | 1.73438 | 2 |
| 59 | .48737 | 2.05182 | .50916 | 1.96402 | .53134 | 1.88205 | .55393 | 1.80529 | .57696 | 1.73321 | 1 |
| 60 | .48773 | 2.05030 | .50953 | 1.96261 | .53171 | 1.88073 | .55431 | 1.80405 | .57735 | 1.73205 | 0 |
| | Cotang | Tang | Cotang | Tang | Cotang | Tang | Cotang | Tang | Cotang | Tang | |
| ′ | 64° | | 63° | | 62° | | 61° | | 60° | | ′ |

## TABLE 2 cont.

Natural Tangents and Cotangents

| ′ | 30° Tang | 30° Cotang | 31° Tang | 31° Cotang | 32° Tang | 32° Cotang | 33° Tang | 33° Cotang | 34° Tang | 34° Cotang | ′ |
|---|---|---|---|---|---|---|---|---|---|---|---|
| 0 | .57735 | 1.73205 | .60086 | 1.66428 | .62487 | 1.60033 | .64941 | 1.53986 | .67451 | 1.48256 | 60 |
| 1 | .57774 | 1.73089 | .60126 | 1.66318 | .62527 | 1.59930 | .64982 | 1.53888 | .67493 | 1.48163 | 59 |
| 2 | .57813 | 1.72973 | .60165 | 1.66209 | .62568 | 1.59826 | .65024 | 1.53791 | .67536 | 1.48070 | 58 |
| 3 | .57851 | 1.72857 | .60205 | 1.66099 | .62608 | 1.59723 | .65065 | 1.53693 | .67578 | 1.47977 | 57 |
| 4 | .57890 | 1.72741 | .60245 | 1.65990 | .62649 | 1.59620 | .65106 | 1.53595 | .67620 | 1.47885 | 56 |
| 5 | .57929 | 1.72625 | .60284 | 1.65881 | .62689 | 1.59517 | .65148 | 1.53497 | .67663 | 1.47792 | 55 |
| 6 | .57968 | 1.72509 | .60324 | 1.65772 | .62730 | 1.59414 | .65189 | 1.53400 | .67705 | 1.47699 | 54 |
| 7 | .58007 | 1.72393 | .60364 | 1.65663 | .62770 | 1.59311 | .65231 | 1.53302 | .67748 | 1.47607 | 53 |
| 8 | .58046 | 1.72278 | .60403 | 1.65554 | .62811 | 1.59208 | .65272 | 1.53205 | .67790 | 1.47514 | 52 |
| 9 | .58085 | 1.72163 | .60443 | 1.65445 | .62852 | 1.59105 | .65314 | 1.53107 | .67832 | 1.47422 | 51 |
| 10 | .58124 | 1.72047 | .60483 | 1.65337 | .62892 | 1.59002 | .65355 | 1.53010 | .67875 | 1.47330 | 50 |
| 11 | .58162 | 1.71932 | .60522 | 1.65228 | .62933 | 1.58900 | .65397 | 1.52913 | .67917 | 1.47238 | 49 |
| 12 | .58201 | 1.71817 | .60562 | 1.65120 | .62973 | 1.58797 | .65438 | 1.52816 | .67960 | 1.47146 | 48 |
| 13 | .58240 | 1.71702 | .60602 | 1.65011 | .63014 | 1.58695 | .65480 | 1.52719 | .68002 | 1.47053 | 47 |
| 14 | .58279 | 1.71588 | .60642 | 1.64903 | .63055 | 1.58593 | .65521 | 1.52622 | .68045 | 1.46962 | 46 |
| 15 | .58318 | 1.71473 | .60681 | 1.64795 | .63095 | 1.58490 | .65563 | 1.52525 | .68088 | 1.46870 | 45 |
| 16 | .58357 | 1.71358 | .60721 | 1.64687 | .63136 | 1.58388 | .65604 | 1.52429 | .68130 | 1.46778 | 44 |
| 17 | .58396 | 1.71244 | .60761 | 1.64579 | .63177 | 1.58286 | .65646 | 1.52332 | .68173 | 1.46686 | 43 |
| 18 | .58435 | 1.71129 | .60801 | 1.64471 | .63217 | 1.58184 | .65688 | 1.52235 | .68215 | 1.46595 | 42 |
| 19 | .58474 | 1.71015 | .60841 | 1.64363 | .63258 | 1.58083 | .65729 | 1.52139 | .68258 | 1.46503 | 41 |
| 20 | .58513 | 1.70901 | .60881 | 1.64256 | .63299 | 1.57981 | .65771 | 1.52043 | .68301 | 1.46411 | 40 |
| 21 | .58552 | 1.70787 | .60921 | 1.64148 | .63340 | 1.57879 | .65813 | 1.51946 | .68343 | 1.46320 | 39 |
| 22 | .58591 | 1.70673 | .60960 | 1.64041 | .63380 | 1.57778 | .65854 | 1.51850 | .68386 | 1.46229 | 38 |
| 23 | .58631 | 1.70560 | .61000 | 1.63934 | .63421 | 1.57676 | .65896 | 1.51754 | .68429 | 1.46137 | 37 |
| 24 | .58670 | 1.70446 | .61040 | 1.63826 | .63462 | 1.57575 | .65938 | 1.51658 | .68471 | 1.46046 | 36 |
| 25 | .58709 | 1.70332 | .61080 | 1.63719 | .63503 | 1.57474 | .65980 | 1.51562 | .68514 | 1.45955 | 35 |
| 26 | .58748 | 1.70219 | .61120 | 1.63612 | .63544 | 1.57372 | .66021 | 1.51466 | .68557 | 1.45864 | 34 |
| 27 | .58787 | 1.70106 | .61160 | 1.63505 | .63584 | 1.57271 | .66063 | 1.51370 | .68600 | 1.45773 | 33 |
| 28 | .58826 | 1.69992 | .61200 | 1.63398 | .63625 | 1.57170 | .66105 | 1.51275 | .68642 | 1.45682 | 32 |
| 29 | .58865 | 1.69879 | .61240 | 1.63292 | .63666 | 1.57069 | .66147 | 1.51179 | .68685 | 1.45592 | 31 |
| 30 | .58905 | 1.69766 | .61280 | 1.63185 | .63707 | 1.56969 | .66189 | 1.51084 | .68728 | 1.45501 | 30 |
| 31 | .58944 | 1.69653 | .61320 | 1.63079 | .63748 | 1.56868 | .66230 | 1.50988 | .68771 | 1.45410 | 29 |
| 32 | .58983 | 1.69541 | .61360 | 1.62972 | .63789 | 1.56767 | .66272 | 1.50893 | .68814 | 1.45320 | 28 |
| 33 | .59022 | 1.69428 | .61400 | 1.62866 | .63830 | 1.56667 | .66314 | 1.50797 | .68857 | 1.45229 | 27 |
| 34 | .59061 | 1.69316 | .61440 | 1.62760 | .63871 | 1.56566 | .66356 | 1.50702 | .68900 | 1.45139 | 26 |
| 35 | .59101 | 1.69203 | .61480 | 1.62654 | .63912 | 1.56466 | .66398 | 1.50607 | .68942 | 1.45049 | 25 |
| 36 | .59140 | 1.69091 | .61520 | 1.62548 | .63953 | 1.56366 | .66440 | 1.50512 | .68985 | 1.44958 | 24 |
| 37 | .59179 | 1.68979 | .61561 | 1.62442 | .63994 | 1.56265 | .66482 | 1.50417 | .69028 | 1.44868 | 23 |
| 38 | .59218 | 1.68866 | .61601 | 1.62336 | .64035 | 1.56165 | .66524 | 1.50322 | .69071 | 1.44778 | 22 |
| 39 | .59258 | 1.68754 | .61641 | 1.62230 | .64076 | 1.56065 | .66566 | 1.50228 | .69114 | 1.44688 | 21 |
| 49 | .59297 | 1.68643 | .61681 | 1.62125 | .64117 | 1.55966 | .66608 | 1.50133 | .69157 | 1.44598 | 20 |
| 41 | .59336 | 1.68531 | .61721 | 1.62019 | .64158 | 1.55866 | .66650 | 1.50038 | .69200 | 1.44508 | 19 |
| 42 | .59376 | 1.68419 | .61761 | 1.61914 | .64199 | 1.55766 | .66692 | 1.49944 | .69243 | 1.44418 | 18 |
| 43 | .59415 | 1.68308 | .61801 | 1.61808 | .64240 | 1.55666 | .66734 | 1.49849 | .69286 | 1.44329 | 17 |
| 44 | .59454 | 1.68196 | .61842 | 1.61703 | .64281 | 1.55567 | .66776 | 1.49755 | .69329 | 1.44239 | 16 |
| 45 | .59494 | 1.68085 | .61882 | 1.61598 | .64322 | 1.55467 | .66818 | 1.49661 | .69372 | 1.44149 | 15 |
| 46 | .59533 | 1.67974 | .61922 | 1.61493 | .64363 | 1.55368 | .66860 | 1.49566 | .69416 | 1.44060 | 14 |
| 47 | .59573 | 1.67863 | .61962 | 1.61388 | .64404 | 1.55269 | .66902 | 1.49472 | .69459 | 1.43970 | 13 |
| 48 | .59612 | 1.67752 | .62003 | 1.61283 | .64446 | 1.55170 | .66944 | 1.49378 | .69502 | 1.43881 | 12 |
| 49 | .59651 | 1.67641 | .62043 | 1.61179 | .64487 | 1.55071 | .66986 | 1.49284 | .69545 | 1.43792 | 11 |
| 50 | .59691 | 1.67530 | .62083 | 1.61074 | .64528 | 1.54972 | .67028 | 1.49190 | .69588 | 1.43703 | 10 |
| 51 | .59730 | 1.67419 | .62124 | 1.60970 | .64569 | 1.54873 | .67071 | 1.49097 | .69631 | 1.43614 | 9 |
| 52 | .59770 | 1.67309 | .62164 | 1.60865 | .64610 | 1.54774 | .67113 | 1.49003 | .69675 | 1.43525 | 8 |
| 53 | .59809 | 1.67198 | .62204 | 1.60761 | .64652 | 1.54675 | .67155 | 1.48909 | .69718 | 1.43436 | 7 |
| 54 | .59849 | 1.67088 | .62245 | 1.60657 | .64693 | 1.54576 | .67197 | 1.48816 | .69761 | 1.43347 | 6 |
| 55 | .59888 | 1.66978 | .62285 | 1.60553 | .64734 | 1.54478 | .67239 | 1.48722 | .69804 | 1.43258 | 5 |
| 56 | .59928 | 1.66867 | .62325 | 1.60449 | .64775 | 1.54379 | .67282 | 1.48629 | .69847 | 1.43169 | 4 |
| 57 | .59967 | 1.66757 | .62366 | 1.60345 | .64817 | 1.54281 | .67324 | 1.48536 | .69891 | 1.43080 | 3 |
| 58 | .60007 | 1.66647 | .62406 | 1.60241 | .64858 | 1.54183 | .67366 | 1.48442 | .69934 | 1.42992 | 2 |
| 59 | .60046 | 1.66538 | .62446 | 1.60137 | .64899 | 1.54085 | .67409 | 1.48349 | .69977 | 1.42903 | 1 |
| 60 | .60086 | 1.66428 | .62487 | 1.60033 | .64941 | 1.53986 | .67451 | 1.48256 | .70021 | 1.42815 | 0 |
| ′ | 59° Cotang | 59° Tang | 58° Cotang | 58° Tang | 57° Cotang | 57° Tang | 56° Cotang | 56° Tang | 55° Cotang | 55° Tang | ′ |

## TABLE 2 cont.

Natural Tangents and Cotangents

| ′ | 35° Tang | 35° Cotang | 36° Tang | 36° Cotang | 37° Tang | 37° Cotang | 38° Tang | 38° Cotang | 39° Tang | 39° Cotang | ′ |
|---|---|---|---|---|---|---|---|---|---|---|---|
| 0 | .70021 | 1.42815 | .72654 | 1.37638 | .75355 | 1.32704 | .78129 | 1.27994 | .80978 | 1.23490 | 60 |
| 1 | .70064 | 1.42726 | .72699 | 1.37554 | .75401 | 1.32624 | .78175 | 1.27917 | .81027 | 1.23416 | 59 |
| 2 | .70107 | 1.42638 | .72743 | 1.37470 | .75447 | 1.32544 | .78222 | 1.27841 | .81075 | 1.23343 | 58 |
| 3 | .70151 | 1.42550 | .72788 | 1.37386 | .75492 | 1.32464 | .78269 | 1.27764 | .81123 | 1.23270 | 57 |
| 4 | .70194 | 1.42462 | .72832 | 1.37302 | .75538 | 1.32384 | .78316 | 1.27688 | .81171 | 1.23196 | 56 |
| 5 | .70238 | 1.42374 | .72877 | 1.37218 | .75584 | 1.32304 | .78363 | 1.27611 | .81220 | 1.23123 | 55 |
| 6 | .70281 | 1.42286 | .72921 | 1.37134 | .75629 | 1.32224 | .78410 | 1.27535 | .81268 | 1.23050 | 54 |
| 7 | .70325 | 1.42198 | .72966 | 1.37050 | .75675 | 1.32144 | .78457 | 1.27458 | .81316 | 1.22977 | 53 |
| 8 | .70368 | 1.42110 | .73010 | 1.36967 | .75721 | 1.32064 | .78504 | 1.27382 | .81364 | 1.22904 | 52 |
| 9 | .70412 | 1.42022 | .73055 | 1.36883 | .75767 | 1.31984 | .78551 | 1.27306 | .81413 | 1.22831 | 51 |
| 10 | .70455 | 1.41934 | .73100 | 1.36800 | .75812 | 1.31904 | .78598 | 1.27230 | .81461 | 1.22758 | 50 |
| 11 | .70499 | 1.41847 | .73144 | 1.36716 | .75858 | 1.31825 | .78645 | 1.27153 | .81510 | 1.22685 | 49 |
| 12 | .70542 | 1.41759 | .73189 | 1.36633 | .75904 | 1.31745 | .78692 | 1.27077 | .81558 | 1.22612 | 48 |
| 13 | .70586 | 1.41672 | .73234 | 1.36549 | .75950 | 1.31666 | .78739 | 1.27001 | .81606 | 1.22539 | 47 |
| 14 | .70629 | 1.41584 | .73278 | 1.36466 | .75996 | 1.31586 | .78786 | 1.26925 | .81655 | 1.22467 | 46 |
| 15 | .70673 | 1.41497 | .73323 | 1.36383 | .76042 | 1.31507 | .78834 | 1.26849 | .81703 | 1.22394 | 45 |
| 16 | .70717 | 1.41409 | .73368 | 1.36300 | .76088 | 1.31427 | .78881 | 1.26774 | .81752 | 1.22321 | 44 |
| 17 | .70760 | 1.41322 | .73413 | 1.36217 | .76134 | 1.31348 | .78928 | 1.26698 | .81800 | 1.22249 | 43 |
| 18 | .70804 | 1.41235 | .73457 | 1.36134 | .76180 | 1.31269 | .78975 | 1.26622 | .81849 | 1.22176 | 42 |
| 19 | .70848 | 1.41148 | .73502 | 1.36051 | .76226 | 1.31190 | .79022 | 1.26546 | .81898 | 1.22104 | 41 |
| 20 | .70891 | 1.41061 | .73547 | 1.35968 | .76272 | 1.31110 | .79070 | 1.26471 | .81946 | 1.22031 | 40 |
| 21 | .70935 | 1.40974 | .73592 | 1.35885 | .76318 | 1.31031 | .79117 | 1.26395 | .81995 | 1.21959 | 39 |
| 22 | .70979 | 1.40887 | .73637 | 1.35802 | .76364 | 1.30952 | .79164 | 1.26319 | .82044 | 1.21886 | 38 |
| 23 | .71023 | 1.40800 | .73681 | 1.35719 | .76410 | 1.30873 | .79212 | 1.26244 | .82092 | 1.21814 | 37 |
| 24 | .71066 | 1.40714 | .73726 | 1.35637 | .76456 | 1.30795 | .79259 | 1.26169 | .82141 | 1.21742 | 36 |
| 25 | .71110 | 1.40627 | .73771 | 1.35554 | .76502 | 1.30716 | .79306 | 1.26093 | .82190 | 1.21670 | 35 |
| 26 | .71154 | 1.40540 | .73816 | 1.35472 | .76548 | 1.30637 | .79354 | 1.26018 | .82238 | 1.21598 | 34 |
| 27 | .71198 | 1.40454 | .73861 | 1.35389 | .76594 | 1.30558 | .79401 | 1.25943 | .82287 | 1.21526 | 33 |
| 28 | .71242 | 1.40367 | .73906 | 1.35307 | .76640 | 1.30480 | .79449 | 1.25867 | .82336 | 1.21454 | 32 |
| 29 | .71285 | 1.40281 | .73951 | 1.35224 | .76686 | 1.30401 | .79496 | 1.25792 | .82385 | 1.21382 | 31 |
| 30 | .71329 | 1.40195 | .73996 | 1.35142 | .76733 | 1.30323 | .79544 | 1.25717 | .82434 | 1.21310 | 30 |
| 31 | .71373 | 1.40109 | .74041 | 1.35060 | .76779 | 1.30244 | .79591 | 1.25642 | .82483 | 1.21238 | 29 |
| 32 | .71417 | 1.40022 | .74086 | 1.34978 | .76825 | 1.30166 | .79639 | 1.25567 | .82531 | 1.21166 | 28 |
| 33 | .71461 | 1.39936 | .74131 | 1.34896 | .76871 | 1.30087 | .79686 | 1.25492 | .82580 | 1.21094 | 27 |
| 34 | .71505 | 1.39850 | .74176 | 1.34814 | .76918 | 1.30009 | .79734 | 1.25417 | .82629 | 1.21023 | 26 |
| 35 | .71549 | 1.39764 | .74221 | 1.34732 | .76964 | 1.29931 | .79781 | 1.25343 | .82678 | 1.20951 | 25 |
| 36 | .71593 | 1.39679 | .74267 | 1.34650 | .77010 | 1.29853 | .79829 | 1.25268 | .82727 | 1.20879 | 24 |
| 37 | .71637 | 1.39593 | .74312 | 1.34568 | .77057 | 1.29775 | .79877 | 1.25193 | .82776 | 1.20808 | 23 |
| 38 | .71681 | 1.39507 | .74357 | 1.34487 | .77103 | 1.29696 | .79924 | 1.25118 | .82825 | 1.20736 | 22 |
| 39 | .71725 | 1.39421 | .74402 | 1.34405 | .77149 | 1.29618 | .79972 | 1.25044 | .82874 | 1.20665 | 21 |
| 40 | .71769 | 1.39336 | .74447 | 1.34323 | .77196 | 1.29541 | .80020 | 1.24969 | .82923 | 1.20593 | 20 |
| 41 | .71813 | 1.39250 | .74492 | 1.34242 | .77242 | 1.29463 | .80067 | 1.24895 | .82972 | 1.20522 | 19 |
| 42 | .71857 | 1.39165 | .74538 | 1.34160 | .77289 | 1.29385 | .80115 | 1.24820 | .83022 | 1.20451 | 18 |
| 43 | .71901 | 1.39079 | .74583 | 1.34079 | .77335 | 1.29307 | .80163 | 1.24746 | .83071 | 1.20379 | 17 |
| 44 | .71946 | 1.38994 | .74628 | 1.33998 | .77382 | 1.29229 | .80211 | 1.24672 | .83120 | 1.20308 | 16 |
| 45 | .71990 | 1.38909 | .74674 | 1.33916 | .77428 | 1.29152 | .80258 | 1.24597 | .83169 | 1.20237 | 15 |
| 46 | .72034 | 1.38824 | .74719 | 1.33835 | .77475 | 1.29074 | .80306 | 1.24523 | .83218 | 1.20166 | 14 |
| 47 | .72078 | 1.38738 | .74764 | 1.33754 | .77521 | 1.28997 | .80354 | 1.24449 | .83268 | 1.20095 | 13 |
| 48 | .72122 | 1.38653 | .74810 | 1.33673 | .77568 | 1.28919 | .80402 | 1.24375 | .83317 | 1.20024 | 12 |
| 49 | .72167 | 1.38568 | .74855 | 1.33592 | .77615 | 1.28842 | .80450 | 1.24301 | .83366 | 1.19953 | 11 |
| 50 | .72211 | 1.38484 | .74900 | 1.33511 | .77661 | 1.28764 | .80498 | 1.24227 | .83415 | 1.19882 | 10 |
| 51 | .72255 | 1.38399 | .74946 | 1.33430 | .77708 | 1.28687 | .80546 | 1.24153 | .83465 | 1.19811 | 9 |
| 52 | .72299 | 1.38314 | .74991 | 1.33349 | .77754 | 1.28610 | .80594 | 1.24079 | .83514 | 1.19740 | 8 |
| 53 | .72344 | 1.38229 | .75037 | 1.33268 | .77801 | 1.28533 | .80642 | 1.24005 | .83564 | 1.19669 | 7 |
| 54 | .72388 | 1.38145 | .75082 | 1.33187 | .77848 | 1.28456 | .80690 | 1.23931 | .83613 | 1.19599 | 6 |
| 55 | .72432 | 1.38060 | .75128 | 1.33107 | .77895 | 1.28379 | .80738 | 1.23858 | .83662 | 1.19528 | 5 |
| 56 | .72477 | 1.37976 | .75173 | 1.33026 | .77941 | 1.28302 | .80786 | 1.23784 | .83712 | 1.19457 | 4 |
| 57 | .72521 | 1.37891 | .75219 | 1.32946 | .77988 | 1.28225 | .80834 | 1.23710 | .83761 | 1.19387 | 3 |
| 58 | .72565 | 1.37807 | .75264 | 1.32865 | .78035 | 1.28148 | .80882 | 1.23637 | .83811 | 1.19316 | 2 |
| 59 | .72610 | 1.37722 | .75310 | 1.32785 | .78082 | 1.28071 | .80930 | 1.23563 | .83860 | 1.19246 | 1 |
| 60 | .72654 | 1.37638 | .75355 | 1.32704 | .78129 | 1.27994 | .80978 | 1.23490 | .83910 | 1.19175 | 0 |
| ′ | Cotang | Tang | Cotang | Tang | Cotang | Tang | Cotang | Tang | Cotang | Tang | ′ |
| | 54° | | 53° | | 52° | | 51° | | 50° | | |

**TABLE 2 cont.**

Natural Tangents and Cotangents

| ′ | 40° | | 41° | | 42° | | 43° | | 44° | | ′ |
|---|---|---|---|---|---|---|---|---|---|---|---|
| | Tang | Cotang | Tang | Cotang | Tang | Cotang | Tang | Cotang | Tang | Cotang | |
| 0 | .83910 | 1.19175 | .86929 | 1.15037 | .90040 | 1.11061 | .93252 | 1.07237 | .96569 | 1.03553 | 60 |
| 1 | .83960 | 1.19105 | .86980 | 1.14969 | .90093 | 1.10996 | .93306 | 1.07174 | .96625 | 1.03493 | 59 |
| 2 | .84009 | 1.19035 | .87031 | 1.14902 | .90146 | 1.10931 | .93360 | 1.07112 | .96681 | 1.03433 | 58 |
| 3 | .84059 | 1.18964 | .87082 | 1.14834 | .90199 | 1.10867 | .93415 | 1.07049 | .96738 | 1.03372 | 57 |
| 4 | .84108 | 1.18894 | .87133 | 1.14767 | .90251 | 1.10802 | .93469 | 1.06987 | .96794 | 1.03312 | 56 |
| 5 | .84158 | 1.18824 | .87184 | 1.14699 | .90304 | 1.10737 | .93524 | 1.06925 | .96850 | 1.03252 | 55 |
| 6 | .84208 | 1.18754 | .87236 | 1.14632 | .90357 | 1.10672 | .93578 | 1.06862 | .96907 | 1.03192 | 54 |
| 7 | .84258 | 1.18684 | .87287 | 1.14565 | .90410 | 1.10607 | .93633 | 1.06800 | .96963 | 1.03132 | 53 |
| 8 | .84307 | 1.18614 | .87338 | 1.14498 | .90463 | 1.10543 | .93688 | 1.06738 | .97020 | 1.03072 | 52 |
| 9 | .84357 | 1.18544 | .87389 | 1.14430 | .90516 | 1.10478 | .93742 | 1.06676 | .97076 | 1.03012 | 51 |
| 10 | .84407 | 1.18474 | .87441 | 1.14363 | .90569 | 1.10414 | .93797 | 1.06613 | .97133 | 1.02952 | 50 |
| 11 | .84457 | 1.18404 | .87492 | 1.14296 | .90621 | 1.10349 | .93852 | 1.06551 | .97189 | 1.02892 | 49 |
| 12 | .84507 | 1.18334 | .87543 | 1.14229 | .90674 | 1.10285 | .93906 | 1.06489 | .97246 | 1.02832 | 48 |
| 13 | .84556 | 1.18264 | .87595 | 1.14162 | .90727 | 1.10220 | .93961 | 1.06427 | .97302 | 1.02772 | 47 |
| 14 | .84606 | 1.18194 | .87646 | 1.14095 | .90781 | 1.10156 | .94016 | 1.06365 | .97359 | 1.02713 | 46 |
| 15 | .84656 | 1.18125 | .87698 | 1.14028 | .90834 | 1.10091 | .94071 | 1.06303 | .97416 | 1.02653 | 45 |
| 16 | .84706 | 1.18055 | .87749 | 1.13961 | .90887 | 1.10027 | .94125 | 1.06241 | .97472 | 1.02593 | 44 |
| 17 | .84756 | 1.17986 | .87801 | 1.13894 | .90940 | 1.09963 | .94180 | 1.06179 | .97529 | 1.02533 | 43 |
| 18 | .84806 | 1.17916 | .87852 | 1.13828 | .90993 | 1.09899 | .94235 | 1.06117 | .97586 | 1.02474 | 42 |
| 19 | .84856 | 1.17846 | .87904 | 1.13761 | .91046 | 1.09834 | .94290 | 1.06056 | .97643 | 1.02414 | 41 |
| 20 | .84906 | 1.17777 | .87955 | 1.13694 | .91099 | 1.09770 | .94345 | 1.05994 | .97700 | 1.02355 | 40 |
| 21 | .84956 | 1.17708 | .88007 | 1.13627 | .91153 | 1.09706 | .94400 | 1.05932 | .97756 | 1.02295 | 39 |
| 22 | .85006 | 1.17638 | .88059 | 1.13561 | .91206 | 1.09642 | .94455 | 1.05870 | .97813 | 1.02236 | 38 |
| 23 | .85057 | 1.17569 | .88110 | 1.13494 | .91259 | 1.09578 | .94510 | 1.05809 | .97870 | 1.02176 | 37 |
| 24 | .85107 | 1.17500 | .88162 | 1.13428 | .91313 | 1.09514 | .94565 | 1.05747 | .97927 | 1.02117 | 36 |
| 25 | .85157 | 1.17430 | .88214 | 1.13361 | .91366 | 1.09450 | .94620 | 1.05685 | .97984 | 1.02057 | 35 |
| 26 | .85207 | 1.17361 | .88265 | 1.13295 | .91419 | 1.09386 | .94676 | 1.05624 | .98041 | 1.01998 | 34 |
| 27 | .85257 | 1.17292 | .88317 | 1.13228 | .91473 | 1.09322 | .94731 | 1.05562 | .98098 | 1.01939 | 33 |
| 28 | .85308 | 1.17223 | .88369 | 1.13162 | .91526 | 1.09258 | .94786 | 1.05501 | .98155 | 1.01879 | 32 |
| 29 | .85358 | 1.17154 | .88421 | 1.13096 | .91580 | 1.09195 | .94841 | 1.05439 | .98213 | 1.01820 | 31 |
| 30 | .85408 | 1.17085 | .88473 | 1.13029 | .91633 | 1.09131 | .94896 | 1.05378 | .98270 | 1.01761 | 30 |
| 31 | .85458 | 1.17016 | .88524 | 1.12963 | .91687 | 1.09067 | .94952 | 1.05317 | .98327 | 1.01702 | 29 |
| 32 | .85509 | 1.16947 | .88576 | 1.12897 | .91740 | 1.09003 | .95007 | 1.05255 | .98384 | 1.01642 | 28 |
| 33 | .85559 | 1.16878 | .88628 | 1.12831 | .91794 | 1.08940 | .95062 | 1.05194 | .98441 | 1.01583 | 27 |
| 34 | .85609 | 1.16809 | .88680 | 1.12765 | .91847 | 1.08876 | .95118 | 1.05133 | .98499 | 1.01524 | 26 |
| 35 | .85660 | 1.16741 | .88732 | 1.12699 | .91901 | 1.08813 | .95173 | 1.05072 | .98556 | 1.01465 | 25 |
| 36 | .85710 | 1.16672 | .88784 | 1.12633 | .91955 | 1.08749 | .95229 | 1.05010 | .98613 | 1.01406 | 24 |
| 37 | .85761 | 1.16603 | .88836 | 1.12567 | .92008 | 1.08686 | .95284 | 1.04949 | .98671 | 1.01347 | 23 |
| 38 | .85811 | 1.16535 | .88888 | 1.12501 | .92062 | 1.08622 | .95340 | 1.04888 | .98728 | 1.01288 | 22 |
| 39 | .85862 | 1.16466 | .88940 | 1.12435 | .92116 | 1.08559 | .95395 | 1.04827 | .98786 | 1.01229 | 21 |
| 40 | .85912 | 1.16398 | .88992 | 1.12369 | .92170 | 1.08496 | .95451 | 1.04766 | .98843 | 1.01170 | 20 |
| 41 | .85963 | 1.16329 | .89045 | 1.12303 | 92224 | 1.08432 | .95506 | 1.04705 | .98901 | 1.01112 | 19 |
| 42 | .86014 | 1.16261 | .89097 | 1.12238 | .92277 | 1.08369 | .95562 | 1.04644 | .98958 | 1.01053 | 18 |
| 43 | .86064 | 1.16192 | .89149 | 1.12172 | .92331 | 1.08306 | .95618 | 1.04583 | .99016 | 1.00994 | 17 |
| 44 | .86115 | 1.16124 | .89201 | 1.12106 | .92385 | 1.08243 | .95673 | 1.04522 | .99073 | 1.00935 | 16 |
| 45 | .86166 | 1.16056 | .89253 | 1.12041 | .92439 | 1.08179 | .95729 | 1.04461 | .99131 | 1.00876 | 15 |
| 46 | .86216 | 1.15987 | .89306 | 1.11975 | .92493 | 1.08116 | .95785 | 1.04401 | .99189 | 1.00818 | 14 |
| 47 | .86267 | 1.15919 | .89358 | 1.11909 | .92547 | 1.08053 | .95841 | 1.04340 | .99247 | 1.00759 | 13 |
| 48 | .86318 | 1.15851 | .89410 | 1.11844 | .92601 | 1.07990 | .95897 | 1.04279 | .99304 | 1.00701 | 12 |
| 49 | .86368 | 1.15783 | .89463 | 1.11778 | .92655 | 1.07927 | .95952 | 1.04218 | .99362 | 1.00642 | 11 |
| 50 | .86419 | 1.15715 | .89515 | 1.11713 | .92709 | 1.07864 | .96008 | 1.04158 | .99420 | 1.00583 | 10 |
| 51 | .86470 | 1.15647 | .89567 | 1.11648 | .92763 | 1.07801 | .96064 | 1.04097 | .99478 | 1.00525 | 9 |
| 52 | .86521 | 1.15579 | .89620 | 1.11582 | .92817 | 1.07738 | .96120 | 1.04036 | .99536 | 1.00467 | 8 |
| 53 | .86572 | 1.15511 | .89672 | 1.11517 | .92872 | 1.07676 | .96176 | 1.03976 | .99594 | 1.00408 | 7 |
| 54 | .86623 | 1.15443 | .89725 | 1.11452 | .92926 | 1.07613 | .96232 | 1.03915 | .99652 | 1.00350 | 6 |
| 55 | .86674 | 1.15375 | .89777 | 1.11387 | .92980 | 1.07550 | .96288 | 1.03855 | .99710 | 1.00291 | 5 |
| 56 | .86725 | 1.15308 | .89830 | 1.11321 | .93034 | 1.07487 | .96344 | 1.03794 | .99768 | 1.00233 | 4 |
| 57 | .86776 | 1.15240 | .89883 | 1.11256 | .93088 | 1.07425 | .96400 | 1.03734 | .99826 | 1.00175 | 3 |
| 58 | .86827 | 1.15172 | .89935 | 1.11191 | .93143 | 1.07362 | .96457 | 1.03674 | .99884 | 1.00116 | 2 |
| 59 | .86878 | 1.15104 | .89988 | 1.11126 | .93197 | 1.07299 | .96513 | 1.03613 | .99942 | 1.00058 | 1 |
| 60 | .86929 | 1.15037 | .90040 | 1.11061 | .93252 | 1.07237 | .96569 | 1.03553 | 1.00000 | 1.00000 | 0 |
| ′ | Cotang | Tang | Cotang | Tang | Cotang | Tang | Cotang | Tang | Cotang | Tang | ′ |
| | 49° | | 48° | | 47° | | 46° | | 45° | | |

# APPENDIX E

## STANDARD STADIA REDUCTION TABLE

### TABLE 1

Vertical Heights

| Minutes | 0° | 1° | 2° | 3° | 4° | 5° | 6° | 7° | 8° | 9° |
|---|---|---|---|---|---|---|---|---|---|---|
| 0... | 0.00 | 1.74 | 3.49 | 5.23 | 6.96 | 8.68 | 10.40 | 12.10 | 13.78 | 15.45 |
| 2... | 0.06 | 1.80 | 3.55 | 5.28 | 7.02 | 8.74 | 10.45 | 12.15 | 13.84 | 15.51 |
| 4... | 0.12 | 1.86 | 3.60 | 5.34 | 7.07 | 8.80 | 10.51 | 12.21 | 13.89 | 15.56 |
| 6... | 0.17 | 1.92 | 3.66 | 5.40 | 7.13 | 8.85 | 10.57 | 12.26 | 13.95 | 15.62 |
| 8... | 0.23 | 1.98 | 3.72 | 5.46 | 7.19 | 8.91 | 10.62 | 12.32 | 14.01 | 15.67 |
| 10... | 0.29 | 2.04 | 3.78 | 5.52 | 7.25 | 8.97 | 10.68 | 12.38 | 14.06 | 15.73 |
| 12... | 0.35 | 2.09 | 3.84 | 5.57 | 7.30 | 9.03 | 10.74 | 12.43 | 14.12 | 15.78 |
| 14... | 0.41 | 2.15 | 3.90 | 5.63 | 7.36 | 9.08 | 10.79 | 12.49 | 14.17 | 15.84 |
| 16... | 0.47 | 2.21 | 3.95 | 5.69 | 7.42 | 9.14 | 10.85 | 12.55 | 14.23 | 15.89 |
| 18... | 0.52 | 2.27 | 4.01 | 5.75 | 7.48 | 9.20 | 10.91 | 12.60 | 14.28 | 15.95 |
| 20... | 0.58 | 2.33 | 4.07 | 5.80 | 7.53 | 9.25 | 10.96 | 12.66 | 14.34 | 16.00 |
| 22... | 0.64 | 2.38 | 4.13 | 5.86 | 7.59 | 9.31 | 11.02 | 12.72 | 14.40 | 16.06 |
| 24... | 0.70 | 2.44 | 4.18 | 5.92 | 7.65 | 9.37 | 11.08 | 12.77 | 14.45 | 16.11 |
| 26... | 0.76 | 2.50 | 4.24 | 5.98 | 7.71 | 9.43 | 11.13 | 12.83 | 14.51 | 16.17 |
| 28... | 0.81 | 2.56 | 4.30 | 6.04 | 7.76 | 9.48 | 11.19 | 12.88 | 14.56 | 16.22 |
| 30... | 0.87 | 2.62 | 4.36 | 6.09 | 7.82 | 9.54 | 11.25 | 12.94 | 14.62 | 16.28 |
| 32... | 0.93 | 2.67 | 4.42 | 6.15 | 7.88 | 9.60 | 11.30 | 13.00 | 14.67 | 16.33 |
| 34... | 0.99 | 2.73 | 4.48 | 6.21 | 7.94 | 9.65 | 11.36 | 13.05 | 14.73 | 16.39 |
| 36... | 1.05 | 2.79 | 4.53 | 6.27 | 7.99 | 9.71 | 11.42 | 13.11 | 14.79 | 16.44 |
| 38... | 1.11 | 2.85 | 4.59 | 6.33 | 8.05 | 9.77 | 11.47 | 13.17 | 14.84 | 16.50 |
| 40... | 1.16 | 2.91 | 4.65 | 6.38 | 8.11 | 9.83 | 11.53 | 13.22 | 14.90 | 16.55 |
| 42... | 1.22 | 2.97 | 4.71 | 6.44 | 8.17 | 9.88 | 11.59 | 13.28 | 14.95 | 16.61 |
| 44... | 1.28 | 3.02 | 4.76 | 6.50 | 8.22 | 9.94 | 11.64 | 13.33 | 15.01 | 16.66 |
| 46... | 1.34 | 3.08 | 4.82 | 6.56 | 8.28 | 10.00 | 11.70 | 13.39 | 15.06 | 16.72 |
| 48... | 1.40 | 3.14 | 4.88 | 6.61 | 8.34 | 10.05 | 11.76 | 13.45 | 15.12 | 16.77 |
| 50... | 1.45 | 3.20 | 4.94 | 6.67 | 8.40 | 10.11 | 11.81 | 13.50 | 15.17 | 16.83 |
| 52... | 1.51 | 3.26 | 4.99 | 6.73 | 8.45 | 10.17 | 11.87 | 13.56 | 15.23 | 16.88 |
| 54... | 1.57 | 3.31 | 5.05 | 6.79 | 8.51 | 10.22 | 11.93 | 13.61 | 15.28 | 16.94 |
| 56... | 1.63 | 3.37 | 5.11 | 6.84 | 8.57 | 10.28 | 11.98 | 13.67 | 15.34 | 16.99 |
| 58... | 1.69 | 3.43 | 5.17 | 6.90 | 8.63 | 10.34 | 12.04 | 13.73 | 15.40 | 17.05 |
| 60... | 1.74 | 3.49 | 5.23 | 6.96 | 8.68 | 10.40 | 12.10 | 13.78 | 15.45 | 17.10 |

Horizontal Corrections

| Dist. | 0° | 1° | 2° | 3° | 4° | 5° | 6° | 7° | 8° | 9° |
|---|---|---|---|---|---|---|---|---|---|---|
| 100.. | 0.0 | 0.0 | 0.1 | 0.3 | 0.5 | 0.8 | 1.1 | 1.5 | 1.9 | 2.5 |
| 200.. | 0.0 | 0.1 | 0.2 | 0.5 | 1.0 | 1.5 | 2.2 | 3.0 | 3.9 | 4.9 |
| 300.. | 0.0 | 0.1 | 0.4 | 0.8 | 1.5 | 2.3 | 3.3 | 4.5 | 5.8 | 7.4 |
| 400.. | 0.0 | 0.1 | 0.5 | 1.1 | 2.0 | 3.0 | 4.4 | 6.0 | 7.8 | 9.8 |
| 500.. | 0.0 | 0.2 | 0.6 | 1.4 | 2.5 | 3.8 | 5.5 | 7.5 | 9.7 | 12.3 |
| 600.. | 0.0 | 0.2 | 0.7 | 1.6 | 2.9 | 4.6 | 6.5 | 8.9 | 11.6 | 14.7 |
| 700.. | 0.0 | 0.2 | 0.8 | 1.9 | 3.4 | 5.3 | 7.6 | 10.4 | 13.6 | 17.2 |
| 800.. | 0.0 | 0.2 | 1.0 | 2.2 | 3.9 | 6.1 | 8.7 | 11.9 | 15.5 | 19.6 |
| 900.. | 0.0 | 0.3 | 1.1 | 2.4 | 4.4 | 6.8 | 0.8 | 13.4 | 17.5 | 22.1 |
| 1000.. | 0.0 | 0.3 | 1.2 | 2.7 | 4.9 | 7.6 | 10.9 | 14.9 | 19.4 | 24.5 |

## TABLE 1 cont.

Vertical Heights

| Minutes | 10° | 11° | 12° | 13° | 14° | 15° | 16° | 17° | 18° | 19° |
|---|---|---|---|---|---|---|---|---|---|---|
| 0... | 17.10 | 18.73 | 20.34 | 21.92 | 23.47 | 25.00 | 26.50 | 27.96 | 29.39 | 30.78 |
| 2... | 17.16 | 18.78 | 20.39 | 21.97 | 23.52 | 25.05 | 26.55 | 28.01 | 29.44 | 30.83 |
| 4... | 17.21 | 18.84 | 20.44 | 22.02 | 23.58 | 25.10 | 26.59 | 28.06 | 29.48 | 30.87 |
| 6... | 17.26 | 18.89 | 20.50 | 22.08 | 23.63 | 25.15 | 26.64 | 28.10 | 29.53 | 30.92 |
| 8... | 17.32 | 18.95 | 20.55 | 22.13 | 23.68 | 25.20 | 26.69 | 28.15 | 29.58 | 30.97 |
| 10... | 17.37 | 19.00 | 20.60 | 22.18 | 23.73 | 25.25 | 26.74 | 28.20 | 29.62 | 31.01 |
| 12... | 17.43 | 19.05 | 20.66 | 22.23 | 23.78 | 25.30 | 26.79 | 28.25 | 29.67 | 31.06 |
| 14... | 17.48 | 19.11 | 20.71 | 22.28 | 23.83 | 25.35 | 26.84 | 28.30 | 29.72 | 31.10 |
| 16... | 17.54 | 19.16 | 20.76 | 22.34 | 23.88 | 25.40 | 26.89 | 28.34 | 29.76 | 31.15 |
| 18... | 17.59 | 19.21 | 20.81 | 22.39 | 23.93 | 25.45 | 26.94 | 28.39 | 29.81 | 31.19 |
| 20... | 17.65 | 19.27 | 20.87 | 22.44 | 23.99 | 25.50 | 26.99 | 28.44 | 29.86 | 31.24 |
| 22... | 17.70 | 19.32 | 20.92 | 22.49 | 24.04 | 25.55 | 27.04 | 28.49 | 29.90 | 31.28 |
| 24... | 17.76 | 19.38 | 20.97 | 22.54 | 24.09 | 25.60 | 27.09 | 28.54 | 29.95 | 31.33 |
| 26... | 17.81 | 19.43 | 21.03 | 22.60 | 24.14 | 25.65 | 27.13 | 28.58 | 30.00 | 31.38 |
| 28... | 17.86 | 19.48 | 21.08 | 22.65 | 24.19 | 25.70 | 27.18 | 28.63 | 30.04 | 31.42 |
| 30... | 17.92 | 19.54 | 21.13 | 22.70 | 24.24 | 25.75 | 27.23 | 28.68 | 30.09 | 31.47 |
| 32... | 17.97 | 19.59 | 21.18 | 22.75 | 24.29 | 25.80 | 27.28 | 28.73 | 30.14 | 31.51 |
| 34... | 18.03 | 19.64 | 21.24 | 22.80 | 24.34 | 25.85 | 27.33 | 28.77 | 30.19 | 31.56 |
| 36... | 18.08 | 19.70 | 21.29 | 22.85 | 24.39 | 25.90 | 27.38 | 28.82 | 30.23 | 31.60 |
| 38... | 18.14 | 19.75 | 21.34 | 22.91 | 24.44 | 25.95 | 27.43 | 28.87 | 30.28 | 31.65 |
| 40... | 18.19 | 19.80 | 21.39 | 22.96 | 24.49 | 26.00 | 27.48 | 28.92 | 30.32 | 31.69 |
| 42... | 18.24 | 19.86 | 21.45 | 23.01 | 24.55 | 26.05 | 27.52 | 28.96 | 30.37 | 31.74 |
| 44... | 18.30 | 19.91 | 21.50 | 23.06 | 24.60 | 26.10 | 27.57 | 29.01 | 30.41 | 31.78 |
| 46... | 18.35 | 19.96 | 21.55 | 23.11 | 24.65 | 26.15 | 27.62 | 29.06 | 30.46 | 31.83 |
| 48... | 18.41 | 20.02 | 21.60 | 23.16 | 24.70 | 26.20 | 27.67 | 29.11 | 30.51 | 31.87 |
| 50... | 18.46 | 20.07 | 21.66 | 23.22 | 24.75 | 26.25 | 27.72 | 29.15 | 30.55 | 31.92 |
| 52... | 18.51 | 20.12 | 21.71 | 23.27 | 24.80 | 26.30 | 27.77 | 29.20 | 30.60 | 31.96 |
| 54... | 18.57 | 20.18 | 21.76 | 23.32 | 24.85 | 26.35 | 27.81 | 29.25 | 30.65 | 32.01 |
| 56... | 18.62 | 20.23 | 21.81 | 23.37 | 24.90 | 26.40 | 27.86 | 29.30 | 30.69 | 32.05 |
| 58... | 18.68 | 20.28 | 21.87 | 23.42 | 24.95 | 26.45 | 27.91 | 29.34 | 30.74 | 32.09 |
| 60... | 18.73 | 20.34 | 21.92 | 23.47 | 25.00 | 26.50 | 27.96 | 29.39 | 30.78 | 32.14 |

Horizontal Corrections

| Dist. | 10° | 11° | 12° | 13° | 14° | 15° | 16° | 17° | 18° | 19° |
|---|---|---|---|---|---|---|---|---|---|---|
| 100.. | 3.0 | 3.6 | 4.3 | 5.1 | 5.9 | 6.7 | 7.6 | 8.5 | 9.5 | 10.6 |
| 200.. | 6.0 | 7.3 | 8.6 | 10.1 | 11.7 | 13.4 | 15.2 | 17.1 | 19.1 | 21.2 |
| 300.. | 9.1 | 10.9 | 13.0 | 15.2 | 17.6 | 20.1 | 22.8 | 25.6 | 28.6 | 31.8 |
| 400.. | 12.1 | 14.6 | 17.3 | 20.2 | 23.4 | 26.8 | 30.4 | 34.2 | 38.2 | 42.4 |
| 500.. | 15.1 | 18.2 | 21.6 | 25.3 | 29.3 | 33.5 | 38.0 | 42.7 | 47.7 | 53.0 |
| 600.. | 18.1 | 21.8 | 25.9 | 30.4 | 35.1 | 40.2 | 45.6 | 51.3 | 57.3 | 63.6 |
| 700.. | 21.1 | 25.5 | 30.2 | 35.4 | 41.0 | 46.9 | 53.2 | 59.8 | 66.8 | 74.2 |
| 800.. | 24.2 | 29.1 | 34.6 | 40.5 | 46.8 | 53.6 | 60.8 | 68.4 | 76.4 | 84.8 |
| 900.. | 27.2 | 32.8 | 38.9 | 45.5 | 52.7 | 60.3 | 68.4 | 76.9 | 85.9 | 95.4 |
| 1000 . | 30.2 | 36.4 | 43.2 | 50.6 | 58.5 | 67.0 | 76.0 | 85.5 | 95.5 | 106.0 |

## TABLE 1 cont.

### Vertical Heights

| Minutes | 20° | 21° | 22° | 23° | 24° | 25° | 26° | 27° | 28° | 29° |
|---|---|---|---|---|---|---|---|---|---|---|
| 0... | 32.14 | 33.46 | 34.73 | 35.97 | 37.16 | 38.30 | 39.40 | 40.45 | 41.45 | 42.40 |
| 2... | 32.18 | 33.50 | 34.77 | 36.01 | 37.20 | 38.34 | 39.44 | 40.49 | 41.48 | 42.43 |
| 4... | 32.23 | 33.54 | 34.82 | 36.05 | 37.23 | 38.38 | 39.47 | 40.52 | 41.52 | 42.46 |
| 6... | 32.27 | 33.59 | 34.86 | 36.09 | 37.27 | 38.41 | 39.51 | 40.55 | 41.55 | 42.49 |
| 8... | 32.32 | 33.63 | 34.90 | 36.13 | 37.31 | 38.45 | 39.54 | 40.59 | 41.58 | 42.53 |
| 10... | 32.36 | 33.67 | 34.94 | 36.17 | 37.35 | 38.49 | 39.58 | 40.62 | 41.61 | 42.56 |
| 12... | 32.41 | 33.72 | 34.98 | 36.21 | 37.39 | 38.53 | 39.61 | 40.66 | 41.65 | 42.59 |
| 14... | 32.45 | 33.76 | 35.02 | 36.25 | 37.43 | 38.56 | 39.65 | 40.69 | 41.68 | 42.62 |
| 16... | 32.49 | 33.80 | 35.07 | 36.29 | 37.47 | 38.60 | 39.69 | 40.72 | 41.71 | 42.65 |
| 18... | 32.54 | 33.84 | 35.11 | 36.33 | 37.51 | 38.64 | 39.72 | 40.76 | 41.74 | 42.68 |
| 20... | 32.58 | 33.89 | 35.15 | 36.37 | 37.54 | 38.67 | 39.76 | 40.79 | 41.77 | 42.71 |
| 22... | 32.63 | 33.93 | 35.19 | 36.41 | 37.58 | 38.71 | 39.79 | 40.82 | 41.81 | 42.74 |
| 24... | 32.67 | 33.97 | 35.23 | 36.45 | 37.62 | 38.75 | 39.83 | 40.86 | 41.84 | 42.77 |
| 26... | 32.72 | 34.01 | 35.27 | 36.49 | 37.66 | 38.78 | 39.86 | 40.89 | 41.87 | 42.80 |
| 28... | 32.76 | 34.06 | 35.31 | 36.53 | 37.70 | 38.82 | 39.90 | 40.92 | 41.90 | 42.83 |
| 30... | 32.80 | 34.10 | 35.36 | 36.57 | 37.74 | 38.86 | 39.93 | 40.96 | 41.93 | 42.86 |
| 32... | 32.85 | 34.14 | 35.40 | 36.61 | 37.77 | 38.89 | 39.97 | 40.99 | 41.97 | 42.89 |
| 34... | 32.89 | 34.18 | 35.44 | 36.65 | 37.81 | 38.93 | 40.00 | 41.02 | 42.00 | 42.92 |
| 36... | 32.93 | 34.23 | 35.48 | 36.69 | 37.85 | 38.97 | 40.04 | 41.06 | 42.03 | 42.95 |
| 38... | 32.98 | 34.27 | 35.52 | 36.73 | 37.89 | 39.00 | 40.07 | 41.09 | 42.06 | 42.98 |
| 40... | 33.02 | 34.31 | 35.56 | 36.77 | 37.93 | 39.04 | 40.11 | 41.12 | 42.09 | 43.01 |
| 42... | 33.07 | 34.35 | 35.60 | 36.80 | 37.96 | 39.08 | 40.14 | 41.16 | 42.12 | 43.04 |
| 44... | 33.11 | 34.40 | 35.64 | 36.84 | 38.00 | 39.11 | 40.18 | 41.19 | 42.15 | 43.07 |
| 46... | 33.15 | 34.44 | 35.68 | 36.88 | 38.04 | 39.15 | 40.21 | 41.22 | 42.19 | 43.10 |
| 48... | 33.20 | 34.48 | 35.72 | 36.92 | 38.08 | 39.18 | 40.24 | 41.26 | 42.22 | 43.13 |
| 50... | 33.24 | 34.52 | 35.76 | 36.96 | 38.11 | 39.22 | 40.28 | 41.29 | 42.25 | 43.16 |
| 52... | 33.28 | 34.57 | 35.80 | 37.00 | 38.15 | 39.26 | 40.31 | 41.32 | 42.28 | 43.18 |
| 54... | 33.33 | 34.61 | 35.85 | 37.04 | 38.19 | 39.29 | 40.35 | 41.35 | 42.31 | 43.21 |
| 56... | 33.37 | 34.65 | 35.89 | 37.08 | 38.23 | 39.33 | 40.38 | 41.39 | 42.34 | 43.24 |
| 58... | 33.41 | 34.69 | 35.93 | 37.12 | 38.26 | 39.36 | 40.42 | 41.42 | 42.37 | 43.27 |
| 60... | 33.46 | 34.73 | 35.97 | 37.16 | 38.30 | 39.40 | 40.45 | 41.45 | 42.40 | 43.30 |

### Horizontal Corrections

| Dist. | 20° | 21° | 22° | 23° | 24° | 25° | 26° | 27° | 28° | 29° |
|---|---|---|---|---|---|---|---|---|---|---|
| 100.. | 11.7 | 12.8 | 14.0 | 15.3 | 16.5 | 17.9 | 19.2 | 20.6 | 22.0 | 23.5 |
| 200.. | 23.4 | 25.7 | 28.1 | 30.5 | 33.1 | 35.7 | 38.4 | 41.2 | 44.1 | 47.0 |
| 300.. | 35.1 | 38.5 | 42.1 | 45.8 | 49.6 | 53.6 | 57.7 | 61.8 | 66.1 | 70.5 |
| 400.. | 46.8 | 51.4 | 56.1 | 61.1 | 66.2 | 71.4 | 76.9 | 82.4 | 88.2 | 94.0 |
| 500.. | 58.5 | 64.2 | 70.2 | 76.4 | 82.7 | 89.3 | 96.1 | 103.1 | 110.2 | 117.5 |
| 600.. | 70.2 | 77.0 | 84.2 | 91.6 | 99.2 | 107.2 | 115.3 | 123.7 | 132.2 | 141.0 |
| 700.. | 81.9 | 89.9 | 98.2 | 106.9 | 115.8 | 125.0 | 134.5 | 144.3 | 154.3 | 164.5 |
| 800.. | 93.6 | 102.7 | 112.2 | 122.2 | 132.3 | 142.9 | 153.8 | 164.9 | 176.3 | 188.0 |
| 900.. | 105.3 | 115.6 | 126.3 | 137.4 | 148.9 | 160.7 | 173.0 | 185.5 | 198.4 | 211.5 |
| 1000.. | 117.0 | 128.4 | 140.3 | 152.7 | 165.4 | 178.6 | 192.2 | 206.1 | 220.4 | 235.0 |

## TABLE 1 cont.

Vertical Heights

| Minutes | 30° | 31° | 32° | 33° | 34° | 35° | 36° | 37° | 38° | 39° |
|---|---|---|---|---|---|---|---|---|---|---|
| 0 | 43.30 | 44.15 | 44.94 | 45.68 | 46.36 | 46.98 | 47.55 | 48.06 | 48.52 | 48.91 |
| 2 | 43.33 | 44.17 | 44.97 | 45.70 | 46.38 | 47.00 | 47.57 | 48.08 | 48.53 | 48.92 |
| 4 | 43.36 | 44.20 | 44.99 | 45.72 | 46.40 | 47.02 | 47.59 | 48.10 | 48.54 | 48.93 |
| 6 | 43.39 | 44.23 | 45.02 | 45.75 | 46.42 | 47.04 | 47.61 | 48.11 | 48.56 | 48.94 |
| 8 | 43.42 | 44.26 | 45.04 | 45.77 | 46.45 | 47.06 | 47.62 | 48.13 | 48.57 | 48.96 |
| 10 | 43.45 | 44.28 | 45.07 | 45.80 | 46.47 | 47.08 | 47.64 | 48.14 | 48.58 | 48.97 |
| 12 | 43.47 | 44.31 | 45.09 | 45.82 | 46.49 | 47.10 | 47.66 | 48.16 | 48.60 | 48.98 |
| 14 | 43.50 | 44.34 | 45.12 | 45.84 | 46.51 | 47.12 | 47.68 | 48.17 | 48.61 | 48.99 |
| 16 | 43.52 | 44.36 | 45.14 | 45.86 | 46.53 | 47.14 | 47.69 | 48.19 | 48.63 | 49.00 |
| 18 | 43.56 | 44.39 | 45.17 | 45.89 | 46.55 | 47.16 | 47.71 | 48.21 | 48.64 | 49.01 |
| 20 | 43.59 | 44.42 | 45.19 | 45.91 | 46.57 | 47.18 | 47.73 | 48.22 | 48.65 | 49.03 |
| 22 | 43.62 | 44.44 | 45.22 | 45.93 | 46.60 | 47.20 | 47.75 | 48.24 | 48.67 | 49.04 |
| 24 | 43.65 | 44.47 | 45.24 | 45.96 | 46.62 | 47.22 | 47.76 | 48.25 | 48.68 | 49.05 |
| 26 | 43.67 | 44.50 | 45.27 | 45.98 | 46.64 | 47.24 | 47.78 | 48.27 | 48.69 | 49.06 |
| 28 | 43.70 | 44.52 | 45.29 | 46.00 | 46.66 | 47.26 | 47.80 | 48.28 | 48.71 | 49.07 |
| 30 | 43.73 | 44.55 | 45.32 | 46.03 | 46.68 | 47.28 | 47.82 | 48.30 | 48.72 | 49.08 |
| 32 | 43.76 | 44.58 | 45.34 | 46.05 | 46.70 | 47.30 | 47.83 | 48.31 | 48.73 | 49.09 |
| 34 | 43.79 | 44.60 | 45.36 | 46.07 | 46.72 | 47.31 | 47.85 | 48.33 | 48.74 | 49.10 |
| 36 | 43.82 | 44.63 | 45.39 | 46.09 | 46.74 | 47.33 | 47.87 | 48.34 | 48.76 | 49.11 |
| 38 | 43.84 | 44.66 | 45.41 | 46.12 | 46.76 | 47.35 | 47.88 | 48.36 | 48.77 | 49.13 |
| 40 | 43.87 | 44.68 | 45.44 | 46.14 | 46.78 | 47.37 | 47.90 | 48.37 | 48.78 | 49.14 |
| 42 | 43.90 | 44.71 | 45.46 | 46.16 | 46.80 | 47.39 | 47.92 | 48.39 | 48.80 | 49.15 |
| 44 | 43.93 | 44.74 | 45.49 | 46.18 | 46.82 | 47.41 | 47.93 | 48.40 | 48.81 | 49.16 |
| 46 | 43.95 | 44.76 | 45.51 | 46.21 | 46.84 | 47.43 | 47.95 | 48.41 | 48.82 | 49.17 |
| 48 | 43.98 | 44.79 | 45.53 | 46.23 | 46.86 | 47.44 | 47.97 | 48.43 | 48.83 | 49.18 |
| 50 | 44.01 | 44.81 | 45.56 | 46.25 | 46.88 | 47.46 | 47.98 | 48.44 | 48.85 | 49.19 |
| 52 | 44.04 | 44.84 | 45.58 | 46.27 | 46.90 | 47.48 | 48.00 | 48.46 | 48.86 | 49.20 |
| 54 | 44.07 | 44.86 | 45.61 | 46.29 | 46.92 | 47.50 | 48.01 | 48.47 | 48.87 | 49.21 |
| 56 | 44.09 | 44.89 | 45.63 | 46.32 | 46.94 | 47.52 | 48.03 | 48.49 | 48.88 | 49.22 |
| 58 | 44.12 | 44.91 | 45.65 | 46.34 | 46.96 | 47.54 | 48.05 | 48.50 | 48.90 | 49.23 |
| 60 | 44.15 | 44.94 | 45.68 | 46.36 | 46.98 | 47.55 | 48.06 | 48.52 | 48.91 | 49.24 |

Horizontal Corrections

| Dist. | 30° 00′ | 30° 30′ | 31° 00′ | 31° 30′ | 32° 00′ | 32° 30′ | 33° 00′ | 33° 30′ | 34° 00′ | 34° 30′ |
|---|---|---|---|---|---|---|---|---|---|---|
| 100 | 25.0 | 25.8 | 26.5 | 27.3 | 28.1 | 28.9 | 29.7 | 30.5 | 31.3 | 32.1 |
| 200 | 50.0 | 51.5 | 53.1 | 54.6 | 56.2 | 57.7 | 59.3 | 60.9 | 62.5 | 64.2 |
| 300 | 75.0 | 77.3 | 79.6 | 81.9 | 84.2 | 86.6 | 89.0 | 91.4 | 93.8 | 96.2 |
| 400 | 100.0 | 103.0 | 106.1 | 109.2 | 112.3 | 115.5 | 118.6 | 121.8 | 125.1 | 128.3 |
| 500 | 125.0 | 128.8 | 132.6 | 136.5 | 140.4 | 144.3 | 148.3 | 152.3 | 156.3 | 160.4 |

| Dist. | 35° 00′ | 35° 30′ | 36° 00′ | 36° 30′ | 37° 00′ | 37° 30′ | 38° 00′ | 38° 30′ | 39° 00′ | 39° 30′ |
|---|---|---|---|---|---|---|---|---|---|---|
| 100 | 32.9 | 33.7 | 34.6 | 35.4 | 36.2 | 37.1 | 37.9 | 38.7 | 39.6 | 40.5 |
| 200 | 65.8 | 67.4 | 69.1 | 70.8 | 72.4 | 74.1 | 75.8 | 77.5 | 79.2 | 80.9 |
| 300 | 98.7 | 101.2 | 103.7 | 106.1 | 108.7 | 111.2 | 113.7 | 116.2 | 118.8 | 121.4 |
| 400 | 131.6 | 134.9 | 138.2 | 141.5 | 144.9 | 148.2 | 151.6 | 155.0 | 158.4 | 161.8 |
| 500 | 164.5 | 168.6 | 172.8 | 176.9 | 181.1 | 185.3 | 189.5 | 193.7 | 198.0 | 202.3 |

# APPENDIX F

## PHOTOGRAPHS, DIAGRAM, AND COMPONENTS OF TRANSIT

### TRANSIT

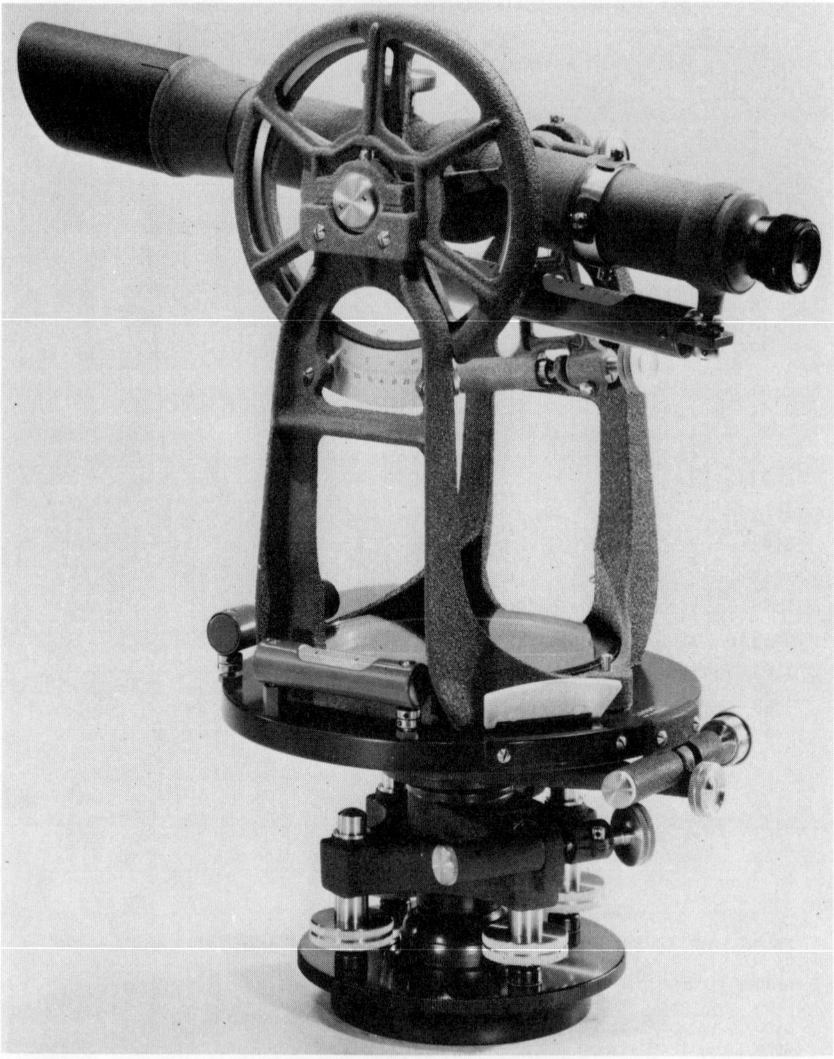

## COMPONENT PARTS OF A PARAGON TRANSIT

**5205-1** Tripod Plate
**2** Leveling Head
**3** Leveling Screw Head
**4** Leveling Screw Stem
**5** Leveling Screw Shoe
**6** Leveling Screw Cap (Left Hand Thread)
**7** Shifting Plate
**8** Half Ball
**9** Half Ball Lock Screw
**10** Outer Center
**11** Vernier Plate Clamp
**12** Vernier Plate Clamp Gib
**13** Vern. Pl. Cl. Screw
**14** Vernier Plate Clamp Screw Pin
**15** Lower Clamp
**16** Lower Clamp Collar
**17** Lower Clamp Collar Mounting Screw
**18** Lower Clamp Gib
**19** Lower Clamp Screw
**20** Lower Clamp Screw Pin
**21** Lower Clamp Tangent Screw
**21A** Clamp Tangent Screw Pivot Pin
**22** Clamp Tangent Screw Tension Screw
**23** Lower Clamp Tangent Screw Plunger
**24** Lower Clamp Tangent Screw Spring
**25** Lower Clamp Tangent Screw Cap
**26** Horizontal Circle
**27** Horizontal Circle Mounting Screw
**28** Horiz. Circle Vernier
**29** Horiz. Circle Vernier Mounting Screw
**30** Horizontal Circle Adjusting Screw
**31** Vernier Plate
**32** Vernier Plate Mounting Screw
**33** Inner Center
**34** Center Nut
**35** Center Nut Lock Screw
**36** Center Cap
**37** Center Spring
**38** Center Ball
**39** Plumb Bob Chain and Hook
**100** Compass Needle
**101** Compass Needle Lifter
**102** Compass Needle Lifter Bushing
**103** Compass Needle Lifter Screw Assembly
**104** Compass Needle Pivot
**105** Compass Ring
**5205-106** Compass Ring Spring
**107** Compass Dial
**108** Compass Cover Glass and Mount
**109** Vernier Cover Glass
**110** Vernier Cover Glass Strap
**110A** Vernier Cover Glass Strap Screw (not shown)
**111** Vernier Reflector
**112** Vern. Reflector Frame and Hinge
**113** Vernier Plate Clamp Tangent Screw
**114** Vern. Pl. Cl. Tangent Screw Spring Box
**115†** Tangent Screw Plunger
**116†** Tangent Screw Spring
**117†** Tangent Screw Cap
†(Not shown but similar to Nos. 23, 24 & 25.) (For Vernier Plate and Telescope Cl. Tang. Screws)
**118** Plate Level Bracket and Posts complete
**119** Plate Level Vial Guard
**120** Plate and Standard Level Adj. Nut
**121** Plate and Standard Level Spring
**122** Plate and Standard Level Post Cap
**123** Plate and Standard Level Vial, Tube and Ends complete
**124** Plate and Standard Level Vial
**125†** Standard Level Post, Fixed
**126†** Standard Level Post, Adjustable
†Not shown.
**127** Declination Adjustm'nt Pinion and Washer
**129** Standard
**130** Trunnion Cap
**131** Trunnion Cap Screw
**132** Trunnion Friction Sc.
**133** Trunnion Bearing Block
**134** Trunnion Bearing Block Adjusting Screw
**135** Trunnion Bearing Block Lock Screw
**200** Tele. Barrel and Axle
**201** Tele. Axle End Cap
**202†** Telescope Clamp
**203†** Telescope Clamp Gib
†Not shown.
**204** Tele. Clamp Screw
**205** Tele. Clamp Screw Pin (Not shown).
**206** Telescope Clamp Tangent Screw
**5205-207** Tele. Level Vial, Tube and Ends complete
**208** Tele. Level Vial only
**209** Telescope Level Tube End Lock Screw
**210** Tele. Level Adjust. Nut
**211** Objective Cap
**212** Objective Lens and Mount
**213** Sunshade
**214** Telescope Draw Tube
**215** Tele. Focusing Lens
**216** Telescope Focusing Lens Mount
**217** Tele. Focusing Lens Mount Lock Screw
**218** Telescope Focusing Lens Lock Ring
**219** Tele. Focusing Pinion, Pinion Head (222) & Screw (223) complete
**220** Tele. Focusing Pinion
**221** Telescope Focusing Pinion Lock Screw
**222** Telescope Focusing Pinion Head
**223** Telescope Focusing Pinion Head Screw
**224** Reticule (See 5097-49)
**225** Reticule Adjust. Screw
**226** Reticule Adjusting Screw Shutter
**227*** Eyepiece Lens I and Mount
**228*** Eyepiece Lens II and Mount
**229*** Eyepiece Tube
**230*** Eyepiece Lens III and Mount
**231*** Eyepiece Focusing Lens and Mount
**232*** Eyepiece Focusing Ring
**233*** Eyepiece Focusing Ring Set Screw
**234*** Eyepiece Cap
**235*** Eyepiece Focus. Sleeve
**236*** Eyepiece Focusing Sleeve Screw
**237** Vertical (Stadia) Circle
**238** Vertical Circle Vernier
**239** Vertical Circle Vernier Posts and Nuts
**240** Vertical Circle Guard
**241** Vert. Cir. Guard Screw
**242** Stadia Index, Horizontal
**243** Stadia Index, Vertical
**244** Stadia Index Lock Sc.
**245** Stadia Index Adjusting Screw
**246** Stadia Index Frame
**247** Stadia Index Frame Sc.
*Nos. 227 to 236 apply to Achromatic Eyepiece only.

# DIAGRAM OF A PARAGON TRANSIT

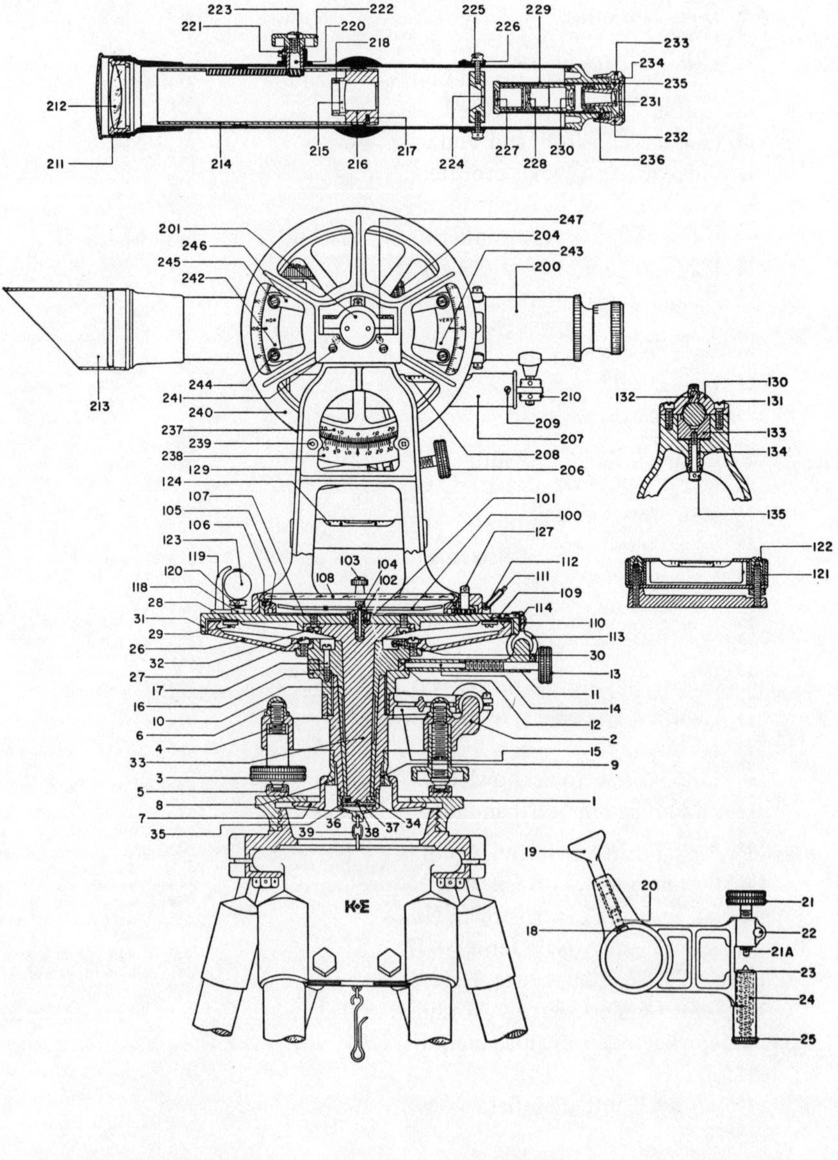

# ACKNOWLEDGMENTS

Assistance of the staff of Tri-County Community College, Murphy, North Carolina is gratefully acknowledged, particularly the assistance of Randall Shields, evening program director, and June Brooks. The author appreciates the time spent by his wife in typing, proofreading, and editorial comment, and also the excellent work of Foster Dionne in making Photographic illustrations.

Rear Admiral Russell C. Brinker encouraged the author to write this book to share Tri-County Community College's teaching approach with others. His valuable advice and editorial comment, which was so generously volunteered, is also appreciated.

**The Delmar Staff**

Source Editor: Mark W. Huth
Associate Editor: Vincent A. De Santis
Photo Editor: Sherry Patnode
Copy Editor: Kathleen Beiswenger

**Classroom Testing**

The material in this textbook was classroom tested by the students in Tri-County Community College, Murphy, North Carolina.

The author wishes to acknowledge the following companies and organizations which have supplied material and information:

Hewlett Packard Co., Loveland, Colorado – figures 7-20, 7-21, 11-13
Kern Instruments Inc., Port Chester, New York – figures 11-12, 11-14
Keuffel and Esser Co., Morristown, New Jersey – Appendix F
Monroe, Morris Plains, New Jersey – figure 7-19
Tellurometer-USA, Hauppauge, New York – figure 11-11
United States Geodetic Survey, Washington, D.C. – figures 4-6, 14-1, 14-9
Wild Heerbrugg Instruments Inc., Farmingdale, New York – figures 11-10, 11-15, 11-16
Zena Company, South Plainfield, New Jersey – 11-9

# INDEX